George Perkins Merrill

A Treatise on Rocks, Rock-Weathering and Soils

George Perkins Merrill

A Treatise on Rocks, Rock-Weathering and Soils

ISBN/EAN: 9783337337018

Printed in Europe, USA, Canada, Australia, Japan

Cover: Foto ©berggeist007 / pixelio.de

More available books at **www.hansebooks.com**

A TREATISE

ON

ROCKS, ROCK-WEATHERING
AND SOILS

BY

GEORGE P. MERRILL

CURATOR OF GEOLOGY IN THE UNITED STATES NATIONAL MUSEUM
OF GEOLOGY IN THE CORCORAN SCIENTIFIC SCHOOL
SCHOOL OF COLUMBIAN UNIVERSITY, WASHINGTON
AUTHOR OF "STONES FOR BUILDING AND DECORATION"

New York
THE MACMILLAN COMPANY
LONDON: MACMILLAN & CO., Ltd.
1897

All rights reserved

"The ruins of an older world are visible in the present structure of our planet; and the strata which now compose our continents have been once beneath the sea, and were formed out of the waste of pre-existing continents. The same forces are still destroying, by chemical decomposition or mechanical violence, even the hardest rocks, and transporting the materials to the sea, where they are spread out, and form strata analogous to those of more ancient date." — HUTTON.

PREFATORY NOTE

IN the work here presented the writer has endeavored to bring together in systematic form the results of several years' study of the phenomena attendant upon rock degeneration and soil formation. Although beginning with a discussion of rocks and rock-forming minerals, the work must be considered in no sense a petrology as this word is commonly used. What is here given relative to the origin, structure, and composition of rock masses is regarded as an essential introduction to the chapters on rock-weathering. The portion dealing with the structure and composition of the resultant materials is an essential corollary to these same chapters.

It is believed that no apology is necessary even in this day of many books for bringing out the present work. The origin, structure, and mineral composition of rocks, particularly the eruptive varieties, are matters which have of late received much attention. In fact, it is to these rocks that the petrologists have devoted their best efforts. Since the introduction of the microscope into petrographic work, there has, however, been very little time devoted to the study of rocks in a weathered condition. The chemists have made analyses, but have disregarded the physical and mineralogical nature of the material analyzed. Other workers have studied the physical properties of rocks decayed, — in the form of soils, — but have in their turn disregarded their mineral and chemical nature. The writer has aimed to bring together here such results obtained by these workers in divers fields as it is believed will be for the mutual benefit of all concerned. The state of comminution reached by rocks during the processes of long-continued, secular decay, and the amount of leaching such have undergone, are certainly of as much practical interest to the agriculturist as of theoretical interest to the geologist.

To the one, these residues are essential to the life and well-being of man through furnishing the soils from whence is derived directly and indirectly the food for life's sustenance; to the other they are but transitory phases in the earth's history, representing the materials from which, through a process of fractional separation by running waters, have been made up the thousands of feet of secondary rocks which to-day occupy so large a portion of its surface.

The very general scheme of classification adopted in the treatment of the unconsolidated clastic materials may at first seem disappointing. It has, however, been the writer's special aim to introduce into this preliminary volume as few new terms as possible, using only those which through years of service have become a part of our language. It is of course possible that in his desire to avoid any possible confusion such as might arise through putting forward a purely tentative classification he has been overcautious.

It is possible, further, that in numerous instances it may appear that too much reliance is placed upon single analyses, particularly in the discussions relating to the character of decomposed material. Regarding this it can only be said that in those instances upon which most reliance is placed, the materials were not merely collected by the author himself, but that he made his own chemical analyses and microscopic determinations as well. It is believed that the fresh and residual materials examined are in each instance as truly representative of the same rock mass, as would be samples of fresh rock collected equal distances apart. In all cases special effort was made to obtain material concerning the lithological identity of which there could be no doubt, and in the majority of cases the residuary matter was collected from positions immediately overlying the still unaltered rock. Where such a procedure was impossible, especial care was exercised to obtain only such as was originally of the same lithological nature as the fresh rock, and which had suffered no contamination from extraneous sources. The fact that stratified rocks are likely to vary so greatly within short distances, and hence that a residual clay cannot be relied upon to represent the residue from rocks of the same nature immediately underlying, will serve to explain in part the author's limiting himself so largely to a discussion of massive eruptive materials. That so little use has been made of other analyses, made in greater detail or by those more skilled in analytical methods, is due to a lack of satisfactory information relative to the mutual association of the fresh and decomposed materials and the mineralogical and physical nature of the residual product.

As will be readily perceived by those at all acquainted with the general literature, the publications of the U. S. Geological Survey, the U. S. National Museum, and the Bulletins of the Geological Society of America have been drawn upon to furnish materials for illustration. The writer is under special obligation to Dr. Milton Whitney of the U. S. Department of Agriculture for many of the mechanical analyses given, and to Mr. L. H. Merrill of the Maine Experiment Station for numerous criticisms and suggestions.

To the late Dr. G. Brown Goode he is indebted for permission to utilize photographs and specimens forming a part of the collections of the National Museum and also for electrotypes of sundry plates and figures in its publications.

GEORGE P. MERRILL.

U. S. NATIONAL MUSEUM, January, 1897.

CONTENTS

PART I

THE CONSTITUENTS, PHYSICAL AND CHEMICAL PROPERTIES, AND MODE OF OCCURRENCE OF ROCKS

		PAGE
I.	INTRODUCTORY: ROCKS DEFINED	1
II.	THE CHEMICAL ELEMENTS CONSTITUTING ROCKS .	4
III.	THE MINERALS CONSTITUTING ROCKS	9
IV.	THE PHYSICAL AND CHEMICAL PROPERTIES OF ROCKS . .	33
	1. The Structure of Rocks, macroscopic and microscopic .	33
	2. The Specific Gravity of Rocks	43
	3. The Chemical Composition of Rocks	44
	4. The Color of Rocks	45
V.	THE MODE OF OCCURRENCE OF ROCKS	49

·PART II

THE KINDS OF ROCKS

GENERALITIES, AND CLASSIFICATION 56

I. IGNEOUS ROCKS: ORIGIN OF, AND CLASSIFICATION; RELATIONSHIP EXISTING BETWEEN PLUTONIC AND EFFUSIVE ROCKS 59
    1. The Granite-Liparite Group 65
    2. The Syenite-Trachyte Group 73
    3. The Foyaite-Phonolite Group 77
    4. The Diorite-Andesite Group 81
    5. The Gabbro-Basalt Group 85
    6. The Theralite-Basanite Group 93

		PAGE
	7. The Peridotite-Limburgite Group	95
	8. The Pyroxenite-Augitite Group	99
	9. The Leucite-Nepheline Rocks	102

II. AQUEOUS ROCKS 105
    1. Rocks formed through Chemical Agencies 105
        (1) Oxides 106
        (2) Carbonates 111
        (3) Silicates 114
        (4) Sulphates 117
        (5) Phosphates 119
        (6) Chlorides 119
        (7) Hydrocarbon Compounds 120
    2. Rocks formed as Sedimentary Deposits 129
        (1) Rocks composed mainly of Inorganic Material . . 131
            (1) The Arenaceous Group: Psammites . . 131
            (2) The Argillaceous Group: Pelites . . . 135
            (3) The Calcareous Group: Calcareous Conglomerate and Breccia 139
            (4) The Volcanic Group: Tuffs 139
        (2) Rocks composed mainly of débris from Plant and Animal Life. Organagenous . . . 141
            (1) The Siliceous Group: Infusorial Earth . 141
            (2) The Calcareous Group: Limestone, Marl, etc. 143
            (3) The Carbonaceous Group: Peat, Lignite, and Coal 148
            (4) The Phosphatic Group 151

III. ÆOLIAN ROCKS 153
    Volcanic Dust; Dune Sands, etc. 153

IV. METAMORPHIC ROCKS 155
    Agencies and Results of Metamorphism and Metasomatosis . 155
    1. Stratified or Bedded 162
        (1) The Crystalline Limestones and Dolomites . 162
    2. Foliated or Schistose 164
        (1) The Gneisses 164
        (2) The Crystalline Schists 168

PART III

THE WEATHERING OF ROCKS

I. THE PRINCIPLES INVOLVED IN ROCK-WEATHERING: Statement of General Problem; Weathering defined; Reference to Authorities and Opinions held 173
   1. Action of the Atmosphere 176
      (1) Nitrogen, Nitric Acid, and Ammonia of the Atmosphere 176
      (2) Carbonic Acid of the Atmosphere . . . 178
      (3) Oxygen of the Atmosphere 180
      (4) Effects of Heat and Cold 180
      (5) Effects of Wind 184
   2. Chemical Action of Water 186
      (1) Oxidation 187
      (2) Deoxidation 187
      (3) Hydration 187
      (4) Solution 189
   3. Mechanical Action of Water and of Ice; Erosion by Water; Daubrée's Experiments; Action of Freezing Water and of Ice 195
   4. Action of Plants and Animals; Effect of Lichens, Mosses, Root Action, Organic Acids, etc.; Solvent Power of Citric Acid; Action of Bacteria; Action of Ants and Termites; Action of Marine Invertebrates; Production of Carbonates . . . 201

II. CONSIDERATION OF SPECIAL CASES 206
   (1) Weathering of Granite, District of Columbia . . . 206
   (2) Weathering of Gneiss, Albemarle County, Virginia . . 214
   (3) Weathering of Elæolite Syenite, Little Rock, Arkansas . 216
   (4) Weathering of Phonolites, Marienfels, Bohemia . . 217
   (5) Weathering of Diabase, Medford, Massachusetts . . 218
   (6) Weathering of Diabase, Venezuela 222
   (7) Weathering of Basalt, Kammar Bull, Bohemia . . 223
   (8) Weathering of Basalt, Haute Loire, France . . 223

	PAGE
(9) Weathering of Diorite, Albemarle County, Virginia . .	224
(10) Weathering of Peridotites and Pyroxenites . . .	225
(a) Serpentine of Harford County, Maryland . . .	226
(b) Soapstones of Albemarle and Fairfax Counties, Virginia	226
(11) Weathering of Clastic Rocks	228
(a) Argillites of Harford County, Maryland . . .	229
(b) Cherts of Missouri and Arkansas	230
(12) Weathering of Limestones, Arkansas	232
(13) Résumé: Importance of Hydration; Loss of Constituents; Relative Durability of Various Minerals; Discussion of Processes involved in Feldspathic Decomposition . .	234

III. THE PHYSICAL MANIFESTATIONS OF WEATHERING . . . 241
(1) Disintegration without Decomposition 241
(2) Weathering influenced by Crystalline Structure . . 243
(3) Weathering influenced by Structure of Rock Masses . . 244
(4) Weathering influenced by Mineral Composition . . 248
(5) Results due to Position 252
(6) Induration on Exposure 254
(7) Changes in Color incidental to Weathering . . . 257
(8) Relative Amount of Material removed in Solution . . 258
(9) Incidental Surface Contours 259
(10) Effacement of Original Characteristics 262
(11) Simplification of Chemical Compounds incidental to Weathering 265
(12) Other Results incidental to Decomposition and Erosion . 266

IV. TIME CONSIDERATIONS 268
(1) Rate of Weathering influenced by Texture . . . 268
(2) Rate of Weathering influenced by Composition . . 269
(3) Rate of Weathering influenced by Humidity . . 270
(4) Rate of Weathering influenced by Position . . . 270
(5) Relative Rapidity of Weathering among Eruptive and Sedimentary Rocks 271
(6) Time Limit of Decay: Post-Cretaceous Weathering of Granite; Weathered Implements of Human Workmanship; Post-Glacial Weathering of Diabase; Post-Jurassic

CONTENTS xv

PAGE

and Post-Pliocene Decay of Rocks of the Sierras; Pre-Palæozoic Weathering of Archæan Rocks . . . 272
(7) Extent of Weathering: In the District of Columbia, Georgia, Missouri, Nicaragua, Brazil, and South Africa . 276
(8) Relative Rapidity of Weathering in Warm and Cold Climates: Opinions hitherto held; Supposed Protective Action of Frost Effects of Forests 278
(9) Difference in Kind of Weathering in Cold and Warm Climates 283
(10) Relative Amounts of Materials lost through Weathering in Hilly and Plains Regions 284

PART IV

TRANSPORTATION AND REDEPOSITION OF ROCK DÉBRIS

1. ACTION OF GRAVITY 286
2. ACTION OF WATER AND ICE 287
3. ACTION OF WIND 292

PART V

THE REGOLITH

I. CLASSIFICATION AND GENERAL DESCRIPTION 299
    1. Sedentary Materials 300
       (1) Residuary Deposits: Residual Sands and Clays; Terra Rossa; Laterite, etc. 301
       (2) Cumulose Deposits: Peat; Muck and Swamp Soils in part; Infusorial Earths 313
    2. Transported Materials 318
       (1) Colluvial Deposits: Talus, Cliff Débris and Material of Avalanches 319
       (2) Alluvial Deposits: Modern Alluvium; Sea-coast Swamps; Loess; Adobe in part; Champlain Clays; Beach Sands and Gravel 320

		PAGE
(3) Æolian Deposits: Wind-blown Sand; Sand Dunes; Volcanic Dust		344
(4) Glacial Deposits: Moraine Material; Eskers; Drumlins, etc.		350
3. The Soil		358
(1) The Chemical Nature of Soils		358
(2) The Mineral Composition of Soils		374
(3) The Physical Condition of Soils		379
(4) The Weight of Soils		382
(5) The Kinds and Classification of Soils		382
(6) The Color of Soils		385
(7) The Age of Soils		387
(8) Soils as Affected by Plant and Animal Life		390

ILLUSTRATIONS

FULL-PAGE PLATES

	FACING PAGE
PLATE 1	*Frontispiece*

Stone Mountain, Georgia. A Residual Boss of Granite. From a photograph by J. K. Hillers.

PLATE 2 33
    Porphyritic and Flow Structures.

PLATE 3 35
    Slaggy and Vesicular Structures.

PLATE 4 38
    Brecciated Structures.

PLATE 5 41
    Microscopic Structures of Rocks.

PLATE 6 65
    Fig. 1. Lithophysæ in Liparite.
    Fig. 2. Cross-section of Stalagmite.
    Fig. 3. Concretionary Aragonite.
    Fig. 4. Pegmatite.

PLATE 7 70
    Fig. 1. Liparite, Nevadite Form.
    Fig. 2. Liparite, Rhyolite Form.
    Fig. 3. Liparite, Obsidian Form.
    Fig. 4. Liparite, Pumiceous Form.

PLATE 8 82
    Fig. 1. Orbicular Diorite.
    Fig. 2. Granite Spheroid.

PLATE 9 107
    Fig. 1. Botryoidal Hematite.
    Fig. 2. Septarian Nodule.

PLATE 10 113
    View in Limestone Cavern.

                                                               FACING PAGE
PLATE 11 130
    Fig. 1. Shell Limestone.
    Fig. 2. Coquina.
    Fig. 3. Crinoidal Limestone.

PLATE 12 143
    Fig. 1. Pisolitic Limestone.
    Fig. 2. Oölitic Limestone.

PLATE 13 164
    Banded and Foliated Gneisses.

PLATE 14 172
    Weathered Granite, District of Columbia. From a photograph
    by George P. Merrill.

PLATE 15 193
    Corroded Limestones.

PLATE 16 199
    Fig. 1. Diorite Boulder split along Joint Planes by Frost.
    Fig. 2. Corroded Surface of Pyroxenic Limestone.
    Fig. 3. Corroded Limestone.

PLATE 17 219
    Weathered Diabase Dike, Medford, Mass. From a photograph by
    G. H. Barton.

PLATE 18 241
    Fig. 1. Exfoliated Granite in the Sierras. From a photograph
    by H. W. Turner.
    Fig. 2. Talus Slopes on Pike's Peak. From a photograph by W.
    H. Jackson.
    Fig. 3. Disintegrated Granite, Ute Pass, Colorado. From a
    photograph by W. H. Jackson.

PLATE 19 248
    Fig. 1. Weathered Schists, Coast of Cape Elizabeth, Maine.
    Fig. 2. Sandstone bored by Bees.
    Fig. 3. Slab of Glaciated Limestone.

PLATE 20 258
    Fig. 1. Weathered Boulder of Oriskany Sandstone.
    Fig. 2. Concentric Weathering in Diabase.
    Fig. 3. Zonal Structure in Weathered Argillite.
    Fig. 4. Weathered Sandstone showing Induration along Joint
    Planes.

ILLUSTRATIONS xix

FACING PAGE

PLATE 21 267
    Fig. 1. Sink-hole near Knoxville, Tennessee. From a photograph by George P. Merrill.
    Fig. 2. Beds of Marble corroded by Meteoric Waters, Pickens County, Georgia.

PLATE 22 285
    Fig. 1. Forest destroyed by Wind-blown Sand. From a photograph by I. C. Russell.
    Fig. 2. Calcareous Conglomerate carved and polished by Wind-blown Sand.
    Fig. 3. Rock being undermined by Wind-blown Sand. After G. K. Gilbert.

PLATE 23 319
    Rock Disintegration and Formation of Talus, Mount Sneffels, Colorado. From a photograph by Whitman Cross.

PLATE 24 345
    Fig. 1. Section of Beds of Leda Clay, Lewiston, Maine. From a photograph by L. H. Merrill.
    Fig. 2. Beds of Volcanic Dust, Reese Creek, Gallatin County, Montana. From a photograph by George P. Merrill.

PLATE 25 357
    Fig. 1. Section of Glacial Till. From a photograph by G. F. Wright.
    Fig. 2. Glaciated Landscape. From a photograph by L. H. Merrill.

Plates 2, 3, 4, 5, 6, 7, 8, 9, 11, 12, 13, 15, 19, and 20, and Fig. 3 Plate 16, and Fig. 2 Plate 22, from specimens in the Geological Department of the United States National Museum.

FIGURES IN TEXT

FIG.   PAGE
1. Augite partially altered into Hornblende 40
2. Mounted Thin Section of Rock 43
3. Microscopic Structure of Muscovite-Biotite Granite, Hallowell, Maine 67
4. Microscopic Structure of Diabase, Weehawken, New Jersey . . 88
5. Microscopic Structure of Peridotite (Porphyritic Lherzolite) . 96
6. Microscopic Structure of Pyroxenite 100

ILLUSTRATIONS

FIG.		PAGE
7.	Microscopic Structure of Oölitic Limestone	112
8.	Pyroxene partially altered into Serpentine	115
9.	Microstructure of Sandstone	131
10.	Section through Lake Basin, showing Bed of Infusorial Earth	142
11.	Microstructure of Oölitic Limestone	144
12.	Microstructure of Fossiliferous Limestone	145
13.	Microstructure of Quartzite	158
14.	Microstructure of Crystalline Limestone	163
15.	Microstructure of Gneiss	165
16.	Microstructure of Quartzite	169
17.	Influence of Joints in the Production of Boulders	244
18.	Exfoliation of Granite, Stone Mountain, Georgia	245
19.	Concentric Exfoliation of Granite, Canada	246
20.	Microstructure of Sandstone, with Large Absorptive Power	269
21.	Microstructure of Diabase, with relatively Little Absorptive Power	269
22.	Flint Implement showing Weathered Surface	274
23.	Sketch showing Pre-Palæozoic Decay of Rocks	276
24.	Diagram showing Direction and Rate of Motion of Soil	287
25.	Diagram showing Flood Plain of River	289
26.	Angular Outlines of Particles in Residual Soil from Gneiss	301
27.	Section across Central Kentucky, showing Inherited Characteristics of Soils	303
28.	Angular Quartz Particles from Decomposed Gneiss	304
29.	Outlines of Kaolinite Crystals and Kaolin Particles	309
30.	Section across Small Lake	314
31.	Talus Slopes	319
32.	Alluvial Plains	323
33.	Outlines of Particles in Chinese Loess	329
34.	Particles washed from Leda Clays	335
35.	Cross-section of Marine Marsh	338
36.	Quartz Granules in Beach Sand	343
37.	Outlines of Particles of Glass in Volcanic Dust	349
38.	Section through Carboniferous Soil	386
39.	Section showing Varying Character of Residual Soil	387
40.	Section through Ant Nest	390
41 and 42.	Sections showing the Effect of Tree Roots in Soil	395

Fig. 1, after G. W. Hawes; 5 and 6, after G. H. Williams; 18 and 22, after Robert Bell; 10, 23, 24, 26, 29, 30, 31, 34, 37, 38, 39, 40, and 41, after Shaler, Twelfth Annual Report United States Geological Survey, 1890–1891.

ROCKS, ROCK-WEATHERING, AND SOILS

PART I

THE CONSTITUENTS, PHYSICAL AND CHEMICAL PROPERTIES, AND MODE OF OCCURRENCE OF ROCKS

I. INTRODUCTORY

A ROCK is a mineral aggregate; more than this, it is an essential portion of the earth's crust, a geological body occupying a more or less well-defined position in the structure of the earth, either in the form of stratified beds, eruptive masses, sheets or dikes, or in that of veins and other chemical deposits of comparatively little importance as regards size and extent. In giving this definition, origin, chemical composition, and state of aggregation of the individual particles are for the time ignored. From a strictly geological standpoint, the beds of loose sand, and even the water of the ocean itself, may be considered as rocks, and either, under favorable circumstances, may undergo a process of induration such as shall be productive of the condition of solidity commonly ascribed to rocks by the popular mind.

In ever-varying conditions as regards compactness, color, texture, and structure, rocks form the entire mass of the globe so far as it is as yet made known to us, with the exception of a scarcely appreciable proportion of organic matter. It is rock which forms the substance of mountain ranges and the vast stretches of valley and plain. It is from the rocks that we gain our food, our fuel, and the supplies of metal which are seemingly so essential to our well-being; we cannot ignore them, even if we would. We borrow from the rocks that which is essential to our life to-day, but when that brief day is ended return it once more, with neither loss nor gain, to its original source.

Those portions of the earth's crust which are available for study comprise at best but a few thousand vertical feet, though from the fact that the stratified rocks have been so extensively thrown out of their original, horizontal position, and again eroded, we are enabled to measure their thickness, and may hence claim to know with a reasonable degree of accuracy the character of the material forming this crust down to a depth of perhaps twenty miles.[1] Throughout all this vast thickness, comprising millions upon millions of cubic feet, in weight far beyond all comprehension, we find a constant recurrence of materials alike in composition and similarity in origin to those upon the immediate surface. There is at times, as noted later, a difference in structure due to metamorphism, between the older, deeper lying portions and those more recent, but the ultimate composition is essentially the same, and all the knowledge thus far gained points to a wonderful unity in nature's methods, and shows with seeming conclusiveness that the geological agencies of the past, the methods by which rocks were made and again destroyed, differed in no essential particular from those in progress to-day. What these processes were, how they operated, and with what results, it shall be our aim to here set forth.

Among the many interesting, and at first thought seemingly unaccountable, things we shall encounter in the progress of our work, not the least is the fact that so large a proportion of natural objects are more or less out of harmony with their surroundings. Throughout life every organic being is in a constant struggle with the elements to preserve that life, fulfil all its functions, and gratify its natural desires. No sooner does life depart than decomposition and disintegration ensue. As with organic beings, so with inorganic substances. Every mass of rock pushed up by the faulting and folding of the earth's crust, exposed by denudation, or erupted as molten matter from the earth's interior, finds almost at once that its various elements, in their existing combinations, are not in har-

[1] The total mean depth of the fossiliferous formations of Europe as stated by Geikie ('Text-book of Geology, p. 675) has been set down as 75,000 feet.

mony with their environment. The summer's heat and winter's cold, the chemical action of atmospheres and acidulated rains, combine their forces; a breaking up ensues, to be succeeded by new combinations and perhaps reconsolidations more in keeping with the then existing circumstances. An intermediate product in all this endless cycle of change, of disintegration and recombination, is a comparatively thin, superficial mantle of loose débris, which, mixed with more or less organic matter, nearly everywhere covers the land, and by its combined chemical and mechanical properties furnishes food and foothold for myriads of plants, and hence, indirectly, sustenance for man and beast as well. In brief, what is commonly known as soil is but disintegrated and more or less decomposed rock material, intermingled, perhaps, with organic matter from plant decay. Such being the case, a study of the processes of rock weathering and the transportation, deposition, and physical properties of the resultant débris, is but a study of the origin of soils on the broadest and most comprehensive basis, and soils themselves may justly be regarded as secondary rocks in a state of incomplete consolidation. Their study belongs, therefore, as legitimately to the realm of geology as does that of any subject relating to rock formation or other phases of the earth's history.

Accepting the above, we will begin our studies by a consideration of (1) the elements which in their single or combined state make up the minerals; (2) the minerals which make up the rocks; (3) the rocks themselves, with particular reference to their mineralogical and chemical natures; (4) the breaking down or degeneration of rocks through processes in part chemical and in part mechanical; and (5) the result of this clasmatic process as manifested in the production of clay, sand, gravel, and incidental soil. There are other points which will be touched upon more briefly, in order to make our work systematic, as the action of wind and water in assorting and redepositing rock débris and tending to reduce the land surface to one general level.

II. THE CHEMICAL ELEMENTS CONSTITUTING ROCKS

Although there are 69 elements now known, but 16 occur in any abundance or form more than an extremely small proportion of the material of the earth's crust. Indeed, of this number probably fully one-half, taken collectively, will not constitute more than 4 or 5% of the earth's crust so far as known. These 16, arranged according to their chemical properties and order of their abundance, are as follows: oxygen, silicon, carbon, sulphur, hydrogen, chlorine, phosphorus, fluorine, aluminum, calcium, magnesium, potassium, sodium, iron, manganese, and barium. The eight more important, with their approximate percentage amounts as given by Roscoe and Schorlemmer,[1] are as below: —

Oxygen	44.0 to 48.7%
Silicon	22.8 to 36.2
Aluminum	9.9 to 6.1
Iron	9.9 to 2.4
Calcium	6.6 to 0.9
Magnesium	2.7 to 0.1
Sodium	2.4 to 2.5
Potassium	1.7 to 3.1

It must not for a moment be imagined, however, that these elements exist for the most part in a *free* or uncombined state: on the contrary, in the majority of cases so great is their affinity for one another that it is only momentarily, or under abnormal conditions, that they are met with at all in this elementary form. Those elements which are most common in the free state, though even these occur more commonly combined with others, are, (1) the gas oxygen, and (2) the solids, carbon, sulphur, and, more rarely, iron. Still more rarely, and under such abnormal conditions, as exist during volcanic eruptions, are found the free gases, hydrogen, chlorine, and fluorine. The gas nitrogen, although so abundant a constituent of the atmosphere,

---

[1] Treatise on Chemistry, Vol. I, p. 55, 1878.

is, as a primary constituent of the earth's crust, almost wholly unknown, and needs no consideration at this stage of our work.

Oxygen, as is well known, is the active, even the aggressive, principle of the atmosphere, of which it constitutes about one-fifth by bulk. Combined with other elements, it is, however, of vastly greater geological importance, being estimated, as noted above, to constitute from 44 to 48.7% of the entire mass of the earth's crust; that is to say, could the earth's crust be once more resolved into its original elements, the oxygen thus liberated would be found very nearly equal to all the other elements taken together. The simpler forms of oxygen compounds are known as oxides, and of these the oxide of hydrogen, water (H_2O), is by far the most common, and, anomalous as it may at first seem, is a true mineral and to be classed as an anhydrous oxide at that. Aside from being so essential to human life, oxygen, as will be noted later, is a very potent factor in the manifold changes which are constantly taking place in the more superficial portions of the earth's crust.

Silicon. — Next to oxygen silicon is the most abundant of the earth's constituents, though it exists only in combination, either as an oxide (SiO_2), or with other elements to form *silicates*. In these two forms it is the predominating constituent in all but the calcareous rocks. As silica (SiO_2), or quartz, it forms one of the most indestructible of natural compounds, and hence is to be found as the prevailing constituent in nearly all sands and soils.

Aluminum is next to oxygen and silicon probably the most important element when regarded from our present standpoint. It occurs mainly in combination with silicon and oxygen, forming an important series of minerals known as aluminous silicates. As a sesquioxide it is well known in the minerals corundum and beauxite.

Iron, although less abundant than either oxygen or silica, occupies a very important place as a rock constituent, owing to the variety of compounds of which it forms a part, as well as to the decided colors which are characteristic of its oxides and of the iron-bearing silicates. The most conspicuous forms of iron on the immediate surface of the earth are the oxides, but which at greater depths, or where the atmosphere has as yet exercised less influence, give way to carbonates, sulphides, and silicates.

Iron, although so common in combination with other elements, occurs but rarely free, owing to its affinity for oxygen. It is possible that far below the surface, beyond the reach of meteoric waters and atmospheric air, it is to be found in a metallic state much more abundantly, but of this we have no other proof than that the specific gravity of the globe, in its entirety, is much greater than that of the most dense minerals which constitute its outer portion. The inference seems unavoidable that at great depths some of these elements exist uncombined, and in a state of greater molecular density than at the surface.

Calcium is a very important element of the earth's crust, although, as we have seen, it has been estimated to compose only about one-sixteenth of its mass. Its most conspicuous form of occurrence is in combination with carbon dioxide, forming the mineral calcite ($CaCO_3$), or the rock limestone. In this form it is slightly soluble in water containing carbonic acid, and hence has become an almost universal ingredient of all natural waters, whence it furnishes the lime necessary for the formation of shells and skeletons of the various tribes of mollusca and corals. In combination with sulphuric acid, calcium forms the rock gypsum. It is also an important constituent of many silicates.

Magnesium is found in combination with carbonic acid as carbonate, forming thus an essential part of the rock dolomite. The bitter taste of sea-water and some mineral waters is due to the presence of salts of magnesia. In combination with silica as a *silicate* it forms an essential part of such rocks as serpentine, soapstone, and talc.

Potassium combined with silica is also an important element in many mineral silicates, as orthoclase, leucite, and nepheline. In smaller amounts it is found in silicates of the mica, amphibole, and pyroxene groups. The following table will serve to show the varying amounts of potash (K_2O) in rocks of various kinds: —

Granite	2.6 to 6.50%
Diorite	0.1 to 2.42%
Basalt	0.058 to 0.50%
Gabbro	0.00 to 0.93%
Limestone	0.10 to 1.22%
Sandstone	0.00 to 3.30%
Slate (fissile argillite)	0.00 to 3.83%

As a chloride, potassium is invariably present in sea-water, and as a nitrate it forms the rare, but valuable mineral nitre, or saltpetre.

Sodium. — The most common and wide-spread form of the element sodium is the compound with chlorine known as sodium chloride (NaCl) or common salt. In this form it is the most abundant of the salts occurring in sea-water, and constitutes also rock masses of no inconsiderable dimensions interstratified with other rocks of the earth's crust. Combined with silica, lime, and alumina, sodium is an important constituent of the soda-lime feldspars, and of numerous other silicate minerals. In the form of carbonate and sulphate it occurs as an incrustation on the surface, or disseminated throughout the soils in poorly drained portions of arid countries, giving rise to the so-called "alkali soils," for which such regions are frequently noted. As a nitrate, sodium occurs in the desert regions of Chili, forming the soda nitre so valuable for fertilizing purposes.

Manganese is, next to iron, the most abundant of the heavy metals, occurring as oxide, carbonate, or in combination with two or more other elements as a silicate.

Barium is found mainly combined with sulphuric acid, to form the mineral barite or heavy spar. It sometimes occurs as a carbonate, and more rarely as a silicate.

Phosphorus, although existing in comparatively insignificant proportions, is nevertheless an important element, though in nature it occurs only in combination with various bases, principally lime, to form phosphates. In this form it is found in the bones of animals, the seeds of plants, and constitutes the essential portions of the minerals apatite and phosphorite. Though small in proportion, phosphorus is a very important constituent of any fertile soils. Its chief source, in the older, crystalline rocks, is the mineral apatite, as noted later. As found in the secondary rocks, as limestones and marls, it is evidently derived from animal remains. (See p. 151.) Analyses have shown that the amount of phosphorus, in the form of phosphoric anhydride (P_2O_5), in rocks rarely exceeds 1%, and usually falls much lower, being most abundant in the basic eruptive rocks, as diorites and gabbros, and most lacking in the siliceous fragmentals, as sandstones and slates. The following table will serve to show the small percentages of this constituent in rocks of various kinds: —

Granite	0.07 to	0.25%
Diorite	0.18 to	1.06%
Basalt	0.03 to	1.18%
Limestone	0.06 to	10.00%
Shale	0.02 to	0.25%
Sandstone	0.00 to	0.1 %

Of the solid elements occurring *free*, or uncombined, carbon is by far the more abundant, being found in the forms known as diamond and graphite, or when quite impure as coal. In combination as a dioxide (CO_2), it forms the well-known carbonic acid gas, which, like oxygen, is a powerful agent in bringing about important changes in the rocks with which it comes in contact. Free sulphur occurs more rarely, being as a rule a product of volcanic activity, or due to the reduction of the sulphides and sulphates of the metal with which it more commonly exists in combination.

III. THE MINERALS CONSTITUTING ROCKS

A rock, as previously stated, is a mineral aggregate. As a rule, the number of mineral species constituting any essential portion of a rock is very small, seldom exceeding three or four. In common crystalline limestones, the only essential constituent is the mineral calcite; granite, on the other hand, is, as a rule, composed of minerals of three or four independent species. As has been elsewhere stated, the mineral composition of rocks in general is greatly simplified by the wide range of conditions, under which the commonest minerals can be formed, thus allowing their presence in rocks of all classes and of whatever origin. Thus quartz, feldspar, mica, the minerals of the hornblende or pyroxene group, can be formed in a mass cooling from a state of fusion; they may be crystallized from solution, or be formed from volatilized products. They are therefore the commonest of minerals and rarely excluded from rocks of any class, since there is no process of rock formation which determines their absence. Moreover, most of the common minerals, like the feldspars, micas, hornblendes, pyroxenes, and the alkaline carbonates, possess the capacity of adapting themselves to a very considerable range of compositions. In the feldspars, for example, the alkalies, lime, soda, or potash may replace each other almost indefinitely, and it is now commonly assumed that true species do not exist, all being but isomorphous admixtures passing into one another by all gradations, and the names albite, oligloclase, anorthite, etc., are to be used only as indicating convenient stopping and starting points in the series. Hornblende or pyroxene, further, may be pure silicates of lime and magnesia, or iron and manganese may partially replace these substances. Lime carbonate may be pure, or magnesia may replace the lime in any proportion. These illustrations are sufficient to indicate the reason of the great simplicity of rock masses as regards their chief constituents, and that whatever may be the composition of a mass within nature's limits, and whatever may be the conditions of its origin, the probabilities are that it will be formed essentially of one or more of a half a dozen minerals in some of their varieties.

But however great the adaptability of these few minerals may be, they are, nevertheless, subject to very definite laws of chemical equivalence. There are elements which they cannot take into their composition, and there are circumstances which retard their formation while other minerals may be crystallizing. In a mass of more or less accidental composition it may, therefore, be expected that other minerals will form in considerable numbers, but minute quantities. It is customary to speak of those minerals which form the chief ingredients of any rock, and which may be regarded as characteristic of any particular variety, as the *essential* constituents, while those which occur in but small quantities, and whose presence or absence does not fundamentally affect its character, are called *accessory* constituents. The accessory mineral which predominates, and which is, as a rule, present in such quantities as to be recognizable by the unaided eye, is the *characterizing* accessory. Thus a biotite granite is a stone composed of the essential minerals quartz and potash feldspar, but in which the accessory mineral biotite occurs in such quantities as to give a definite character to the rock. The minerals of rocks may also be conveniently divided into two groups, according as they are products of the first consolidation of the mass or of subsequent changes. This is the system here adopted. We thus have:—

(1) The original or primary constituents, those which formed upon its first consolidation. All the essential constituents are original, but, on the other hand, all the original constituents are not essential. Thus, in granite, quartz and orthoclase are both original and essential, while beryl and zircon or apatite, though original, are not essential.

(2) The secondary constituents are those which result from changes in a rock subsequent to its first consolidation, changes which are due in great part to the chemical action of percolating water. Such are the calcite, chalcedony, quartz, and zeolite deposits which form in the druses and amygdaloidal cavities of traps and other rocks.

Below is given a list of the more important rock-forming minerals, arranged as above indicated. Although these are sufficiently described as regards their chemical and crystallographic properties in any of the mineralogies, it has seemed advisable to devote some space here to a reconsideration of those most prominent as rock constituents, in order that the individual characteristics of the rocks of which they form a part may be better understood. In passing them in review we will also note briefly the characteristic alteration and decomposition products to which they give rise, though the cause of such changes must be left for another chapter.

A. Original Minerals.

1. Quartz.
2. The Feldspars.
   2 a. Orthoclase.
   2 b. Microcline.
   2 c. Albite.
   2 d. Oligoclase.
   2 e. Andesite.
   2 f. Labradorite.
   2 g. Bytownite.
   2 h. Anorthite.
3. The Amphiboles.
   3 a. Hornblende.
   3 b. Tremolite.
   3 c. Actinolite.
   3 d. Arvedsonite.
   3 e. Glaucophane.
   3 f. Smaragdite.
4. The Monoclinic Pyroxenes.
   4 a. Malacolite.
   4 b. Diallage.
   4 c. Augite.
   4 d. Acmite.
   4 e. Ægerite.
5. The Rhombic Pyroxenes.
   5 a. Enstatite (Bronzite).
   5 b. Hypersthene.
6. The Micas.
   6 a. Muscovite.
   6 b. Biotite.
   6 c. Phlogopite.
7. Calcite (and Aragonite).
8. Dolomite.
9. Gypsum.
10. Olivine.
11. Garnet.
12. Epidote.
13. Zoisite.
14. Andalusite.
15. Staurolite.
16. Scapolite.
17. Elæolite and Nepheline.
18. Leucite.
19. Sodalite.
20. Hauyn (nosean).
21. Apatite.
22. Menaccanite.
23. Magnetite.
24. Hematite.
25. Chromite.
26. Halite (common salt).
27. Fluorite.
28. Graphite.
29. Carbon.
30. Pyrite.

B. Secondary Minerals.

1. Quartz.
   1 a. Chalcedony.
   1 b. Opal.
   1 c. Tridymite.

2. Albite.
3. The Amphiboles.
   3 a. Hornblende.
   3 b. Tremolite.
   3 c. Actinolite.
   3 d. Uralite.
4. Muscovite (Sericite).
5. The Chlorites.
   5 a. Jefferisite.
   5 b. Ripidolite.
   5 c. Penninite.
   5 d. Prochlorite.
6. Calcite (and aragonite).
7. Wollastonite.
8. Scapolite.
9. Garnet.
10. Epidote.
11. Zoisite.
12. Serpentine.
13. Talc.
14. Glauconite.
15. Kaolin.

16. The Zeolites.
   16 a. Pectolite.
   16 b. Laumontite.
   16 c. Phrenite.
   16 d. Thomsonite.
   16 e. Natrolite.
   16 f. Analcite.
   16 g. Datolite.
   16 h. Chabazite.
   16 i. Stilbite.
   16 k. Heulandite.
   16 l. Phillipsite.
   16 m. Ptilolite.
   16 n. Mordenite.
   16 o. Harmotome.
17. Hematite.
18. Limonite.
19. Göthite.
20. Turgite.
21. Pyrite.
22. Marcasite.

Quartz. —*Composition:* Pure silica, SiO_2; specific gravity 2.6; hardness, 7.[1]

This is one of the commonest and most widely distributed minerals of the earth's crust, and forms an essential constituent in a variety of eruptive and sedimentary rocks, such as granite,

---

[1] For convenience in determining minerals, the "scale of hardness" given below has been adopted by mineralogists. By means of it one is enabled to designate the comparative hardness of minerals with ease and definiteness. Thus, in saying that serpentine has a hardness equal to 4, is meant that it is of the same hardness as the mineral fluorite, and can therefore be cut with a knife, but less readily than calcite or marble.

1. *Talc:* Easily scratched with the thumbnail.
2. *Gypsum:* Can be scratched by the thumbnail.
3. *Calcite:* Not scratched by the thumbnail, but easily cut with a knife.
4. *Fluorite:* Can be cut with a knife, but less easily than calcite.
5. *Apatite:* Can be cut with a knife, but only with difficulty.
6. *Orthoclase feldspar:* Can be cut with a knife only with great difficulty and on thin edges.
7. *Quartz:* Cannot be cut with a knife; scratches glass.
8. *Topaz:* Will scratch quartz.
9. *Corundum:* Will scratch topaz.
10. *Diamond:* Will scratch corundum.

quartz porphyry, liparite, gneiss, mica schist, quartzite, and sandstones. In the granites, gneisses, and schists it occurs in the form of irregular granules destitute of crystal outlines. In the quartz porphyries and liparites it is found as a porphyritic constituent, usually with well-defined crystal outlines, which may however have become more or less obliterated through the corrosive action of a molten magma. (See Fig. 3, Pl. 5.) In the secondary rocks, quartzite and sandstone, the quartz occurs as more or less rounded or irregularly angular grains without crystal outlines, except it may be through a secondary deposition of silica, as explained on p. 158. Quartz is the hardest and most indestructible of the common constituents, and hence when rocks containing it decompose and their débris becomes exposed to combined chemical and mechanical agencies, it remains unaltered to the very last, forming the chief constituent of beds of sand and gravel, which in turn may become transformed into sandstones, quartzites, or conglomerates.

Quartz is usually easily recognized, either under the microscope or by the unaided eye, by its clear, colorless appearance, irregular, glass-like fracture, — having no true cleavage, — hardness, and insolubility in any acids but hydrofluoric. Under the microscope it appears in clear, pellucid grains, often highly charged with minute cavities filled with liquid and gaseous carbonic acid, the latter like the bubble in a spirit level, dancing about from side to side of its minute chamber as though endowed with life. In other cases the cavity may be filled with a saline solution from which has separated out a minute cube of common salt.

As a secondary constituent quartz occurs, filling veins and cracks in other rocks, and in the impure crypto-crystalline and amorphous forms known as chalcedony, chert, flint, opal, hyalite, and agate is found as an infiltration product in the cavities of many trappean rocks, in lenticular and oval concretionary masses in limestones, and replacing the organic matter of wood and other organisms. The name *tridymite* is given to a quartz occurring in minute, usually microscopic, tablets in cavities in volcanic rocks, particularly the more acid varieties. (See further on p. 71.)

The Feldspars. — Hardness, 5 to 7; specific gravity, 2.5 to 2.8. The feldspars are essentially anhydrous silicates of aluminum, with varying amounts of lime, potash, or soda, and rarely barium. They have in common the characteristics of two easy cleavages inclined to one another at an angle of 90°, or nearly 90°; close relationship in optical properties; similarity in colors, which vary from clear and transparent through white, yellowish pink, and red; more rarely greenish, and often opaque through impurities or decomposition; and lastly, a constant intergradation in composition, as already noted on p. 9.

Nine varieties of feldspar are commonly recognized, which on crystallographic grounds are divided into two groups: the first, crystallizing in the monoclinic system, including orthoclase and hyalophane; and the second, crystallizing in the triclinic system, including microcline, anorthoclase, and the albite-anorthite series albite, oligoclase, andesine, labradorite, and anorthite.

The Monoclinic Feldspars: Orthoclase (Sanidin), Potash Feldspars. — *Composition:* $K_2Al_2Si_6O_{16}$ = silicia, 64.7%; alumina, 18.4%; potash, 16.9%.

This is one of the commonest and most abundant of feldspars, and forms an essential constituent of the acid rocks, such as granite, gneiss, syenite, and the orthoclase and quartzose porphyries; more rarely it occurs as an accessory in the more basic eruptives. Under the name *sanidin* is included the clear glassy variety of orthoclase occurring in tertiary and modern lavas, such as trachyte, phonolite, and the liparites.

Among the older rocks orthoclase not infrequently occurs in very coarse pegmatitic crystallizations with quartz and mica, and is quarried for utilization in pottery manufacture. As a rock constituent the potash feldspars are of primary importance, imparting by their preponderance, not merely color and important structural features, but on their decomposition yielding up the alkali potash, valuable for plant food, and the mineral kaolin so essential for porcelain ware, or in its impure state, as clay for pottery and brick making. In the thin sections, under the microscope, the orthoclase of the older rocks is, as a rule, found to be quite opaque, or at least muddy, through impurities or incipient kaolinization. In many eruptives it has been one of the first minerals to separate out from the molten magma, and shows, therefore, more or less well-defined crystallographic boundaries — is *idiomorphic,* to use a more technical term. A well-defined zonal structure is frequently observed, which is due to interrupted periods of growth, and not infrequently to a gradual change in the character of the magma, whereby the outer zones are more or less translucent or opaque from impurities. Twin structure is very common after what is known as the Carlsbad law, and when the crystals are of sufficient size is easily recognized by the unequal reflection of the light from the two sides of a crystal on a cleavage surface.

The Triclinic Feldspars. — The chemical relationship existing between the triclinic feldspars is shown in the following table: —

	SiO_2	Al_2O_3	K_2O	Na_2O	CaO
Microcline	65.00%	18.00%	17.00%	....	....
Albite	68.00	20.00	....	12.00%	~~12.00%~~
Oligoclase	62.00	24.00	....	9.00	5.00
Labradorite	53.00	30.00	....	4.00	13.00
Anorthite	43.00	37.00	....	....	20.00

Considering only the last four of these, as arranged, it will be noted that they become gradually poorer in the acid element silicia, and richer in alumina and other bases; that is, they become more basic. Also that albite carries some 12 % of soda and no lime; that oligoclase carries 9 % of soda and 5 % of lime; labradorite but 4 % of soda and 13 % of lime, while anorthite, the most basic of all, has no soda, and carries 20 % of lime. They have hence come to be known, respectively, as soda feldspar, soda-lime feldspar, lime-soda feldspar, and lime feldspar. As a matter of fact, however, these varieties all grade into one another, through the replacing power of the various elements, and are regarded, not as true species, but rather as isomorphous admixtures, forming what is known as the *albite-anorthite series*.

Their distinction, either in hand specimens by the unaided eye, or in thin sections by the microscope, is a matter of considerable difficulty, and as in addition to other characteristics they have in common two eminent cleavages occurring at oblique angles, it has become customary to group all under he general term of *plagioclase*, a name derived from two

Greek words signifying oblique and fracture. We can then treat of the subject under the heads of (1) microline and (2) plagioclase.

(1) **Microcline (Triclinic Potash Feldspar).** — As a rock constituent, this feldspar is in every way nearly, if not quite, identical with orthoclase, from which it can be distinguished only in thin sections under the microscope. Its composition, manner of occurrence, and associations are those of orthoclase, and need not be repeated here. Anorthoclase is a triclinic soda-potash feldspar of a form closely resembling that of orthoclase and which for all present purposes may be regarded as orthoclase in which soda replaces a considerable proportion of the potash.

(2) **The Plagioclases.** — With the exception of albite the plagioclases are all prominent and essential constituents of the basic eruptives. As a rule they are recognizable only as feldspars by the unaided eye, and recourse must be had to the microscope or to chemical tests for their final determination. Examined in thin sections and by polarized light, they almost invariably show a beautiful parallel banding in light and dark colors, which is due to multiple twinning, the alternate bands becoming light and dark in turn as the stage of the microscope is revolved. When the crystals are of sufficient size, this twinning is sometimes evident in the form of fine straight, parallel bands, or striæ, but in rock masses, as already noted, recourse must be made to microscopic methods. In form the plagioclase of effusive rocks is most frequently slender and elongated, *lath-shaped*, as commonly described, and often with very perfect crystal outlines. In the norites and gabbros, they are often short and stout, imparting a granular character to the rock. They occur frequently in crystals of two or more generations, of which the earlier formed are usually the largest and best developed. The common forms are described in detail below: —

(1) *Albite*, or soda feldspar, occurs as an original constituent in many granites in company with orthoclase; it is also found in gneiss, the crystalline schists, and not infrequently in diorite, phonolite, trachyte, and other eruptives. (2) *Oligoclase*, a soda-lime feldspar, occurs like albite in the acid eruptives like granite and quartz porphyry, but is also a common constituent of diorite, and the younger eruptives such as trachyte, the ande-

sites, and more rarely of the diabases. It is also a constituent of many gneisses. (3) *Labradorite*, or lime-soda feldspar, is a prominent constituent of the basic eruptives of all geological ages, such as the norites, diabases, diorites, and basalts. Andesine and bytownite are closely allied varieties of similar habit, the first being a trifle more acid, and the second more basic than labradorite. (4) *Anorthite*, or lime feldspar, is also a prominent and important constituent of the basic eruptives, and has been found in meteorites and terrestrial peridotites.

On account of their abundance and wide distribution, as well as on account of the character of their decomposition products, the feldspars are to be considered as among the most important of rock constituents. As it is from the débris of the older feldspathic rocks that have been made up a large proportion of all the sedimentaries of more recent date, so too it may be claimed that from the decomposition of this feldspathic constituent has been derived a large share of the salts of potash, lime, and soda, as well as aluminous silicates which form so essential a portion of the soils. The method of feldspathic decomposition as commonly understood is given on p. 237.

This decomposition usually manifests itself by a whitening of the mass, accompanied by opacity and a general softening, whereby it falls away to loose powder unless confined. As seen in thin sections under the microscope, the decomposition goes on most rapidly along lines of cleavage, naturally attacking the outer portions first, so that the crystals show fresh unaltered cores surrounded by opaque and "muddy" borders. In cases where the feldspars carry iron this usually makes its presence known by a reddening or browning of the mass, due to oxidation. In presence of abundant carbonic acid, the liberated iron may enter into combination as a carbonate and the color remain unchanged.

Daubree, who submitted feldspathic fragments to trituration in revolving cylinders of stone and iron, found that in all such cases not merely were the particles worn down to the condition of fine silt, but that there was an actual decomposition, whereby a certain proportion of the alkalies in the form of soluble silicates were formed in the water with which the cylinders were partially filled. When the trituration was carried on in iron cylinders, a certain amount of iron oxides were formed which combined with the silica of the alkaline silicate, leaving the alkali itself free. As in nearly all decomposing rocks there exists more or less of iron oxides from decomposing ferruginous minerals, it is not impossible that a similar reaction is there going on.

The production of kaolin through feldspathic decomposition has become so well recognized that it is customary to speak of this form of decomposition as *kaolinization*, a term which we shall have frequent cause to use as we proceed.

It should be noted that orthoclase, though so frequently found muddied and impure, apparently in an advanced stage of decomposition, does not in reality decompose so readily as the plagioclase (soda-lime) varieties. This fact has been noted by Lemberg,[1] who states that the apparent decomposition may be due to physical causes, as disintegration, inclusions of some easily decomposable silicate, or to originally water-filled cavities whose contents have been absorbed through the formation of secondary hydrous silicates.

Leucite. — *Composition:* Silica, 55.0 %; alumina, 23.5 %; potash, 21.5 %.

Leucite occurs as an original and essential constituent of many volcanic rocks, such as leucitophyre, leucotephrite, and leucitite. More rarely it occurs in trachyte. It is a common associate of nepheline in recent lavas, and has been found associated with elæolite in the elæolite syenites of Hot Springs, Arkansas. When well developed it shows polyhedral, garnet-like outlines.

Leucite as a rock constituent is not an abundant mineral except in rare instances. Its chief interest, from our present standpoint, lies in its high percentage of potash which must become available as plant food on decomposition. Leucite is a common constituent of certain Vesuvian lavas, and it is not improbable that this fact may account in part for the well-known fertility of the soils of this region, though naturally climatic influence has much to do.

Nepheline (Elæolite). — These names are given to what are varietal forms of one and the same mineral. In composition they are silicates of alumina, soda, and potash of the formula $(NaK)_2Al_2Si_2O_8 =$ silica, 41.24; alumina, 35.26; potash, 6.46; soda, 17.04.

[1] Zeit. Deut. Geol. Gesellschaft, 35, 1883.

Nepheline occurs in Tertiary and post-Tertiary eruptive rocks, and is an essential constituent of phonolite, tephrite, and nephelinite. Secondary nepheline has been found in the ejected volcanic blocks found in the lava of Mount Somma. The variety elæolite occurs only in older rocks, and is an essential constituent of elæolite syenite. Cancrinite is a yellowish granular mineral, in some cases apparently resulting from the alteration of elæolite, with which it occurs.

Both nepheline and elæolite gelatinize readily with hydrochloric acid, and the powdered rock when treated on a glass slide with this acid yields abundant microscopic cubes of sodium chloride. This is one of the easiest of microchemical tests for the determination of the mineral. Nepheline occurs as a rule in well-defined short and stout hexagonal prisms, which in longitudinal sections show up as short, colorless rectangular areas extinguishing parallel with the sides of the prism. Elæolite differs in being more opaque and occurring in less well-defined, more granular forms. When occurring in sufficient abundance in a rock mass it is readily recognized by its characteristic greasy appearance. The mineral undergoes a ready alteration, giving rise to zeolitic minerals and on ultimate decomposition through weathering, yielding a rich and fertile soil.

The Amphiboles. — *Composition:* Two principal varieties are recognized. (1) Non-aluminous, consisting mainly of the meta-silicates of magnesium and calcium, with 55 to 59% of silica, 21 to 27% of magnesia, 11 to 15% of lime, and small proportions of protoxides of iron and manganese. Under this head are included the white, gray, and pale green, often fibrous forms, as tremolite, actinolite, and asbestos. (2) Aluminous, containing silica, 40 to 51%; magnesia, 10 to 23%; alumina, 6 to 14%; lime, 10 to 13%; ferrous and ferric oxides, 12 to 20%. Here are included the dark green, brown, and black varieties.

The aluminous variety, common hornblende, is an original and essential constituent of diorite, and of many varieties of granite, gneiss, syenite, schist, andesite, and trachyte, and is also present as a secondary constituent in many rocks, resulting from the molecular alteration of the augite. The non-aluminous varieties occur in gneiss, crystalline limestone, and other metamorphic rocks.

By the unaided eye, or by means of blowpipe tests, it is often impossible to distinguish the minerals of this group from the pyroxenes. In the thin sections this distinction is, however, a matter of comparative ease, basal sections showing not merely a greater development of prismatic faces, but also cleavages cutting at angles of 66° and 124° instead of nearly at right angles, as in the latter. Green fibrous hornblendes frequently result from the molecular alteration of augite, and all varieties are susceptible of alteration into chloritic and ferruginous products with the separation of calcite. In the recent lavas it is a common occurrence to find the hornblendes surrounded by a black border, or wholly changed by corrosion of the molten magma into an aggregate of small black opaque granules, which in certain instances have been proven to be augites.

On decomposing, the amphiboles give rise to ferruginous and aluminous or magnesian products, as do the pyroxenes, next to be described. With the darker colored varieties, the decomposition begins with hydration and the peroxidation of the iron along lines of cleavage and fracture, whereby the crystal becomes riddled with corroded areas filled with the liberated iron in the form of hydrated sesquioxide.

When the disintegration is complete, the whole mass is converted into an ochre-brown, earthy substance. These chemical changes are indicated in the following analysis of I. fresh, and II. decomposed hornblende from Haavi on Fillefjeld, Norway:[1]—

	I		II
Silica	45.37	......	40.32
Alumina	14.81	......	17.49
Iron protoxide	8.74	Iron peroxide	18.26
Manganese	1.50	......	2.14
Lime	14.91	......	5.37
Magnesia	14.33	......	9.23
Water	....	......	8.00
	99.66		100.81

The most striking features of the above analyses are (1) the complete conversion of the protoxides into sesquioxides, (2) the loss in lime and magnesia which have presumably

[1] Bischof's Chemical Geology, Vol. II, p. 354.

been carried away in the form of carbonates, and (3) the assumption of 8% of water. As the dark aluminous and ferruginous hornblendes are among the commonest and most wide-spread of minerals, it is apparent from the above that they may have an important bearing upon the color and physical qualities of the residual clays; to which they thus give rise. The peroxidation of the iron gives yellow, brown, or red colors, while the hydrated aluminous silicate (clay) imparts tenacity. The final product of such decomposition is, then, a ferruginous clay.

The Pyroxenes. — The rock-forming pyroxenes are divided upon crystallographic grounds into two groups, the one orthorhombic in crystallization, and the other monoclinic. All varieties, when in good crystalline form, show in basal sections an octagonal outline bounded by prismatic and pinacoidal faces and with a well-defined cleavage parallel with the prism faces. Chemically they are silicates of magnesia and iron with lime and alumina in varying proportions. They are hard, tough minerals and have an important bearing upon the physical properties of the rocks of which they form a part. Their distribution, in some of their varieties, is almost universal, being found in metamorphic and eruptive rocks of almost every class and every age.

The Monoclinic Pyroxenes. — Two principal varieties are recognized. (1) Pyroxenes containing little or no alumina, and composed of silica, 45.95 to 55.6%; lime, 21.06 to 25.9%; magnesia, 13.08 to 18.5%, with sometimes varying quantities of iron oxides and water. Under this head are included the lighter colored varieties, *malacolite, sahlite,* and *diallage.* (2) Pyroxenes containing alumina, and composed of silica, 49.40 to 51.50%; alumina, 6.15 to 6.70%; magnesia, 13.06 to 17.69%; lime, 21.88 to 23.80%; iron oxides, 0.35 to 7.83%, with sometimes small quantities of soda and water. Under this head are included the darker varieties, augite and leucaugite.

The lighter colored, non-aluminous varieties, malacolite and sahlite, are common in mica and hornblendic schists, gneiss, and granite, though not always in sufficient abundance to be noticeable to the naked eye. The foliated variety, diallage, is an essential constituent of the rock gabbro, and is also common in peridotites. The darker colored, aluminous variety, augite, is an essential constituent of diabase and basalt, and also occurs in many syenites, andesites, and other eruptive rocks.

In the thin sections the monoclinic pyroxenes are usually readily recognized by their nearly rectangular cleavages on basal sections (see Fig. 1), lack of pleochroism, and high extinction angles on sections parallel to the clinopinacoids. The aluminous varieties undergo alteration into chloritic and ferruginous products, while the non-aluminous give rise to serpentine, either process being attended by the separation of free calcite.

Ægerine and acmite are soda-bearing pyroxenes corresponding to the formula $Na_2OFe_2O_3 4SiO_2$. They are less abundant than the above-mentioned varieties, and so far as yet described seem to be confined mainly to the elæolite syenites.

The Orthorhombic Pyroxenes. — These are essentially silicates of magnesia and iron, the latter replacing the former in varying proportions up to as high as 25 %. Two principal varieties are recognized, the distinction being founded mainly upon their optical properties which seem to be affected very largely by the percentages of iron. *Enstatite* is the theoretically pure magnesian silicate of the formula $MgSiO_3$, but which, as a matter of fact, usually contains from 2 to 10 % or more of iron. The highly ferruginous varieties are known as *bronzite*, from their bronze-like lustre. *Hypersthene* differs from enstatite in being strongly pleochroic in thin sections, and it contains from 10 to 25 % of ferruginous oxide.

Both enstatite and hypersthene are common constituents of basic igneous rocks, such as the gabbros, norites, and peridotites. Enstatite is a common constituent of meteorites, occurring not infrequently in peculiar fan-shaped, radiating masses not greatly unlike certain organic forms for which they were once mistaken. Both forms are liable to alteration, giving rise to serpentinous pseudomorphs to which the name *bastite* has been given, and to talcose and chloritic products. The general character of the decomposition products of the pyroxenes, as well as the methods by which the decomposition progresses, are in every way similar to those of the amphiboles, and need not be further dwelt upon here.

The Micas. — There are several species of mica which are prominent as rock constituents, the more important being the white variety, muscovite, and the dark variety, biotite. Both occur, as a rule, in thin, platy forms, splitting readily into thin, elastic folia, which in crystalline form are hexagonal in outline. The folia are often bent and distorted, and the mineral not infrequently undergoes alteration into a chloritic or sericitic product. The micas exercise an important influence upon the rocks containing them, on both color and structural grounds. Other things being equal, the muscovite-bearing rocks are lighter in color than those carrying biotite. If the mica plates are arranged in definite planes, the rock assumes a schistose structure and splits more or less readily into sheets — an important feature from an economic standpoint. Muscovite, or potash mica, a silicate of alumina and potash, is a constituent of many granites, gneisses, and schists, but is rarely met with in other rocks, and is wholly wanting in the basic eruptives. Another white or nearly colorless mica is *sericite*, a silvery white, or greenish, hydrous, secondary constituent of metamorphic schists, or occurring as an alteration product from feldspar: paragonite and margarite are other hydrous micas, confined mainly to the schists and to veins. Lepidolite, a lithia mica of a white or faint pink color, is frequently found in pegmatitic veins in the older rocks.

Biotite, the black iron mica, is a silicate of alumina, iron, and magnesia, and is much more general in its distribution than is muscovite, occurring in both eruptive and metamorphic rocks of all kinds and of all ages. It undergoes alteration into chloritic and ferruginous products and is often an important feature in hastening rock disintegration. Other black micas, sometimes distinguishable from biotite only by chemical means, are lepidomelane and houghtonite. A pearl gray potash mica phlogopite is an important constituent of many limestones, as in northern New York and adjacent portions of Canada.

All micas, owing to their eminently fissile structure, allow the ready percolation of moisture, and hence, though in themselves of difficult solubility, are elements of weakness in any stone of which they may form a part. The characteristic form of decomposition begins as in other silicate minerals, with hydration. This in the dark varieties is accompanied by a higher oxidation of the iron. The folia gradually lose their elasticity and crumble away, the bases being removed in solution as before. The complete decomposition of the micas is, however, brought about very slowly, and almost any granitic soil, however thoroughly decomposed, will, on washing, show small flakes of the mineral still remaining. However rusty, too, these may appear, a little hydrochloric acid cleans them up, showing remnant shreds still fresh and readily recognizable. For some unexplained reason those granitic rocks containing a considerable proportion of white mica are almost invariably more friable and easily disintegrated than those containing biotite.

Olivine (Chrysolite, Peridote). — *Composition:* Silicate of iron and magnesia, $(MgFe)_2 SiO_4$.

This is an essential constituent of basalt, dunite, limburgite, lherzolite, and pikrite, and a prominent ingredient of many lavas, diabases, gabbros, and other igneous rocks. It also occurs occasionally in metamorphic rocks and is a constituent of most meteorites. Olivine is subject to extensive alteration, becoming changed by hydration into serpentine or talcose and chloritic products, with the separation of free iron oxides. Under the microscope olivine is as a rule easily recognized by its lack of cleavage and brilliant polarization colors. It occurs in well-defined crystals and also in irregular grains, either singly or grouped in peculiar clusters to which the name *polysomatic* has been applied by Tschermak. The serpentinous alteration takes place along the irregular curvilinear lines of fracture, and under favorable conditions continues until the transformation is complete. The following analyses by Holland, as quoted in Teall's British Petrography, illustrate the simplicity of the chemical changes which here take place: —

	I	II	III
SiO_2	41.32%	42.72%	43.48%
Al_2O_3	0.28	0.06	
Fe_2O_3	2.30	2.25	
CrO	0.05	Trace	
MgO	54.09	42.52	43.48
H_2O	0.20	13.39	13.04
	98.93%	100.94%	100.00%

I. Olivine, Snarum, Norway. II. Serpentine derived from the same. III. The theoretical composition of serpentine.

Aside from the assumption of some 13 % of water, the principal change, as will be noted, is a loss in magnesia which as a rule separates out as a carbonate. The iron, which existed as protoxide, is further oxidized and crystallizes out along lines of fracture as magnetite or hematite, or in the hydrous sesquioxide form known as limonite. Through decomposition, a portion or all of the silica may be set free as opal or chalcedony, the magnesia going over to the condition of carbonate, and the iron passing into various hydrated oxide forms such as are most stable under the existing circumstances.

Epidote. — *Composition:* Silica, 37.83 %; alumina, 22.63 %; iron oxides, 15.98 %; lime 23.27 %; water 2.05 %.

This mineral is a common constituent of many granites, gneisses, and schists, especially the hornblendic varieties. It is particularly abundant, however, as a secondary constituent in basic eruptives, where it results from the alteration of the original ferromagnesian constituents such as the augites, hornblendes, or micas. It is the presence of this mineral or a secondary chlorite that gives the characteristic color to many of the so-called *greenstones* (altered basalts, diabases, diorites, etc.).

The name *piedmontite* is given to a red manganese epidote, which has been found in certain Japanese schists and has also, in sparing amounts, been observed by Professor Haworth,[1] in the quartz porphyries of Missouri, and a few foreign porphyrites. Zoisite is a closely related mineral crystallizing in the orthorhombic system and relatively poorer in iron and richer in alumina than is epidote. It is chiefly characteristic of the crystalline schists, though sometimes found in granitic rocks, intergrown with common epidote as has been noted in Maryland, by Keyes.[2]

Allanite, or orthite, as it is sometimes called, is closely allied to common epidote, but contains cerium and other of the more rare alkaline earths. In the form of brown acicular crystals it is a common constituent of New England granites and has recently been described in a granite porphyry near Ilchester, Maryland, where it occurs enclosed as a nucleus in the ordinary epidote.

Calcite (Calcium Carbonate). — *Composition:* $CaCO_3$ = Carbon dioxide, 44 %; lime, 56 %. Hardness, 3.

[1] American Geologist, Vol. I, p. 365.
[2] 15th Ann. Rep. U. S. Geol. Survey, 1890-94.

This is an original constituent of many secondary rocks, such as limestone, ophiolite, and calcareous shales. It is the essential constituent of most marbles, of stalactites, travertine, and calc-sinter. The shells of foraminifera, brachiopods, crustaceans, and many lamellibranchs and gasteropods are also of this material. As a secondary constituent, resulting from the decomposition of other minerals, it occurs almost universally, filling wholly or in part cavities in rocks of all ages, such as granite, gneiss, syenite, diabase, diorite, liparite, trachyte, andesite, and basalt.

The effervescence of the mineral when treated with a dilute acid furnishes the most ready means for its detection. Under the microscope it appears as colorless grains with faint iridescent polarization, and is best recognized by its cleavage and characteristic twinning lines as shown in the figure on p. 163. Being soluble in carbonated waters, it is liable to complete removal, or leaves only its impurities behind as a mark of its decay.

Aragonite. — *Composition:* $CaCO_3$ = Carbon dioxide, 44 %; lime, 56 %.

This mineral has the same chemical composition as calcite, but differs in its crystalline form and specific gravity. It occurs with beds of gypsum and veins of ore, and also in stalactitic and stalagmitic forms. In small quantities it occurs as a secondary product in many trap rocks and basalts, and is the substance of which the shells of many gasteropod and lamellibranch molluscs are composed.

The mineral occurs nearly always in clustered aggregates of radiating, divergent needles, and is distinguished from calcite by its crystallization and cleavage. As a rock constituent it is comparatively unimportant, but frequently occurs as a decomposition product in basic eruptives. This form of calcium carbonate, as long ago pointed out by Sorby, is less stable than calcite, and in many instances where the substance has first crystallized in the orthorhombic form aragonite, it is found to have undergone a molecular alteration into calcite.

Dolomite.—*Composition:* $(CaMg)CaO_3$ = Calcium carbonate, 54.35 %; magnesium carbonate, 45.65 %. Hardness, 3.5–4.

This mineral, like calcite, is a wide-spread constituent of rocks, and not infrequently forms extensive masses which are of value as sources of building material. It is distinguishable from calcite by its greater hardness, higher specific gravity, and in being but slightly acted on by acetic or dilute hydrochloric acid. In itself the mineral is less susceptible to atmospheric influence than calcite, yielding much less readily to decomposing agencies of a purely chemical nature. Nevertheless, Roth[1] has shown that in the weathering of dolomitic limestones the magnesia is sometimes removed by leaching, in greater proportional quantities than the more soluble lime carbonate.

Apatite. — *Composition:* Phosphate of lime. Hardness, 5.

Apatite is an almost universal constituent of eruptive rocks, both acid and basic, though as a rule present only in microscopic proportions. In the granular limestones, schists, and other metamorphic and vein rocks it sometimes occurs in large crystals or massive forms in such abundance as to be of value as a source of mineral phosphate for fertilizing purposes. In the thin sections the apatites of eruptive rocks are as a rule colorless, and without evident cleavage, though presenting good crystallographic forms. Rarely the mineral is pleochroic in red or brown or bluish colors. If a drop of an acid solution of ammonium molybdate be placed upon an apatite crystal in an uncovered slide, the mineral will be slowly dissolved and minute crystals of phosphomolybdate of ammonium be contemporaneously deposited. The process is an easy one, readily performed while the slide is still on the stage, and forms one of the most interesting and accurate of the many microchemical tests. Though present in but small amounts, apatite is an important constituent, since it is the only common rock constituent containing the valuable element phosphorus.

THE IRON ORES

Under this head we may conveniently treat the several forms of iron oxides which commonly occur as rock constituents, and which from their opacity in even the thinnest sections, and similarly in crystallographic outline, are separable with difficulty by optical tests alone.

Magnetite. — *Composition:* $FeO + Fe_2O_3 =$ iron sesquioxide, 68.97 %; iron protoxide, 31.03 %.

This is a wide-spread and almost universal constituent of

---

[1] Chemische u. Allgemeine Geologie.

eruptive rocks, occurring as a rule in the form of scattering, small, and rather inconspicuous granules, which under the microscope are characterized by a complete opacity and bluish lustre. When of sufficient size to be distinguished by the unaided eye, magnetite is easily recognized by its brilliant lustre, weight, and its property of being readily attracted by the magnet. It is as a rule one of the first minerals to separate out from the molten magma, and hence presents good crystal outlines in which octahedral forms prevail. Skeleton forms of great beauty are not infrequent. Magnetite also occurs as a constituent of metamorphic rocks and is sometimes found in large beds, constituting a valuable ore of iron. Under continual alternations of heat and cold, moisture and dryness, it slowly decomposes, giving rise to hydrated sesquioxides which impart color, but no valuable qualities, to the resultant sands and clays.

Menaccanite (Ilmenite or Titanic Iron). — *Composition:* $(TiFe)_2O_3$, a mixture in varying proportions of the oxides of iron and titanium.

This, like magnetite, occurs in scattering granules as an original constituent of many eruptive rocks, and also in micaceous lamellar and vein-like masses in other rocks. Under the microscope it shows, by incident light, a brownish rather than bluish lustre, but is best recognized by its characteristic alteration products, which are whitish, gummy, and opaque. The name *leucoxene* was given by Gumbel to the final product of this alteration. This form of iron ore is extremely refractory to atmospheric agencies and is to be found scarcely, if any, changed in the residuary materials resulting from the breaking down of the rocks in which it originated.

Hematite (Specular Iron Ore.) — *Composition:* Anhydrous sesquioxide of iron, Fe_2O_3 = iron, 70.9 %; oxygen, 30.20 %. H = 5.5–6.5.

This mineral occurs in varying proportions and under varying conditions in rocks of all ages. In the form of minute scales of a blood-red color, it is found not infrequently in granitic and other eruptive rocks. It occurs, also, in large beds, forming a valuable ore of iron. In the amorphous condition, it may form the cementing constituent of sandstones, and is the cause of the red color of many rocks, both clastic and metamorphic, and of soils as well. The usual coloring constituent is, however, limonite or turgite, as noted below. The specular and massive forms are best recognized by opacity, brilliant, black, metallic lustre, and red streak.

Limonite (Brown Hematite). — *Composition:* Hydrous sesquioxide of iron, $H_6Fe_2O_6 + Fe_2O_3 =$ iron sesquioxide, 85.6 %; water, 14.4 %. H = 5–5.5.

This is a common constituent of rocks of all ages, but is as a rule wholly secondary, resulting from the decomposition of ferruginous silicates, sulphides, and anhydrous oxides. As a coloring constituent it is even more abundant than hematite, and like it forms a valuable ore of iron. (See p. 107.) Turgite ($Fe_4H_2O_7$) in the form of a brilliant red ochreous material is also a common constituent of soils and clays resulting from the decomposition of siliceous rocks, and is presumably, like limonite, a product of the spontaneous hydration of the iron salts thus set free. (See further under Color of Soils, p. 385.)

Pyrite (Iron Pyrites). — *Composition:* Iron disulphide, FeS_2 = iron, 46.7 %; sulphur, 53.3 %. H = 6–6.5.

Two principal forms of iron disulphide occur in nature, alike in chemical composition, but differing in forms of crystallization and in density. The one is common pyrites which crystallizes in the isometric system, and is easily recognized by its strong brassy yellow color and hardness. Its usual form of occurrence is that of cubes, the corners and edges of which may be more or less modified by secondary planes, and in concretionary masses. The second form *marcasite*, also called *gray*, *white*, or *cockscomb* pyrites, is of lighter color, inferior hardness and density, and crystallizes in the orthorhombic system. Its most common form of occurrence is that of irregular concretionary masses.

Both forms of pyrite are susceptible to oxidation when exposed to atmospheric agencies, though of the two the pyrite proper is much the more refractory.

Mr. A. P. Brown has shown[1] that in this form of the compound a large proportion of the iron exists in a *ferric* condition while in marcasite it is *ferrous*. In other words, marcasite is an unsaturated compound, and hence unstable. This readily explains the relatively more rapid decomposition of the latter mineral. There is also a difference in the character of the products arising from the decomposition of the two compounds,

[1] Proc. American Philos. Soc., Vol. XXXIII, 1894, p. 225.

pyrite yielding, as a rule, limonite and free sulphur, while marcasite, under the same conditions, yields ferrous sulphate, though when decomposing under water, it may also yield much limonite. The sulphate of iron, resulting from pyritiferous decomposition, is, if present in quantity, injurious to plant growth. This fact was well illustrated some years ago on the west front of the National Museum at Washington. Several large masses of iron sulphide, too large for exhibition within the building, were placed here upon a floor of cement bordered by a narrow strip of lawn. Under the oxidizing influence of rain and air the sulphide became slowly converted into sulphate which was washed down upon the cement and thence into the soil, which it so poisoned as to kill the roots and necessitate an entire resodding.

The experiments of Prichard,[1] however, showed that the presence of a small amount of sulphate of iron in a soil *may*, under certain conditions, be beneficial, in that it serves to prevent the loss of ammonia in rapidly decomposing materials. In processes involving slow decomposition, its antiseptic qualities render it of doubtful value.

Chlorite (Viridite). — Under the general name *chlorite* are included several minerals occurring in fibres and folia, closely resembling the micas, from which they differ in their large percentage of water, and in their folia being inelastic. The three principal varieties recognized are, *ripidolite*, *penninite*, and *prochlorite*, any one of which may occur as the essential constituent of a chlorite schist. Chlorite as a secondary product often results from and entirely replaces the pyroxene, hornblende, or mica in rocks of various kinds, and also occurs filling wholly or in part the amygdaloidal cavities of trap rocks. In this form it is frequently visible only with the microscope, and owing to the difficulties in the way of an exact determination of its mineral species is sometimes called *viridite*. It is this mineral which gives the green color to a large share of the more or less altered eruptives, like the diabases and diorites, the "greenstones" of the older geologists.

Serpentine. — *Composition:* A hydrous silicate of magnesium corresponding to the formula $H_4Mg_3Si_2O_9=$ silica, 44.1 %; magnesia, 43.0 %; and water, 12.9 %.

The prevailing color is green, though often spotted and

[1] Ann. de Chemie et Physique, 1892.

streaked; hence the name from the Latin *serpentinus*, a serpent. It has a somewhat greasy lustre and may be cut with a knife, having a hardness of about 4 of the scale. The mineral is always secondary, resulting mainly from the hydration of pure magnesian or lime magnesian silicates. (See further on p. 115.)

Glauconite. — This name is given to a somewhat variable compound consisting essentially of silica, iron, alumina, and water, with smaller amounts of potash, and incidentally lime, magnesia, and soda. The prevailing color is green, and as it occurs in single granules or granular aggregates, it is commonly known as *greensand*. It is always a secondary mineral, and has been formed and is still forming on many shallow sea-bottoms which receive fine sediments derived from the breaking down of siliceous crystalline rocks. (See under Greensand Marl, p. 133.)

The Zeolites. — Under this head are grouped a number of minerals alike in being hydrous silicates of alumina with varying percentages of lime, potash, and soda. They are altogether secondary minerals, resulting from chemical changes taking place in pre-existing rocks, and indicate not infrequently the first or deep-seated stages of rock decay. In a more or less perfect condition they have been assumed to occur in soils, having been derived from the rocks, or, as is contended by some authorities, having formed during the process of rock decomposition or in the soil itself. It is possible that those constituents of a soil which on analysis are found to be "soluble" as the term is ordinarily used, may, in part at least, have existed as zeolites. Hence their consideration in this connection is of importance.

Out of the 22 species of minerals classified as zeolites by Dana in this "System of Mineralogy" there are but 11 which, on account of their abundance or chemical composition, need consideration here. The theoretical composition of these, as indicated from a comparison of several to many analyses, is shown in the accompanying table. In addition to the true zeolites are included several other hydrous silicates closely related, both as regards chemical composition and mode of occurrence, and which, in our present discussion, cannot well be excluded.

	Silica (SiO$_2$)	Alumina (Al$_2$O$_3$)	Lime (CaO)	Barium (BaO)	Potash (K$_2$O)	Soda (Na$_2$O)	Water (H$_2$O)
Ptilolite . . .	70.0	11.9	4.4	....	2.4	0.8	10.5
Mordenite . .	67.2	11.4	2.1	....	3.5	2.3	13.5
Heulandite . .	59.2	16.8	9.2	....	....	....	14.8
Phillipsite . .	48.8	20.7	7.6	....	6.4	....	16.5
Harmotome . .	47.1	16.0	....	20.6	2.1	....	14.1
Stilbite . . .	57.4	16.3	7.7	....	....	1.4	47.2
Laumontite . .	51.1	21.7	11.9	....	....	....	15.3
Chabazite . .	47.2	20.0	5.5	....	....	6.1	21.2
Analcite . . .	54.5	23.2	....	....	....	14.1	8.2
Natrolite . .	47.4	26.8	....	....	....	16.3	9.5
Thomsonite . .	36.9	31.4	11.5	....	....	6.4	13.8
Prehnite . . .	43.7	24.8	27.1	....	....	....	4.4
Apophyllite . .	53.7	....	25.0	....	5.2	....	16.1

PLATE 2

1

2

Fig. 1. Quartz porphyry showing porphyritic structure.
Fig. 2. Quartz porphyry showing flow structure.

IV. THE PHYSICAL AND CHEMICAL PROPERTIES OF ROCKS

1. STRUCTURE

In considering the structure of rocks it will facilitate matters to do so under two heads: (1) the macroscopic (or megascopic) structures, or structures visible to the unaided eye (*macros*, from Greek word μαχρος, signifying large); and (2) microscope structures, or those visible only with the aid of the microscope.

1. **Macroscopic Structures.** — From a structural standpoint all rocks may be classified sufficiently close for present purposes, under the heads of: (1) *Crystalline*, (2) *vitreous* or *glassy*, (3) *colloidal*, and (4) *clastic* or *fragmental*. Of the first of these, ordinary granite or crystalline marbles are good types, being made up wholly of crystal aggregates, without interstitial amorphous or fragmental material. The term *crystalline granular*, or *granular crystalline*, is often applied to such as have a distinctly granular structure, as do many of the granitic rocks. *Vitreous* or *glassy* structures are found only among igneous rocks, and are due always to a cooling of the molten magma too rapidly for the production of crystals. Obviously, as the rate of cooling in rock masses must be extremely variable, so we find all intermediate stages between the completely glassy and the crystalline forms. To these intermediate stages such names as *felsitic* and *microlitic* are given, names the precise meaning of which will be stated under the head of microscopic structures. Rocks originating as chemical deposits, and which have since undergone no structural changes, often present a jelly or glue like structure known as *colloidal*. Such are exemplified in the flints from the English chalk cliffs, the siliceous sinters from the Yellowstone National Park, and by various other forms of silica, as opal, agate, etc., and occasionally by serpentines.

A *clastic* or *fragmental* structure is found only in secondary rocks, and is the result of a breaking down or disintegration of pre-existing rocks, and a reconsolidation of their particles without crystallization. There are many minor points of structure, some of which are common to all of the primary groups above mentioned, while others are limited to one or more. Rocks which are made up of distinct grains, whether crystalline or fragmental, are spoken of as *granular;* when the structure becomes too fine and dense for macroscopic determination it is spoken of as *compact*, though there is no reason why the term should not equally well be applied to the coarser grained rocks in which the individual grains are closely cohering without interstices. The term *massive* is applied to such igneous rocks as show no signs of bedding or stratification, while limestones, sandstones, and such other rocks as are arranged in more or less parallel layers are described as *stratified*. (See Fig. 1, Pl. 13.) The name *foliated* or *schistose* is given to a rock in which the arrangement of the constituent minerals in parallel planes is sufficiently marked to cause it to split in this direction more readily than in any other. Not infrequently the quartzes or feldspars occur in lens-shaped forms about which curve the hornblende or mica folia as shown in Fig. 2, Pl. 13. As explained elsewhere, this structure may be due to original deposition or may be secondary. In eruptive rocks a *fluidal* or *fluxion* structure is not uncommon, as shown in Fig. 2, Pl. 2, and is due to the onward flowing of the mass while gradually cooling and passing into a solid state. Eruptive magmas at the time of their extrusion contain more or less moisture, which, being highly heated, expands whenever sufficient force is developed to overcome the pressure of the overlying mass. In this way are formed innumerable cavities or bubbles, comparable to the cavities caused by carbonic acid from the yeast in well-raised bread. Such cavities are called *vesicles*, and the rocks containing them are *vesicular* (Fig. 2, Pl. 3). By the subsequent action of percolating waters these cavities may become filled with a variety of secondary minerals, among which chalcedony, epidote, calcite, and various zeolites are not uncommon. Such refilled cavities are called *amygdules*, from the Greek word αμυγδαλον, an almond, in allusion to their shape, and the rocks containing them are therefore described as *amygdaloidal*. The upper part of a lava flow not infrequently cools in peculiar ropy

PLATE 3

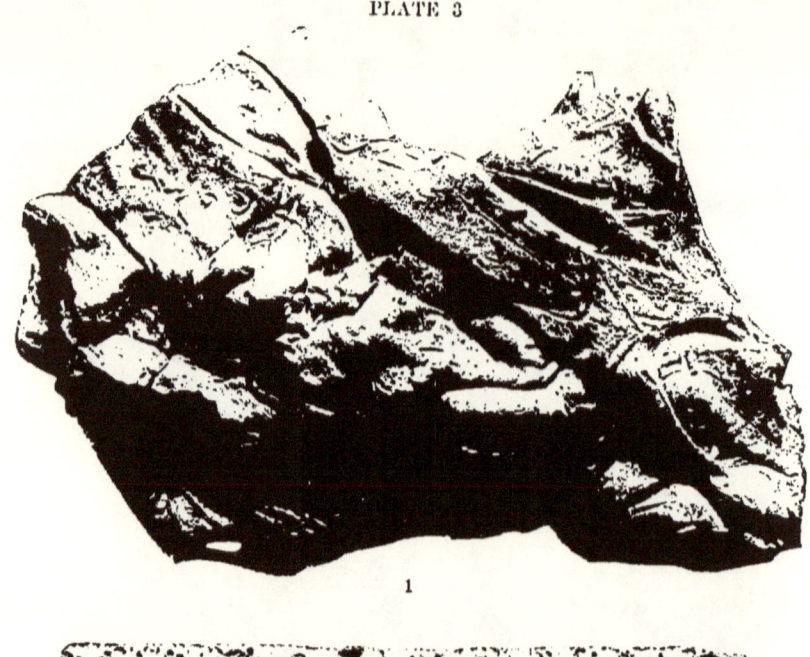

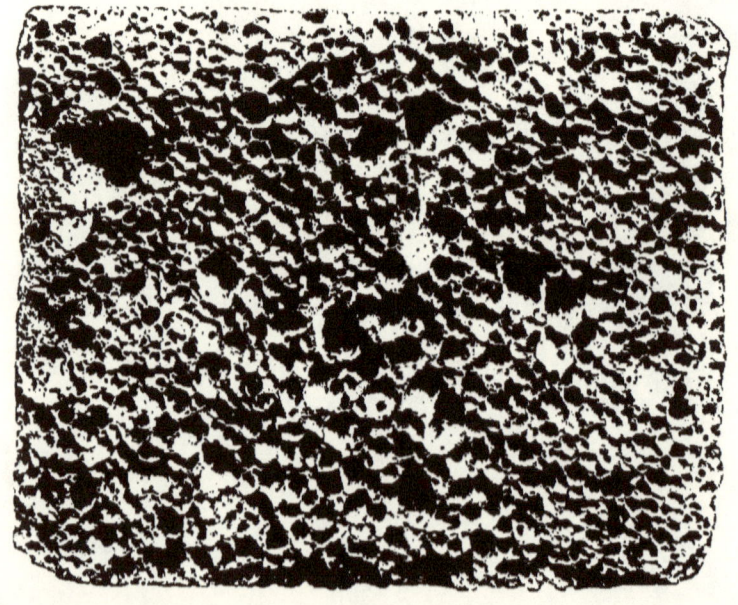

Fig. 1. Basalt showing slaggy structure.   Fig. 2. Basalt showing vesicular structure.

forms like the slag from a smelting furnace. Such forms are known as *slaggy*. (See Fig. 1, Pl. 3.)

When a rock consists of a compact, glassy, or fine and evenly crystalline ground-mass, throughout which are scattered larger crystals, usually of feldspar, the structure is said to be *porphyritic* (Fig. 1, Pl. 2). This structure is quite common in granite, but is not particularly noticeable, owing to the slight contrast in color between the larger crystals and the finer ground-mass. It is most noticeable in such effusive eruptives as the quartz porphyries, in which, as is the case with some of those of eastern Massachusetts, the ground-mass is exceedingly dense and compact and of a black or red color, while the large feldspar crystals are white and stand out in very marked contrasts. This structure is so striking in appearance that rocks possessing it in any marked degree are popularly called porphyries, whatever may be their mineral composition. The term *porphyry* is said to have been originally applied to certain kinds of igneous rocks of a reddish or purple color, such as the celebrated red porphyry or "roseo antico" of Egypt. The word is now used by the best authorities almost wholly in its adjective sense, since any rock may possess this structure whatever its origin or composition may be.

Glassy rocks on cooling sometimes have developed in them a series of concentric cracks whereby the rock on a broken surface shows numerous rounded or globular bodies with an onion-like shell. This structure, which may be visible only with a microscope, is known as *perlitic*. It is not uncommon in glassy forms of Hungarian trachytes.

Glassy and felsitic eruptives, particularly of the liparite and quartz porphyry groups, frequently show spherulitic masses of all sizes, from microscopic to several inches or even feet in diameter, usually with a well-defined radiating structure and which are due to incipient crystallization. Such are known as *spherulites*, and hence rocks in which they occur are described as *spherulitic*.[1]

A concretionary structure is not infrequently developed in rocks either as a primary structure or as due to segregating processes acting subsequent to the formation of the rocks in

[1] The structure and origin of these forms has been worked out in detail by Whitman Cross. Bull. Philosophical Society of Washington, Vol. XI, 1891, pp. 411–462.

which they are found. Many of the forms thus developed are peculiarly deceptive, and it may not be out of place to enter into a discussion of their nature and origin with some detail.

On genetic grounds we may divide such forms, as intimated above, into two groups: (*A*) Primary concretions, formed contemporaneously with the rocks in which they are found, and (*B*) secondary concretions, or those which are due to segregating influences acting subsequent to the formation of the rocks of which they now form a part. All are due to that peculiar and little understood tendency which atoms or molecules of like nature so often manifest in concreting or gathering in amorphous masses or concentric layers about some foreign body which serves as a primary point of attachment. The extreme development of this tendency is seen in crystallization, of which we may perhaps regard this first form of concretionary structure as incipient stages. Under primary concretions may be included the flint and chalcedonic nodules found in chalk and the older limestones, the material of which was in part without doubt derived from the siliceous remains of diatoms and sponges. Such sometimes occur in the form of lenticular nodules with or without an appreciable concentric structure and lying in parallel layers or beds, sometimes continuous for long distances. Clay iron stone, an impure carbonate of iron, occurs characteristically in this form. These latter often crack on drying and consequent shrinkage, the cracks extending from within outward. In these cracks calcite is subsequently deposited, whereby the nodule is divided up into septa of a white or yellowish color. On being cut and polished, these often form beautiful and unique objects. To such the name *septarian nodule* is commonly given. (See Fig. 2, Pl. 9.) The carbonate of lime in inland lakes and seas may not infrequently become deposited in the form of thin pellicles about a minute, perhaps microscopic nucleus, forming small, spherical bodies which, when ultimately consolidated into beds, give rise to the oölitic and pisolitic limestones. (See p. 143.) All primary concretions are not, however, chemical deposits; but, rather, aggregates of mineral particles in a finely fragmental condition.

Such are the clay concretions which are found in the beds of streams and lakes, and which may not so closely simulate animal forms as to be very misleading. The manner in which concretions of this nature are formed was shown in a very interesting manner a few years ago during the process of the work of filling in the so-called Potomac flats, on the river front at Washington, District of Columbia. For the double purpose of raising the flats and deepening the channel, gigantic pumps were employed which raised the sediment from the river bottom in the form of a very thin mud and forced it through iron pipes to the flats, where it flowed out, spreading quietly over the surface. The material of this mud was mainly fine siliceous sand and clay intermingled with occasional freshwater shells and plant débris. As this mud flowed quietly from the mouth of the pipe and spread out over the surface, the clayey particles began immediately to separate from the siliceous sand in the form of concretionary balls, and in the course of a very short time these would grow to be several inches in diameter. Such, owing to the rapidity of their formation, contained a large amount of sand and shells, though clayey matter predominated.

In crystalline rocks concretionary structure is rarely developed. Cases such as shown on Plate 8 are quite unique, and in the case of the orbicular diorite of the greatest interest on account of the beauty of the stone and its adaptability for small ornamentation.

Concretionary structure of a secondary nature may be developed through the process of weathering. Thus, by the oxidizing action of meteoric waters percolating through a porous sand or sandstone, included nodules of iron disulphide (pyrite) may be converted into an oxide which gradually segregates in zones about the original nodule. This oxide, by its cementing action, binds the grains together in the form of a hard crust, leaving the central portion, formerly filled by pyrite, either empty or occupied by loose sand.[1] A zonal banding or shelly structure closely simulating concretionary structure is common in rocks more or less weathered and decomposed, but which is due not to original deposition or crystallization of mineral matter about a centre, but rather to the weathering of jointed blocks, the various chemical agencies acting from without inward.

A botryoidal structure is not infrequent among rocks and minerals of chemical origin. It is, as a rule, confined to such

[1] See On the Formation of Sandstone Concretions, Proceedings U. S. National Museum, Vol. XVII, pp. 87, 88.

as are amorphous or radiating crystalline aggregates of a single mineral, such as chalcedony or the hematite iron ores. (See Fig. 1, Pl. 9.)

A brecciated structure, produced by the presence of angular fragments in a finer ground, is of common occurrence among fragmental rocks (the breccias), but is more rare among the crystallines. It is sometimes produced in volcanic rocks by the imbedding in the still pasty magma of angular fragments of previously consolidated material, as shown in Fig. 2, Pl. 4. Columnar structure, though comparatively common as the structure of a geological body, is rarely developed among the constituents of the rock itself. The columnar structure of many lavas and dike rocks has already been alluded to: occasionally the mineral constituents of some secondary rocks are arranged after this manner. A *cavernous* or *cellular* structure is not infrequently developed through the removal by solution of some constituent or the weathering out of a fossil. As an original structure it occurs in many rocks of chemical origin as the stalagmitic deposits in caves, travertines, etc.

A *laminated* or banded structure, due to the arrangement of the constituents in parallel layers or bands, is common in rocks of sedimentary origin, particularly in sandstones and shales.

2. Microscopic Structures. — Many, if not indeed the majority, of rocks are so fine grained and compact that little of their mineral nature or structural features can be learned from examination by the unaided eye. This difficulty made itself apparent very early in the history of geological science, and to it is perhaps due, more than to any other single cause, the apparent crudities and fallacies of the early workers. As long ago as 1663, the microscope had been to some extent utilized for the examination of minerals; but its application to the study of rocks remained long unrecognized, though early in the present century Cordier and others utilized it in the study of rocks in a pulverized condition. It was not until about 1850, when the subject was taken up by H. Clifton Sorby of England, that the possibility of studying rocks *in thin sections* under the microscope began to be appreciated. Even then the idea failed to bear its legitimate fruits until transplanted to German soils, where, under the fostering care of Professor Zirkel of Leipzig, it soon began to yield an abundant harvest; and to-day the branch of the science of geology known as microscopical pe-

PLATE 4

FIG. 1. Chert breccia cemented by zinc blende.
FIG. 2. Felsite breccia formed of felsitic fragments embedded in a matrix of the same composition.

trography holds a prominent place in all the leading universities, both domestic and foreign. The efficiency of the method is based upon the fact that every crystallized mineral has certain definite optical properties; *i.e.* when cut in such a way as to allow the light to pass through it, will act upon this light in a manner sufficiently characteristic to enable one working with an instrument combining the properties of a microscope and stauroscope to ascertain at least to what crystalline system it belongs, and in most cases by studying also the crystal outlines and lines of cleavage the mineral species as well. To enter upon a detailed description of the method by which this is done would be out of place here, since it involves the polarization of light and other subjects which must be studied elsewhere. The reader is referred to any authoritative work on the subject of light, and to Professor J. P. Idding's translation of Professor Rosenbusch's work on optical mineralogy.[1]

This method of study is of value, not merely as an aid in determining the mineralogical composition of a rock, but also, and what is often of more importance, its structure and the various changes which have taken place in it since its first consolidation. Rocks are not the definite and unchangeable mineral compounds they were once considered, but are rather ever-varying aggregates of minerals, which, even in themselves undergo structural and chemical changes almost without number. It is a common matter to find rock masses which may have had originally the mineral composition and structure of diabase, but which now are mere aggregates of secondary products, such as chlorite, epidote, iron oxides, and kaolin, with perhaps scarcely a trace of the unaltered original constituents; yet the rock mass retains its geological identity, and to the naked eye shows little, if any, sign of the changes that have gone on. These and other changes are in part chemical and in part structural or molecular. A very common mineral transformation in basic rocks is that from augite to hornblende. This takes place merely through a molecular readjustment of the particles, whereby the augite, with its gray or brown colors and rectangular cleavages, passes by uralitic stages over into a green hornblende, a mineral of the same chemical composition, but of different crystallographic form. This transformation in

[1] Microscopic Physiography of Rock-making Minerals, Wiley & Son, New York. See also Professor A. Harkers' Petrology for Students.

its incompleted state is shown in the accompanying figure, in which the central, nearly colorless portion with rectangular cleavage represents the original augite, while the outer dotted portion with cleavage lines cutting at sharp and obtuse angles is the secondary hornblende. This change is due to slow and gradual pressure exerted through unknown periods of time upon the rock masses, and the final result is the production of a rock of entirely different type and structure from that which originally cooled from the molten magma. The change such as above described is further alluded to in the chapter on metamorphism.

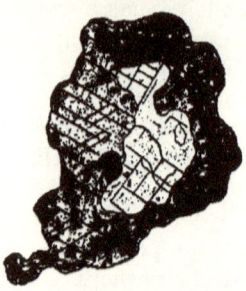

FIG. 1.— Augite partially altered into hornblende.

This science of microscopic petrography, as it is technically called, has also been productive of equally important results in other lines. As an instance of this may be mentioned the discovery that the structural features of a rock are dependent, not upon its chemical composition or geological age, but upon the conditions under which it cooled from a molten magma, portions of the same rock varying all the way from holocrystalline granular through porphyritic to glassy forms. To this fact allusion has already been made.

The general subject of the microscopic structure of rocks of various kinds, will be discussed more fully in describing the rocks themselves. Nevertheless, as in describing these structures it has become necessary to use sundry technical terms, it will be well to refer to them briefly here.

When a rock is made up wholly of crystalline matter, it is spoken of as *holocrystalline;* when, however, it shows interstitial glassy or felsitic matter, it is *hypocrystalline.* Rocks wholly without crystalline secretions are *amorphous.* The glassy, or felsitic matter' occupying the interstices of the other constituents is spoken of as the *base.* This base, together with the microlites and smaller crystallizations of the second generation, is called the *ground-mass;* such may be made up of *microlites —* small needle-like crystals imperfectly developed — when it is called *microlitic,* or of a dense aggregate of quartzose, felspathic and other materials, when it is known as *felsitic.* The larger crystals developed in a glassy, felsitic, microlitic, or finely

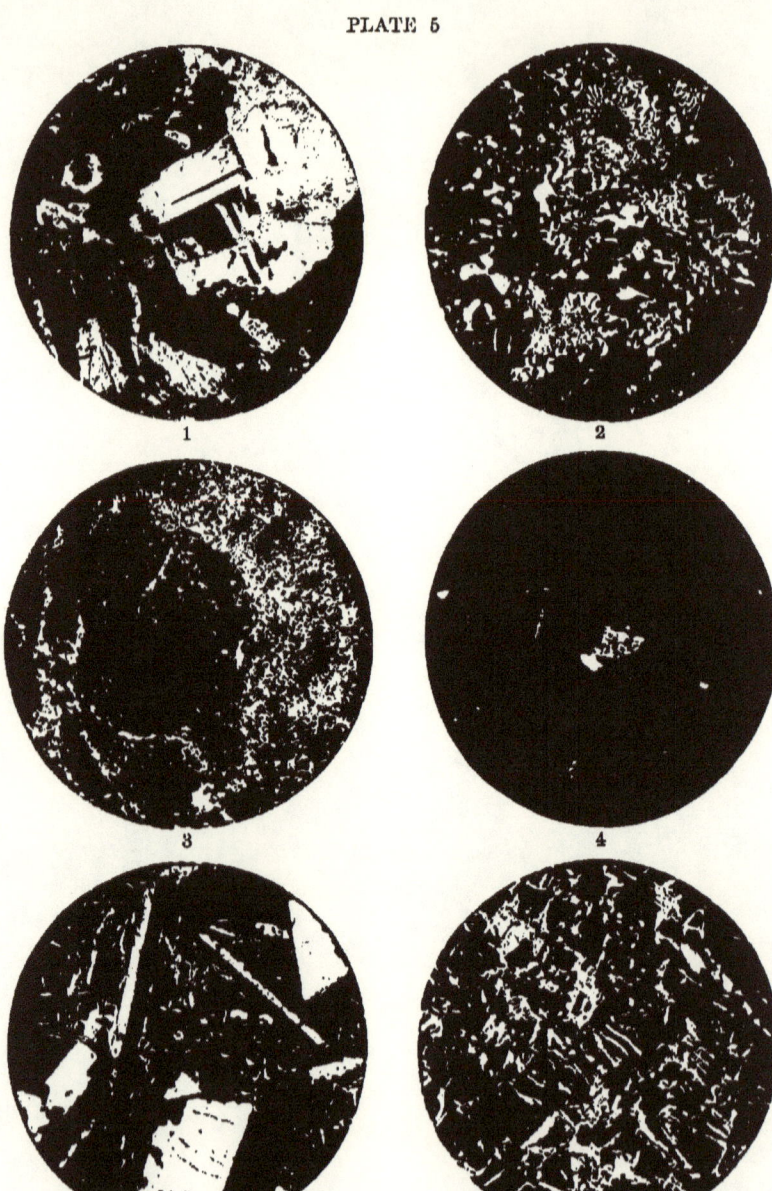

PLATE 5

FIG. 1. Microstructure of granite.
FIG. 2. Microstructure of micropegmatite.
FIG. 3. Microstructure of quartz porphyry.
FIG. 4. Microstructure of porphyritic obsidian.
FIG. 5. Microstructure of trachyte.
FIG. 6. Microstructure of serpentine.

granular *microcrystalline* ground-mass are called *phenocrysts.* When a mineral in a rock shows good crystal outlines, having been uninfluenced in its growth by the proximity of other minerals, it is called *idiomorphic:* when, however, its outline is due not to crystallographic forces, but to interference — to the action of external forces — it is *allotriomorphic.* Many rocks show indications of two or more periods of crystallization, whereby minerals of the same species may be developed. Thus in a molten magma the augites may begin to form under such conditions that for some time their growth is unimpeded and they take on large and well-developed forms. After a time, owing to changed conditions, their growth is stopped, and the rock solidifies with a new crop of smaller and less perfectly developed forms. It is customary to speak of such a mineral as occurring in crystals of two generations. In the case above described, the first developed form the porphyritic constituents, the *phenocrysts*, while the latter formed are a part of the ground-mass. Vitreous or glassy rocks not infrequently show, under the microscope, minute, hair-like or rod-shaped forms, representing the first stages of crystallization, but in which the process was arrested before they were sufficiently developed to render possible an accurate determination of their mineral nature. Such are termed *crystallites;* those in drop-shaped or globular forms being called *globulites*, the rod-shaped ones *belonites*, and the twisted, hair-like forms *trichites*.

The wide variation in microstructure in rocks of essentially the same chemical composition, but which have cooled under the varying conditions indicated above, is shown in Figs. 1 to 4 of Pl. 5, Fig. 1 being a holocrystalline type, and Fig. 4 one almost completely glassy, the first being a deep-seated rock, and the last a surface lava flow. Intermediate structures are often produced through a beginning of crystallization at certain depths below the surface, after which, and while a portion of the magma was still fluid, it was pushed upward toward the surface, or brought under such other conditions as resulted in a more rapid cooling, the final result being a glassy, or microcrystalline rock with scattering porphyritic crystals, or *phenocrysts*. It has not infrequently happened that, subsequent to the formation of these earliest products of crystallization, a second elevation of temperatures has taken place whereby the magma has eaten into or corroded them, as is the case with the quartz crystal shown in the centre of Fig. 3 of Pl. 5.

Inasmuch as this study by the microscope involves the preparation of thin sections, a brief description of the methods pursued may well be given here. The fact that a chip of rock, however dense, can, without breaking, be ground so thin as to be transparent, may at first seem strange, but in reality it is readily accomplished. The work requires only patience and the skill which comes from practice. A small chip of the rock, about the size of a nickel five-cent piece, is broken off with a hammer, care being taken to get it as thin as possible without fracturing. One side of this is then ground flat and smooth by rubbing it in water and emery on a smooth, cast-iron plate. Toward the close of the process fine flour of emery should be used, as the final surface must be very smooth and free from scratches. This chip is then cemented smooth side down on a piece of ordinary double-thick window glass, a convenient size being about 2 x 1 inches, the cementing material being Canada balsam which has been evaporated to the extent that, when cold, it is sufficiently hard to hold firmly, is not at all sticky, but yet is not so hard as to be brittle. The exact degree can only be learned by experience; a hardness such as to be barely indented by the thumb nail will be found about right. This operation of cementing will be best done by means of a thin iron plate laid horizontally on a support and heated not too hot by a lamp beneath. The glass with the balsam upon it is heated to the right temperature, the balsam being fluid and free from bubbles. The rock chip, heated sufficiently to expel all moisture, is then pressed firmly into the balsam, in such a way as to exclude air bubbles, and brought within as close contact with the glass as possible. It is then removed from the iron plate and allowed to cool, when the grinding process is resumed, the glass plate serving merely as support for the film of stone and something for the fingers to hold by. Being transparent, the worker can see just how the grinding is progressing without continually stopping to examine. When sufficiently thin, — usually from $\frac{1}{400}$ to $\frac{1}{800}$ of an inch, — the film is remounted as follows: While on the thick glass on which it was ground, it is thoroughly washed with a brush — an ordinary tooth-brush serves well — to get rid of all particles of emery and other dirt that may adhere. It is then washed in alcohol to get rid of the old hard balsam, which is usually quite dirty from mud produced in grinding. Fresh mounting slips and clean cover glasses being ready, the first is laid upon the warm iron plate with a couple of drops of fresh balsam in the centre, and allowed to heat until it just begins to smoke. Care must here be exercised, as, if heated too much, the balsam becomes hard and brittle, and if too little, the mount is sticky from the balsam which constantly oozes from under the cover. The thick glass, with its film of stone still adhering, is likewise laid upon the warm iron plate, and a drop of fresh balsam placed upon the film. This is then gently heated, and the cover-glass, first warmed, gently laid upon it — one edge placed in position and lowered gradually in such a manner as to force out any accidental air bubbles, being finally pressed flat down against the stone film. The film itself, if sufficiently warmed, no longer adheres to the thick glass, and may be removed to the clean slip for its final mounting. This is best accomplished by taking up the thick glass by means of a pair of forceps and pushing cover-glass and film together, with a needle point set in a handle, off into the balsam on the new slide. The cover-glass here serves merely as a support for the thin film during the process of transferring. Without it there is danger of breakage. When fairly transferred, the new slide is removed from the hot plate, the cover pressed close down against the film, adjusted in proper position and allowed to cool. The superfluous balsam may be then removed with a hot knife and the section finally washed in alcohol. Thus completed, it forms the "*thin section*" of the petrologist.

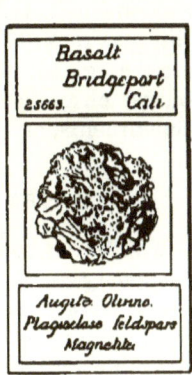

FIG. 2. — Mounted thin section of rock.

2. THE SPECIFIC GRAVITY OF ROCKS

The term *specific gravity* is used to designate the weight of any substance when compared with an equal volume of distilled water at a temperature of $4°$ C. This property is therefore dependent upon the specific gravity of its various constituents and their relative proportions. The exact or true specific gravity of a rock may be obscured by its structure. Thus an obsidian pumice will float upon water, buoyed up by the air contained in its innumerable vesicles, while a compact obsidian of precisely the same chemical composition will sink almost instantly. This property of any subject is spoken of as its apparent specific gravity in distinction from the actual comparative weight, bulk for bulk, of its constituent parts, which could in the case of a pumice be obtained only by finely pulverizing so as to admit the water into all its pores. Inasmuch as the structural peculiarities of any igneous rock — as will be noted later — are dependent upon the condition under which it cooled, it is instructive to notice that a crystalline aggregate has a higher specific gravity, *i.e.* a greater weight, bulk for bulk, than does a glassy, non-crystalline rock of the same chemical composition. The property is therefore dependent upon chemical (and consequently mineral) composition and structure, and as a very general rule it may be said that among the siliceous rocks those which contain the largest amount of silica are the lightest, while those with a comparatively small amount, but which are correspondingly rich in iron, lime, and magnesian constituents, are proportionately heavy.

3. THE CHEMICAL COMPOSITION OF ROCKS

This varies naturally with their mineral composition. It is customary to speak of sedimentary rocks as calcareous, siliceous, ferruginous, or argillaceous, accordingly as lime, silica, iron oxides, or clayey matter are prominent constituents. Among eruptive rocks it is customary to speak of those showing, on analysis, upwards of 60 % silica as acidic, and those showing less than 50 %, but rich in iron, lime, and magnesian constituents, as basic. The extremes, as will be noted, are represented by the rocks of the granite and peridotite groups.

A series illustrating the above-mentioned properties may be arranged as below. With the eruptive rocks only the silica percentages are here given. The results of the complete chemical analysis of each variety are given further on, in the pages devoted to their description.

THE CHEMICAL COMPOSITION OF ROCKS

(1) STRATIFIED ROCKS

Kind	Specific Gravity	Composition
Calcareous:		
Compact limestone	} 2.6 to 2.8	Carbonate of lime.
Crystalline limestone		
Compact dolomite	} 2.8 to 2.95	Carbonate of lime and magnesia.
Crystalline dolomite		
Siliceous:		
Gneiss	2.6 to 2.7	Same as granite.
Siliceous sandstone	2.6	Mainly silica.
Schist	2.6 to 2.8	60 to 80 per cent silica.
Argillaceous:		
Clay slate (argillite)	2.5	Mainly silicate of aluminum.

(2) ERUPTIVE ROCKS

Kind	Specific Gravity	Per cent Silica
Acidic group:		
Granite	2.58 to 2.73	77.65 to 62.00
Liparite	2.53 to 2.70	76.06 to 67.61
Obsidian	2.26 to 2.41	82.80 to 71.19
Obsidian pumice	Floats on water.	82.80 to 71.19
Intermediate group:		
Syenite	2.73 to 2.86	72.30 to 54.65
Trachyte	2.70 to 2.80	64.00 to 60.00
Hyalotrachyte	2.40 to 2.50	64.00 to 60.00
Andesite	2.54 to 2.70	66.75 to 54.73
Basic group:		
Diabase	2.66 to 2.88	50.00 to 48.00
Basalt	2.90 to 3.10	50.59 to 40.74
Peridotite	3.22 to 3.29	42.65 to 33.73
Peridotite (iron rich)	3.80	23.00
Peridotite (meteorite)	3.51	37.70

4. THE COLOR OF ROCKS

The color of a rock is dependent upon a variety of circumstances, but which may all be generalized under the heads of mineral and chemical composition and physical condition. Iron and carbon, in some of their forms, are the common coloring substances and the only ones that need be considered here. The yellow, brown, and red colors, common to fragmental rocks, are due almost wholly to free oxides of iron. The gray, green, dull brown, and even black colors of crystalline rocks are due to the presence of free iron oxides or to the prevalence of silicate minerals rich in iron, as augite, hornblende, or black mica. Rarely copper, manganese, and other metallic oxides than those of iron are present in sufficient abundance to impart their characteristic hues. As a rule, a white or light gray color denotes an absence of an appreciable amount of iron in any of its forms. The amber, bluish and black colors of many rocks, particularly the limestones and slates, are due to the prevalence of carbonaceous matter.

Among siliceous crystalline rocks the more basic, like those of the diabase, diorite, or basalt groups, are as a rule of a darker color than the acid varieties, the color being due to the fine grain and predominance of dark iron-magnesian silicates, such as hornblende, augite, or black mica, or their chloritic alteration products. The red or pink color sometimes occurring in granitic rocks is due to the predominance of red or pink feldspars, which in their turn owe their color to the presence of iron.

Among feldspar-bearing rocks the color is not infrequently due to the physical condition of this important constituent. Thus in many rocks like the norite of Keeseville (New York), and the Quincy (Massachusetts) granite, the dark color is largely due to the fact that the feldspar is clear and glassy, allowing the light rays to penetrate and become absorbed. The beautiful chatoyant play of colors sometimes shown by labradorite-bearing rocks like those of northern New York and of Norway is apparently due to a separation of the individual crystals along cleavage lines, into thin, transparent plates which reflect and partially polarize the light which would otherwise penetrate and become absorbed. Through weathering, such feldspars undergo a further physical change, becoming soft and porous, and no longer allowing the light to penetrate, but wholly reflecting it and causing the stone to appear white. These white feldspars, as has been very neatly expressed by the late Dr. Hawes, bear the same relation to the glassy forms as does the foam of the sea to the water itself, the difference in color being in both cases due to the changed physical condition. Indeed, the color of rocks, as may be imagined, is not constant, but liable to change under varying conditions, particularly those of exposure. Rocks black with carbonaceous matter will fade to almost whiteness on prolonged exposure, owing to the bleaching out of the coloring materials. Rocks rich in magnetite or free iron oxides, protoxide carbonates, or sulphides, or in highly ferruginous silicate minerals, are likewise liable to a change of color, becoming yellowish, red, or brown, through oxidation of the ferruginous constituents. (See p. 257.) Translucent, nearly colorless rocks or minerals, as those made up of crystals of calcite or selenite, will on exposure become nearly opaque and snow-white, owing to purely physical causes, as already noted in the case of the feldspars. (See further in chapter on weathering.)

The cause of the color variations in certain rocks and minerals is, however, a matter concerning which it will not do, as yet, to speak too decidedly. Analysis of a mineral may show the presence of metallic oxides, but it does not necessarily follow that whatever color the mineral may have is due or in any way related to these oxides. Thus the writer has shown [1] that the onyx marbles (travertines) of Arizona and Mexico may vary from pure white to green, and from yellow through brown to red, without appreciable change in the actual amounts of iron, though there may be a change in the form of combination. In the white and green varieties the iron exists as a carbonate; in the yellow, red, and brown varieties as a more or less hydrated sesquioxide. Certain dark amber and bright rose-colored varieties from California, and the Californian Peninsula, show, however, no iron or other of the usual metallic coloring constituents, but burn perfectly white when submitted to high temperatures and yield volatile organic compounds. The fact that serpentines so frequently contain small traces of chromium, early gave rise to the opinion that it was to this element that was due the characteristic green color of the mineral. The writer has elsewhere [2] described serpentines of a beautiful oil yellow and deep green color which, however, contain not a trace of chromium or manganese, but only iron, which in this case is in combination as a silicate. (See p. 114.)

These color characteristics are of greater importance than

[1] Annual Report U. S. National Museum, 1893, p. 558.
[2] On the Serpentine of Montville, New Jersey, Proc. U. S. National Museum, 1888, p. 105.

may at first appear, particularly from an economic standpoint. One of the first essentials in a rock designed for architectural use should be permanency of color. Deleterious changes are particularly liable to occur in stone taken from below the water level, where, protected from oxidation, or from variations in temperature. Certain of the Ohio sandstones are of a blue-gray color below the water level, but buff above, where the included iron sulphides and protoxide carbonates have been acted upon by oxidation. The student should early make himself acquainted with these characteristics, as in the field it is as a rule only the more or less weathered surfaces that present themselves for inspection. This subject is again referred to in the chapter on rock weathering.

Lustre as a property of rocks does not, owing to their complex nature, possess the same value as a determinative characteristic as among minerals. Certain of the more compact and homogeneous varieties possess lustres which may be described as *vitreous, greasy, pearly, metallic,* or *iridescent.*

The meaning of such terms is sufficiently evident, and the subject need not be further dwelt upon here. The fracture, or manner of breaking of any rock, is dependent more upon structure than upon chemical or mineralogical composition. Many fine and evenly grained crystalline or fragmental rocks break with smooth, even surfaces, and are described as having a *straight* or *even* fracture. Others break with shell-like concave and convex surfaces, and are said to have a *conchoidal* fracture. Still others are *splintery, hackly,* or *shaly,* words the meaning of which is sufficiently evident without their being described in detail.

V. THE MODE OF OCCURRENCE OF ROCKS

It is ordinarily assumed that the earth owes its present form to its having originated from a mass of incandescent vapor, and to have passed, by gradual cooling and consequent condensation, from gaseous through pasty or fluidal, and all intermediate stages, to its present condition. This, in brief, is the hypothesis of Kant, and which seems most readily to account for the facts as we now know them. As to the character of the rock masses resulting from this primary cooling, we know but little. Reasoning from analogy, it seems safe to assume that they resembled the slags from a smelting furnace, or some form of modern lavas, more nearly than any other rock masses of which we have knowledge. Whatever may have been their nature, they have long since been obscured by rocks of secondary origin, or become so altered through dynamic and incidental chemical agencies as to be no longer recognizable.

The oldest rocks of which we now have knowledge belong to the group of gneisses and crystalline schists. They are as a rule highly siliceous rocks, though not infrequently including considerable thicknesses of crystalline limestone. They contain no traces of what can be referred beyond doubt to an organic origin, though from their banded or foliated structure, so closely simulating bedding, they have in the past, as a rule, been considered metamorphic rocks; that is to say, rocks laid down as sediments and crystallized by the complex processes comprehended under the term *metamorphism*. Rocks of this type, according to Dana, first appeared in North America in the wide V-shaped area extending from Labrador southwesterly to the Great Lake, and thence northwesterly to the Arctic regions. This area has since been added to by the folding and crumpling processes incident to the formation of the Appalachian and Rocky Mountain systems. Concerning the geographical distribution of these rocks, as they now appear exposed, we have little to say here. They seem to form, as has been stated, the actual floor of the continents upon which all later deposits have been laid down, and through which and into which have been extruded and intruded the great variety of igneous rocks which form so conspicuous a feature in many a mountainous region. In order to properly understand that which is to follow, we may well devote a little space here to a consideration of the manner in which these rock masses occur, so far as exposed to investigation.

Several varieties of igneous rocks, and particularly the granitic types, occur not infrequently in the form of immense oval or rounded masses, protruded into overlying materials which dip away on all sides; such forms are ordinarily designated as *bosses*. (Pl. 1.) It is a form common to granite, gabbros, norites, etc. A *laccolite*[1] is a somewhat similar form due to the welling up of a magma through a comparatively small vent, but which, instead of coming to the surface, spread out laterally into dome-shaped masses between the sheets of the overlying strata. When the intruded matter has been so forced into or between overlying bedded rocks as to appear like more or less regularly defined beds, they are known as *sheets* or *sills*. Such, as a rule, may be distinguished from superficial lava flows by their like condition of compactness along both upper and lower contacts, surface streams being more or less vesicular along the upper portions, owing to the expansion of their included moisture. The name *dike* is given to an eruptive mass of varying width included between well-defined walls, and occupying a fissure or fault in previously consolidated rocks. Such are inclined at all angles with the horizon, and are usually of very moderate width, but may extend for miles. The dikes in any one region will frequently be found to belong to one or more well-defined systems, each system occupying fissures essentially parallel with one another. Any one dike may remain comparatively uniform in width for long distances, excepting when split up into smaller dikes. At times, dikes may be traced to the parent mass — a boss or laccolite — from which they radiate with more or less regularity, being in such cases widest at the start, and gradually

---

[1] It is to be regretted that this name in its present form has been so generally adopted by geologists, since its termination, *ite*, should indicate a kind of rock, whereas, in fact, it but denotes a form of occurrence. Laccolith would be preferable.

thinning out to, it may be, mere knife-like edges. The name *volcanic neck* or plug is given to the cylindrical mass which results from the congealing of that portion of the lava which remains in the volcanic vent when eruption ceases. Through the erosion of the matter composing the cone of a volcano, such are sometimes left, owing to their superior hardness, forming thus a very striking feature of the landscape. The general name *lava* is applied to any igneous rock, regardless of geological age or mineral composition, which has been poured out on the surface of the earth in a molten condition. Such are characterized, as a rule, by less perfect crystallization and a more slaggy and vesicular structure than the deep-seated rocks. A columnar jointing, due to cooling, is by no means uncommon, particularly among basaltic lavas, although it is by no means confined to them.

But a comparatively small proportion of the rocks composing the superficial portions of the earth's crust — the portions with which we are more or less familiar — are eruptive. They are rather what are known as *secondary* rocks ; that is to say, they are rocks made over from these so-called *primary* rocks, which we have been just discussing, by processes which will be described later.

Any rock mass, be it eruptive or otherwise, lying exposed at or near the surface of the ground finds itself subjected to a multitude of disintegrating and decomposing agencies, such as will be described more in detail under the head of rock weathering. Leached and decomposed by meteoric waters, disintegrated by heat and frost, or the mechanical action of waves and currents, the rock masses slowly succumb, their materials being gradually removed in solution, or as débris mechanically transported by every wind, rain, or running stream, down the slopes into the valleys, and from the valleys into the seas. This débris, in various stages of coarseness and fineness, and to which we give the name of bowlders, gravel, sand, and silt, undergoes by these transporting agencies a system of assorting more or less complete, and is carried to distances dependent upon its weight and the force of the transporting agent. It requires no geological or other special training to enable one to understand that the force being the same, the finer and lighter materials will be carried the farthest, and that all must be deposited when the force shall be expended. Consider, then, for purpose of illustration, a stream flowing out from a mountainous region and emptying itself into a lake. Materials falling by gravity from the mountain slopes, or washed by spasmodic rains into the stream, are transported certain distances, according to the strength of the current. For our present purposes, it is sufficient to consider only those portions which are transported quite to the mouth of the stream and dumped into the lake. But as the water leaves its narrow channel and spreads out into the lake, there is an almost instant diminution of the force of its current, and consequent carrying power. As a result, it begins to deposit its load, the coarsest and heaviest first, and the finer materials further out from the shore, the very finest, an impalpable silt it may be, remaining suspended until the very last. There will thus be formed on the bottom of the lake or sea, whichever it may be, a bed, or series of beds of varying thickness, of gravel, sand, and clay, the coarsest at the bottom and nearest the shore, and the finest and last the most remote.

But the streams emptying into the lake vary from time to time in their carrying capacity, and the action of the waves in the sea itself, together with the salts dissolved therein, exert a modifying action, whereby this process of *sedimentation*, as it is called, may not be quite so simple as it first appears.[1] Enough has, however, been said to show that beds of detritus laid down in this manner must occur in approximately horizontal layers, and that the layers may vary greatly in the coarseness and fineness of their materials, as well as in their mineral character. But there are still other processes of sedimentation than the purely mechanical methods described above. All natural waters contain more or less mineral matter, of which lime is the more abundant. Through the secreting power of marine animals, this lime is taken up in the form of a carbonate to form shells and calcareous skeletons of molluscs, corals, and other forms of marine life. On the death of the secreting animal, the calcareous material is left to accumulate in a more or less fragmental condition, forming thus the material of the coral islands, and to a considerable extent the beds of limestone the world over. I have said to a considerable extent, for the reason that it is doubtful if many of our limestones are of purely animal origin; in many a true chemical precipitation plays a not unim-

[1] See Conditions of Sedimentary Deposition, by Bailey Willis, Journal of Geology, 1893, p. 476.

portant part. This is especially true of the oölitic varieties, and the fact is readily apparent when we come to study such in detail. Consider a shallow sea-bottom on which are gradually accumulating in a finely divided condition the fragmental remains of calcareous organisms of any kind. By the undulatory action of the waves these are kept in almost constant motion, though it may be but gently rolling from side to side. Owing to evaporation, or a too rapid accumulation of the lime for it to be abstracted by the lime-secreting animals, the water becomes supercharged with this constituent, which is then precipitated in the form of a thin pellicle around the most available nucleus, in this case the grains of calcareous sand upon the bottom. Thus are gradually built up beds of no inconsiderable thickness, such as the well-known Carboniferous oölitic limestones of Indiana and Kentucky. The microscopic structure of stones of this class is shown in Fig. 7 on p. 112. Rocks which are laid down in the manner we have just described, whether composed of inorganic particles or fragmental materials from marine and fresh water organisms, are designated as *sedimentary*. They occur in more or less well-defined *beds* or *strata*, and hence are spoken of as *bedded* or *stratified*. Owing to the fact that they have in most cases been deposited in comparatively shallow water, they retain not infrequently the superficial markings made upon them by waves and other agencies prior to their final consolidation.

Deposits laid down as above described naturally lie approximately horizontally where not subsequently disturbed by earth movements. The earth's crust, however, is by no means in a state of stable equilibrium, but, being subjected to continuous stress or compressive force, is often broken, crushed, or folded, and crumpled to an extraordinary degree. The name *fault* is applied to the profound fractures made by these movements, and which, inclined at various angles to the horizon, may extend for miles. Usually the rocks on one side of a fault will be found to have sunk down, while those of the other remain stationary or are raised, producing thus an inequality of surface that may assume mountainous proportions. Most mountain ranges, in fact, are due to a combination of faulting and folding processes. It not infrequently happens that the masses of rock, sliding over one another along a line of fault, produce smooth or striated and often highly polished surfaces, to which the name *slickensides* is given. Such are particularly noticeable among serpentinous rocks, being apparently due to motion generated in the mass by increase in bulk incident to its conversion into serpentine.[1] The name *vein* is given to rock masses of chemical origin, deposited along previously existing fractures which may or may not be true faults. By some authorities the name is also made to include the smaller injections of igneous rocks. Such are here classed under the head of dikes, though it must be understood that it is not in all cases possible to state to which of the two classes an occurrence is to be referred. It is customary to divide the veins into two classes: (1) the mineral veins, in which the materials have been deposited from aqueous solution or sublimation between the walls of a fissure; and (2) segregation veins, in which the component materials have crystallized or segregated out of the still unconsolidated, pasty, or colloidal rock. It is not in all cases possible to decide to which of the two classes a vein may belong, but as a rule the mineral (or fissure) veins are separated by sharp and well-defined walls from the country rock, and show a comb or banded structure. The segregation type is less distinctly marked, the vein material being welded to the enclosing rock, or seemingly passing into it by gentle gradations.

The unconsolidated materials, as sands and gravels, occur not only in regularly bedded or stratified forms, but also in hillocks and ridges to which special terms are applied. The loose material washed down the mountain slopes by ephemeral streams, and deposited at the mouth of gorges, not infrequently assumes the form of "a conical mass of low slope descending equally in all directions from the point of issue." To such forms Gilbert has given the name of *alluvial cones*. The material of these cones, as described, varies in size from the finest powder to angular rocks weighing many tons. It exhibits no regular bedding or stratification, but coarse and fine débris are mingled in endless variety. There is a well-marked gradation, however, to be seen as one travels from the apex of a cone toward its periphery. At the apex it is composed mostly of coarse, angular material, with fine silt-like clays filling the interspaces, while toward the periphery the fine material predominates. The name *talus* is given to the accumulations of

[1] See On the Serpentine of Montville, New Jersey, Proc. U. S. National Museum, 1888, p. 105.

débris at the foot of rocky cliffs, and which are composed of angular fragments, large and small, which have fallen from the cliffs above. The name *dune* is given to the rounded hills of wind-blown sand common in arid regions and on windy shores. Such are naturally of moderately fine and quite uniformly assorted materials. In form and position they are ever changing, like drifts of snow, but are usually much steeper on the leeward than on the windward sides. The character of the material of which they are composed is most commonly siliceous sand.

The names *kame, esker, osar,* or *horseback* are given to ridges and mounds of sand and gravel deposited by the melting ice of the glacial epoch. The materials are as a rule well rounded, and as deposited usually show rude lines of stratification. Such, as described, vary greatly in breadth and height, some being 400 to 500 feet broad at the base and from 25 to 60 feet in height. *Drumlin* is the name given to the peculiar low, gently and smoothly sloping lenticular hills composed of unassorted glacial débris, and which are common in eastern Massachusetts and other glacial regions. The general name *moraine* includes the heterogeneous materials brought down by glaciers and ultimately deposited in undulating hills and ridges on their final disappearance. (See further under The Regolith, p. 299.)

PART II

THE KINDS OF ROCKS

"Some rin up hill and down dale knapping the chucky stones to pieces wi hammers like sae many road-makers run daft. They say it is to see how the warld was made." — *St. Ronan's Well.*

REFERENCE has already been made to the fact that but sixteen out of the sixty-nine known elements enter into the composition of the earth's crust in other than comparatively minute quantities. Also to the equally important fact that the combination of these elements as represented in not above a score of well-known mineral species go to make up the essential portion of nearly all rock masses. Nevertheless, owing to the variety of forms under which these rock masses occur, the varying forces or conditions under which they originated, or the proportional quantities of the various minerals which they may contain, we find numerous and widely varying types of rocks, a satisfactory consideration of which necessitates first some attempt at systematic classification. We may say at the outset, however, that rock species, in the sense in which the word is used in mineralogy and zoölogy, scarcely exist. It is true we may have, and particularly among igneous rocks, certain forms which on casual inspection, or indeed on close inspection, with regard only to limited geographical areas, seem to possess an individuality of their own sufficient to entitle them to being considered as true species. Yet, when we come to compare these with others, to take into account their physical and chemical composition, their structure and mode of occurrence, and above all to consider how any rock varies within its own mass, and the still greater variation which may have been produced through alteration, we shall see that one form grades into another almost without limit, that, indeed, no two are exactly alike, and that, were we to attempt any hard and sharp lines of discrimination, our species-making would practically resolve itself into an enumeration of individual occurrences, or specimens. This fact will become apparent as we proceed, and further remarks on the subject may well be deferred until we come to a discussion of individual groups. Indeed, in the present, transitional state of knowledge regarding the chemical and mineralogical composition of rocks, their structural features, and methods of origin, no scheme of classification can be advanced that will prove satisfactory in all its details. The older systems, which were made to answer before the introduction of the microscope into geological science, are now known to be founded upon what were in part false, and what have proven to be wholly inadequate, data. This is especially true in regard to eruptive rocks. The time that has elapsed since this introduction has been too short for the evolution of a perfectly satisfactory system; many have been proposed, but all have been found lacking in some essential particulars. To enter upon a discussion of the merits and demerits of the various schemes would obviously be out of place here, and the student is referred to the published writings of Naumann, Senft, Von Cotta, Richtofen, Vogelsang, Zirkel, Rosenbusch, Michel-Levy, Credner, Jukes Brown, and Geikie, as well as those of the American geologists, Dana,[1] Wadsworth,[2] and Iddings.[3] In the scheme here presented the writer has aimed to simplify matters so far as is consistent with observed facts, and has not hesitated to adopt or reject any such portions of systems proposed by others as have seemed desirable.

All the rocks forming any essential part of the earth's crust are here grouped under four main heads, the distinctions being based upon their origin and structure. Each of the main divisions is again divided into groups or families, the distinctions being based mainly upon mineral and chemical composition, structure, and mode of occurrence. We thus have: —

I. **Igneous Rocks:** *Eruptive.* — Rocks which have been brought up from below in a molten condition, and which owe their present structural peculiarities to variations in conditions of solidification and composition. Having as a rule two

[1] On Some Points in Lithology, Am. Jour. of Science, Vol. XVI, 1878, pp. 335 and 431.
[2] On the Classification of Rocks, Bull. Mus. Comp. Zoöl. Howard College, No. 13, Vol. V ; also Lithological Studies.
[3] The Origin of Igneous Rocks, Bull. Philosophical Society of Washington, 1892.

or more essential constituents. In structure massive, crystalline, or glassy, or in certain altered forms, colloidal.

II. **Aqueous Rocks.** — Rocks formed mainly through the agency of water, as (*A*) chemical precipitates or as (*B*) sedimentary beds. Having one or many essential constituents. In structure laminated or bedded; crystalline, colloidal, or fragmental; never glassy.

III. **Æolian Rocks.** — Rocks formed from wind-drifted materials. In structure irregularly bedded; fragmental.

IV. **Metamorphic Rocks.** — Rocks changed from their original condition through dynamic or chemical agencies and which may have been in part of aqueous, æolian, or of igneous origin. Having one or many essential constituents. In structure bedded, schistose or foliated, and crystalline.

I. ROCKS FORMED THROUGH IGNEOUS AGENCIES. ERUPTIVE

This group includes all those rocks which having once been in a state of igneous fusion have been forced upward and intruded into the overlying rocks in the form of bosses, laccolites, dikes, and sheets, or poured out upon the surface as lavas.

Concerning the source of eruptive rocks we are yet in ignorance. By many they have been supposed to represent portions of the still unconsolidated interior of the earth. The great variety of igneous rocks, the wide variation in chemical composition as well as the apparent independence of closely adjacent volcanoes, both in the matters of time of eruption and character of erupted material, seem, however, to show that they come not from a common reservoir, but from isolated and comparatively small areas where, for reasons not now well understood, previously solidified rock masses have been so highly heated as to become pasty or liquid; and then, through their own expansion, or that of included vapors, or by compressive forces generated in the earth's crust, forced upward into the positions they now occupy. The origin of igneous rocks belongs as yet largely to the realm of speculation. We must here confine ourselves more to their mineral and chemical nature, general physical properties, and the conditions under which they occur.

Consider, then, a mass of molten rock material, — to which the term *magma* may be conveniently applied, — and which by the processes of eruption is forced upward toward the surface, and let us first dwell briefly upon the forms assumed by this magma on cooling under the various conditions in which it finds itself. It is obvious at the start that we can have actually to do with but a comparatively limited portion of the products of any eruption. If the molten material is poured out upon the surface and there remains for our inspection to-day, it is a necessary consequence that the deeper-lying portions are obscured. If, on the other hand, the superficial portions have been removed by erosion so as to expose the deeply lying parts, we have only these for study and observation. It is rare indeed that erosion has so acted on any one rock mass as to expose superficial and deep-seated portions alike. In the older regions, — those of greatest geological antiquity, — erosion, either glacial or otherwise, has not infrequently removed more or less completely the superficial parts and left for our inspection those portions of a magma that at the time of eruption never reached the surface, but cooled, it may be, under thousands of feet of superincumbent matter. Such rocks are as a rule more highly crystalline than those which in the newer, less eroded portions, flowed out upon the surface like our modern lavas. Hence it is that from a very early period it has been found convenient, for purposes of discussion, to divide the eruptive rocks into two general groups: first, the *intrusive* or *plutonic* rocks ; and second, the *effusive*, or *volcanic* rocks.

Although this classification has not been strictly adhered to in the present work, a few words descriptive of the essential distinctions between plutonic and effusive rocks will not be out of place, since such distinctions, particularly in eroded regions, afford the only criteria for discrimination as to the original conditions under which a rock mass has been formed, and hence are of value in the field.

As a general rule, it may be said that the structural features of an eruptive rock depend upon the conditions under which a magma has cooled, although undoubtedly the amount of included vapor of water may exert a powerful influence. As Professor J. P. Iddings has well expressed it, "the chemical differences of igneous rocks are the result of a chemical differentiation of a general magma, and the structure of a rock is dependent upon the physical conditions attending its eruption and solidification." Now it is at once apparent that the greater the depth below the surface at which a magma undergoes solidification, or the greater its mass, the slower, more gradual, will be that solidification, and hence the more complete and coarser will be the crystallization. Hence the strictly plutonic rocks are always holocrystalline. And, inasmuch as the weight of the superincumbent matter has been such as to prevent the expansion of included vapors to form steam cavities, so these rocks are never vesicular or pumiceous, but compact and granular throughout. In cases where a plutonic rock has been voided upward to fill a pre-existing rift in the form of a dike, those portions of the magma coming in contact with the cold walls on either hand will cool most quickly. Hence a dike is as a rule most coarsely crystalline near the centre, becoming finer grained and perhaps microcrystalline or even glassy at the immediate contact. These two phenomena often afford the only means of determining whether a rock mass occurring in the form of a sheet parallel with the stratification, between sedimentary beds, is an intrusive or a contemporaneous lava flow; whether it was injected as we now find it between two previously existing beds; or whether, as a lava flow, it was poured out over the lower, first formed, after which the second was laid down upon its surface. If formed as an intrusive sheet, we may expect to find the rock more dense along both contacts, in addition to which there may, very probably, be more or less contact metamorphism on the sedimentary beds from the action of the hot intruded material. If poured out as a lava, on the other hand, contact metamorphism and the dense, fine-grained portions will be limited to the lower contacts, while, provided there had been no great amount of erosion between the time of the pouring out of the molten mass as a surface flow and the deposition of the newer sediments, the upper portions will be less dense, perhaps even vesicular, scoriaceous, and glassy, while the sediments themselves, having been laid down on cold consolidated material, remain wholly unchanged. Such means of discrimination have been of the greatest value in ascertaining the relative ages of portions of the Triassic sandstones and associated traps in the eastern United States.

The lava flows, cooling so much more rapidly than the plutonic rocks, owing to their exposed position and relief from pressure, often show but incipient forms of crystallization, or are quite glasslike, as is the case with the obsidians of the Yellowstone Park and elsewhere. Chemically these are identical with granite, but they have cooled too quickly for the forces of crystallization to act. Owing, further, to the expansive force of the included vapor of water, — a constituent of all lavas, — these surface flows are not infrequently so filled with cavities as to be quite pumiceous. The pumice purchased at the drug-stores is but the froth from a lava which, had it cooled under greater pressure, might have given us a granite.

A common feature of the effusive or volcanic rocks is a flow structure, sometimes visible only with the microscope, and which is due to a flowing movement of the magma while undergoing consolidation. (See Fig. 2, Pl. 2.) The characteristic structure of effusive rocks is porphyritic, instead of granular, and represents two distinct phases of cooling and crystallization: (1) an intratellurial period, marked by the crystallization of certain constituents while the magma, still buried in the depths of the earth, was cooling very gradually, and (2) an effusive period, marked by the final consolidation of the material on or near the surface. As this final cooling was much the more rapid, the ultimate product is a glassy, felsitic, or sometimes holocrystalline ground-mass, enclosing the porphyritic minerals, or phenocrysts, formed during the first or intratellurial stage.[1] Naturally the deeper-lying portions of an effusive mass,.those forming the under or lower portions of deep lava streams, will be under conditions essentially similar to plutonic magmas, and may cool so slowly as to become holocrystalline. It is, moreover, obvious that, could we trace any superficial mass of erupted material back to its original deep-seated source, we would pass gradually from the volcanic to the plutonic type without at any one point being able to indicate the line of separation. Hence it is that in the laboratory it is not always possible, from the examination of the hand specimen or thin section only, to determine to which of the two classes it may belong. We can easily discriminate between the extremes, but there is a wide intermediate zone where any such attempts are impracticable, as indeed they are unnecessary.[2]

[1] Whitman Cross has shown that there are exceptions to this rule. See, The Laccolitic Mountain Groups of Colorado, 14th Ann. Rep. U. S. Geol. Survey, pp. 231-235.

[2] Intermediate between these plutonic and effusive types is still a third phase of prevailing holocrystalline porphyritic structure, and which, owing to the fact that such have thus far been found only in dikes, it has been proposed to group under the head of *dike rocks* (gangesteine). Since such are but local phases of plutonic magmas, which have been left to cool and crystallize between narrow walls, instead of poured out upon the surface, such a subdivision seems scarcely called for and as tending to still further confuse that which is already sadly confounded. The same may be said with reference to the now prevailing tendency to give varietal names to every phase of magmatic differentiation, and which has resulted already in such monstrosities of nomenclature as ouachitite, monchiquite, yogoite, and absarokite.

Owing to a false impression which formerly prevailed relative to the nature of the Palæozoic effusives and those of Mesozoic, Tertiary, and more recent times, dissimilar names have, in very many instances, been applied to rocks which in other respects than that of geological age are essentially one and the same. Thus the name *andesite* is given to a rock in every respect similar to *porphyrite*, with the possible exception of a slight amount of devitrification the latter may have undergone owing to its greater geological antiquity.

The name *rhyolite* likewise includes rocks with the structure and composition of the older quartz porphyries, and though intended by Richthofen to include only certain comparatively modern acid lavas, has been shown by the late Dr. Williams[1] to be equally applicable to the pre-Cambrian lavas of the South Mountain region of Pennsylvania. These and other names have, however, become too firmly engrafted upon the literature to be too hastily set aside, and may well be retained here.

The following table will serve to show the relationship, so

Intrusive or Plutonic		Effusive or Volcanic	
		Palæovolcanic	Neovolcanic
Acid 65%–75% SiO₂	Granites	Quartz porphyries . .	Liparites (rhyolites)
Intermediate 55% to 65% SiO₂	Syenites	Quartz-free porphyries	Trachytes
	Nepheline syenites (Foyaites)	Phonolites	Phonolites
	Diorites	Porphyrites	Andesites
Basic 40% to 55% SiO₂	Gabbros, norites, and diabases	Melaphyrs and augite porphyrites	Basalts
	Theralites . . .	(Not known)	Thephrites and basanites
	Peridotites . . .	Picrite porphyrites .	Limburgites
	Pyroxenites . . .	(Not known) . . .	Augitites
	(Not known) . .	(Not known) . . .	Leucite rocks
	(Not known) . .	(Not known) . . .	Nepheline rocks
	(Not known) . .	(Not known) . . .	Melilite rocks

far as known, which exists between the plutonic rocks and their effusive equivalents of whatever age. Thus the palæo-

[1] Am. Jour. of Science, Vol. XLIV, p. 482, 1892.

volcanic equivalents of the syenites are the quartz-free porphyries, and the neovolcanic equivalents, the trachytes. The terms *acid*, *intermediate*, and *basic*, as used, have reference to the percentage amounts of silica, both free and combined, contained by the representatives of the several groups. Rocks which, like some of the peridotites, carry even less than 40 % of silica are sometimes spoken of as ultra basic.

The researches of the past few years have made it apparently evident that eruptive rocks are to be satisfactorily studied only when considered in their geographical as well as geological relationships; that is to say, the eruptives of any particular region must be considered with reference to their genetic relation to others of the same region; such a relationship as is suggested by regarding them all as but varying phases of a process of differentiation from a common magma.

That such a relationship in many cases exists has apparently been conclusively demonstrated by the work of Iddings[1] in the Yellowstone Park, J. F. Williams[2] in Arkansas, Pirsson[3] in Montana, and Brögger[4] in Norway. The attempt at correlation of local types with those of a somewhat similar nature at a distance is interesting and instructive, as showing on the whole a remarkable unity in nature's methods;·but we must never lose sight of the fact that each eruptive centre, throughout periods of activity interrupted it may be by thousands of years, works out its own results according to local conditions which may or may not harmonize with those at distant points. It is possible to conceive that, could all the rocks of any successive periods of eruption from a single centre be once more relegated to a common magma, such might, in its entirety, be an exact equivalent of others in remote portions of the globe. The consolidated results from the cooling of extruded portions of this magma may, however, show ever-varying differences due to local conditions. In short, eruptive rocks must be considered by geographic groups and with reference to magmas.

Attempts at a satisfactory classification on other grounds must prove invariably futile and tend only to retard, rather than to promote, the science.

[1] Bull. Philos. Soc. of Washington, XII, 1892.
[2] Ann. Rep. Geol. Survey of Arkansas, Vol. II, 1890.
[3] Bull. Geol. Soc. of America, Vol. VI, 1895.
[4] Die Eruptivgesteine des Kristianiagebiete, Christiana, Norway, 1894.

PLATE 6

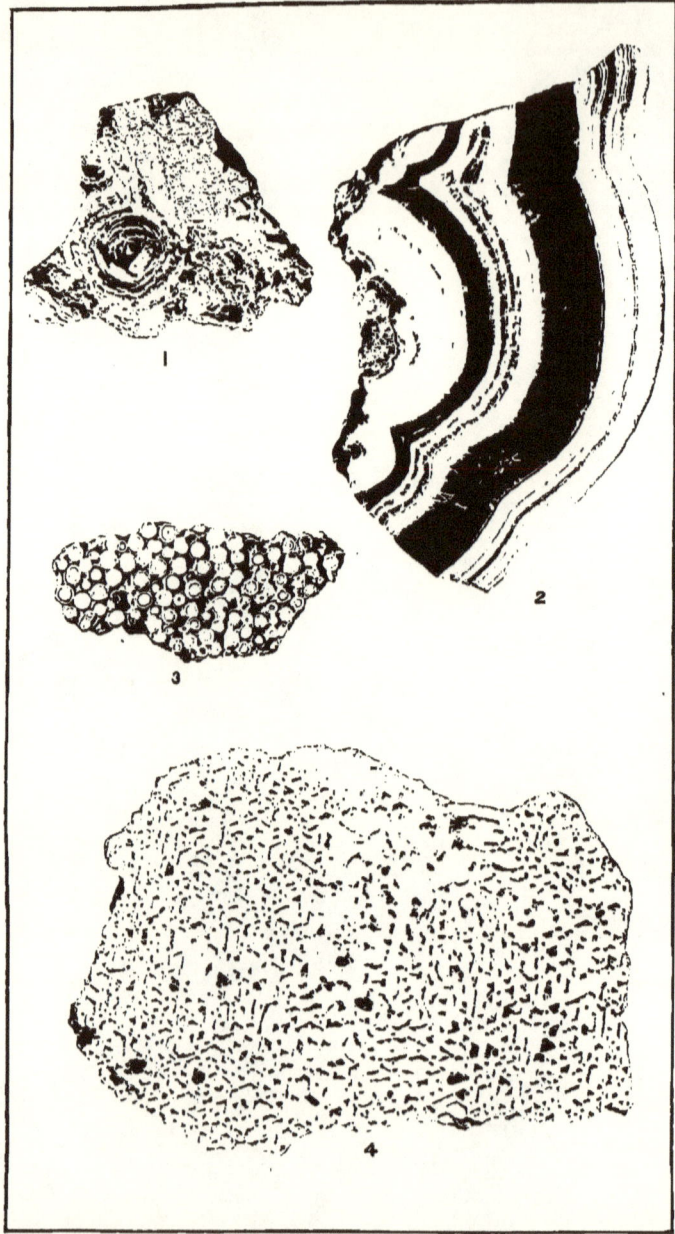

FIG. 1. Lithophysæ in liparite.
FIG. 2. Cross-section of stalagmite.
FIG. 3. Concretionary aragonite.
FIG. 4. Pegmatite.

In the following pages the rocks are discussed in groups, each group comprising all those rocks having essentially the same chemical composition, but differing (1) in degree of crystallization, (2) in mode of occurrence, and (3) in geological age. In all, there is, within certain limits, a considerable range in mineral composition, or at least in the relative proportion of the various essential constituents.

1. THE GRANITE-LIPARITE GROUP

This group includes the most acid of all eruptive rocks; that is, those which on analysis are found to yield the highest percentages of silica. Their chief essential constituents are quartz and potash feldspars, while the more basic ferruginous minerals are in quantities proportionately small. The group includes a deep-seated or plutonic type, granite, and two effusive or volcanic types, quartz porphyry, and liparite or rhyolite. They may be described in detail as below: —

(1) THE GRANITES

Granite, from the Latin "granum," a grain, in allusion to the granular structure.

Mineral Composition. — The essential constituents of granite are quartz and a potash feldspar (either orthoclase or microcline), and plagioclase. Nearly always one or more minerals of the mica, hornblende, or pyroxene group are present, and in small, usually microscopic forms, the accessories magnetite, apatite, and zircon; more rarely occur sphene, beryl, topaz, tourmaline, garnet, epidote, allanite, fluorite, and pyrite. Delesse[1] has made the following determination of the relative proportion of the various constituents in two well-known granites: —

Egyptian Red Granite	Parts	Porphyritic Granite, Vosges	Parts
Red orthoclase	43	White orthoclase	28
White albite	9	Reddish oligoclase	7
Gray quartz	44	Gray quartz	59
Black mica	4	Mica	6
Total	100	Total	100

[1] Prestwich, Chemical and Physical Geology, Vol. I, p. 42.

Chemical Composition. — A general idea of the varying character of these rocks may be gained from the following analyses: —

Kinds and Localities	SiO$_2$	Al$_2$O$_3$	FeO Fe$_2$O$_3$	CaO	MgO	K$_2$O	Na$_2$O
Biotite granite, near Dublin, Ireland	73.0	13.64	2.44	1.84	2.11	4.21	3.53
Biotite granite, Silesia	73.13	12.49	2.58	2.40	0.27	4.13	2.01
Biotite granite, Raleigh, North Carolina	69.28	17.44	2.30	2.30	0.27	2.76	3.64
Hornblende granite, Salt Lake, Utah	71.78	14.75	1.94[1]	2.36	0.71	4.89	3.12
Hornblende granite, Sauk Rapids, Minnesota	64.13	21.01		6.90	1.26	1.22	3.31
Gneissoid biotite granite, District of Columbia	69.33	14.33	3.60	3.21	2.44	2.67	2.70
Hornblende mica granite, Syene, Egypt	68.18	16.20	4.10	1.75	0.48	0.48	2.88

Although the mineral apatite is so universally a constituent of granitic rocks, yet it occurs in such small quantities as to be quite overlooked in the ordinary methods of analysis. Such tests as have been made show that the amount of phosphoric acid (P$_2$O$_5$) contained by rocks of this class rarely exceeds 0.2 % and may fall as low as 0.05 %. Small as is the amount, it is nevertheless probable that it was from just such minute quantities in granites and the more basic eruptives, that was derived the main supply of phosphates existing in soils.

Structure. — The granites are holocrystalline granular rocks. As a rule none of the essential constituents show perfect crystal outlines, though the feldspathic minerals are often quite perfectly formed. The quartz has always been the last mineral to solidify, and hence occurs only as irregular granules occupying the interspaces. It is remarkable from its carrying innumerable cavities filled with liquid and gaseous carbonic acid or with saline matter. So minute are these cavities that it has been estimated by Sorby that from one to ten thousand millions could be contained in a single cubic inch of space. The microscopic structure of a mica granite from Maine is shown in Fig. 3 and in Fig. 1, Pl. 5.

---

[1] Yielded also 1.09 % manganese oxide.

The rocks vary in texture almost indefinitely, presenting all gradations from fine evenly granular rocks to coarsely porphyritic forms in which the feldspars, which are the only constituents porphyritically developed, are several inches or feet in length.

Concretionary forms are rare. A variety from Craftsburg, Vermont, is unique on account of the numerous concretionary masses of black mica it carries.

Colors. — The prevailing color is some shade of gray, though greenish, yellowish, pink, to deep red, are not uncommon.

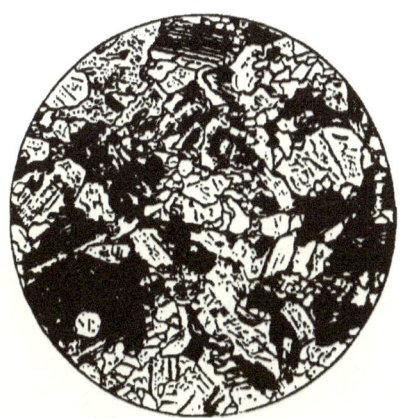

FIG. 3. — Microstructure of muscovite-biotite granite, Hallowell, Maine.

The various hues are due to the color of the prevailing feldspar and the abundance and kind of the accessory minerals. Granites in which muscovite is the prevailing mica, are nearly always very light gray in color. The dark gray varieties are due largely to abundant black mica or hornblende, the greenish and pink or red colors to the prevailing greenish, pink, or red feldspars.

Classification and Nomenclature. — Several varieties are commonly recognized and designated by names dependent upon the predominating accessory mineral. We thus have (1) *muscovite granite*, (2) *biotite granite* or *granitite*, (3) *biotite-muscovite granite*, (4) *hornblende granite*, (5) *hornblende-biotite granite*, and more rarely (6) *pyroxene* (7) *tourmaline* and (8) *epidote granite*. The name *protogine* has been given to a granite in which the mica is in part or wholly replaced by talc. The name is not very generally used.

Graphic granite, or *pegmatite*, is a granitic rock consisting essentially of quartz and orthoclase so crystallized together in long parallel columns or shells that a cross-section bears a crude resemblance to Hebrew writing. (See Fig. 4, Pl. 6.) *Aplit* is a name used by the Germans for a granite very poor in mica and consisting essentially of quartz and feldspar only.

The names *granitell* and *binary granite* have also been used to designate rocks of this class. *Greisen* is a name applied to a quartz-mica rock, with accessory topaz, occurring associated with the tin ores of Saxony and regarded as a granite metamorphosed by exhalations of fluoric acid. *Luxullianite* and *Trowlesworthite* are local names given to tourmaline or tourmaline-fluorite granitic rocks occurring at Luxullian and Trowlesworth, in Cornwall, England. The name *Unakite* has been given to an epidotic granite with pink feldspars occurring in the Unaka Mountains in western North Carolina and eastern Tennessee.

The name *granite porphyry* is made to include a class of rocks placed by Professor Rosenbusch under the head of "gangesteine," or dike rocks, and differing from the true granites mainly in structural features. They consist in their typical forms of orthoclase feldspars and quartzes porphyritically developed in a finer holocrystalline aggregate of the minerals common to the granite group.

The granites are among the most wide-spread and commonest of rocks, and are of great economic importance for structural and monumental work. In the United States they are to be found mainly in the Appalachian region and from the front range of the Rocky Mountains westward to the Pacific coast.

Geological Age and Mode of Occurrence. — The granites are massive rocks, occurring most frequently associated with the older and lower rocks of the earth's crust, sometimes interstratified with metamorphic rocks or forming the central portions of mountain chains. They are not, as once supposed, the oldest of rocks, but occur frequently in eruptive masses or bosses invading rocks of all ages up to late Mesozoic or Tertiary times. Thus Professor Whitney considered the eruptive granites of the Sierra Nevada to be Jurassic. Zirkel divides the granites described in the reports of the 40th Parallel Survey into three groups : (1) Those of Jurassic age ; (2) those of Palæozoic age, and (3) those of Archæan age. The granites of the eastern United States, on the other hand, have, in times past, been regarded as mainly Archæan, though Dr. Wadsworth has shown that the Quincy, Massachusetts, stone is an eruptive rock of late Primordial or more recent age, while Professor Hitchcock regards the eruptive granites of Vermont as having been protruded during Silurian or perhaps Devonian times.

(2) THE QUARTZ PORPHYRIES

Composition. — The mineral and chemical composition of the quartz porphyries is essentially the same as that of the granites, from which they differ mainly in structure. Their essential constituents are quartz and feldspar, with accessory black mica or hornblende in very small quantities; other accessories present, as a rule only in microscopic quantities, are magnetite, pyrite, hematite, and epidote.

Structure. — The prevailing structure is porphyritic. (Fig. 1, Pl. 2.) To the unaided eye they present a very dense and compact ground-mass of uniform reddish, brown, black, gray, or yellowish color, through which are scattered clear glassy crystals of quartz alone, or of quartz and feldspar together. The quartz differs from the quartz of granites in that here it was the first mineral to separate out on cooling, and hence has taken on a more perfect crystalline form; the crystal outlines of the feldspar are also well defined. Under the microscope the ground-mass in the typical porphyry is found to consist of a dense *felsitic*, almost irresolvable substance, which chemical analysis shows to be also a mixture of quartzose and feldspathic material. The porphyritic quartzes show frequently a marked corrosive action from the molten magma, the mineral having again been partially dissolved after its first crystallization. (Fig. 3, Pl. 5.) This difference in structure in rocks of the same chemical composition is believed to be due wholly to the different circumstances under which the two rocks have solidified from a molten magma. The structure of the ground-mass is not always felsitic, but may vary from a glass, as in the pitchstones of Meissen, Isle of Arran, and the Lake Lugano region, through spherulitic, micropegmatitic, and porphyritic to perfectly microcrystalline forms as in the microgranites. This difference in structure may be best understood by reference to Plate 5, which shows the microscopic structure of (1) granite from Sullivan, Hancock County, Maine, (2) micropegmatite from Mount Desert, Maine, and (3) a quartz porphyry from Fairfield, Pennsylvania. Marked fluidal structure is common. (See Pl. 2, Fig. 2.)

Colors. — The colors of the ground-mass, as above noted, vary through reddish, brownish gray to black and sometimes yellowish or green. The porphyritic feldspars vary from red, pink, and yellow to snow-white, and often present a beautiful contrast with the ground-mass, forming a desirable stone for ornamental purposes.

Classification and Nomenclature. — Owing to the very slight development of the accessory minerals, mica, hornblende, etc., it has been found impossible to adopt the system of classification and nomenclature used with the granites and other rocks. Vogelsang's classification as modified by Rosenbusch is based upon the structure of the ground-mass as revealed by the microscope. It is as follows: —

Ground-mass holocrystalline granular . . . . Micro-granite.
Ground-mass holocrystalline, but formed of quartz and feldspar aggregates, rather than district crystals . . . . . . Granophyr.
Ground-mass felsitic . . . . Felsophyr.
Ground-mass glassy . . . . Vitrophyr.

Intermediate forms are designated by a combination of the names, as *granofelsophyr, felsovitrophyr*, etc. The name *felsite* is often given to rocks of this group in which the porphyritic constituents are wholly lacking. The names *felstone* and *petrosilex* are also common, though gradually going out of use. *Elvanite* is a Cornish miner's term and too indefinite to be of great value. *Eurite*, now little used, applies to felsitic forms. The name *felsite pitchstone* or *retinite* has been given to a glassy form with pitch-like lustre, such as occurs in dikes cutting the old red sandstone on the Isle of Arran. *Kugel porphyry* is a name given by German writers to varieties showing spheroids with a radiating or concentric structure. *Micropegmatite* is the term not infrequently applied to such as show under the microscope a pegmatitic structure. (Fig. 2, Pl. 5.) Various popular names, as *leopardite* and *toadstone*, are sometimes applied to such as show a spotted or spherulitic structure.

(3) THE LIPARITES

Mineral Composition. — These rocks may be regarded as the younger equivalents of the quartz porphyries, or the volcanic equivalents of the granites, having essentially the same mineral and chemical composition. The prevailing feldspar is the clear glassy variety of orthoclase known as sanidin; quartz occurs in quite perfect crystal forms often more or less corroded by the molten magmas, as in the porphyries, and in the minute, six-

PLATE 7

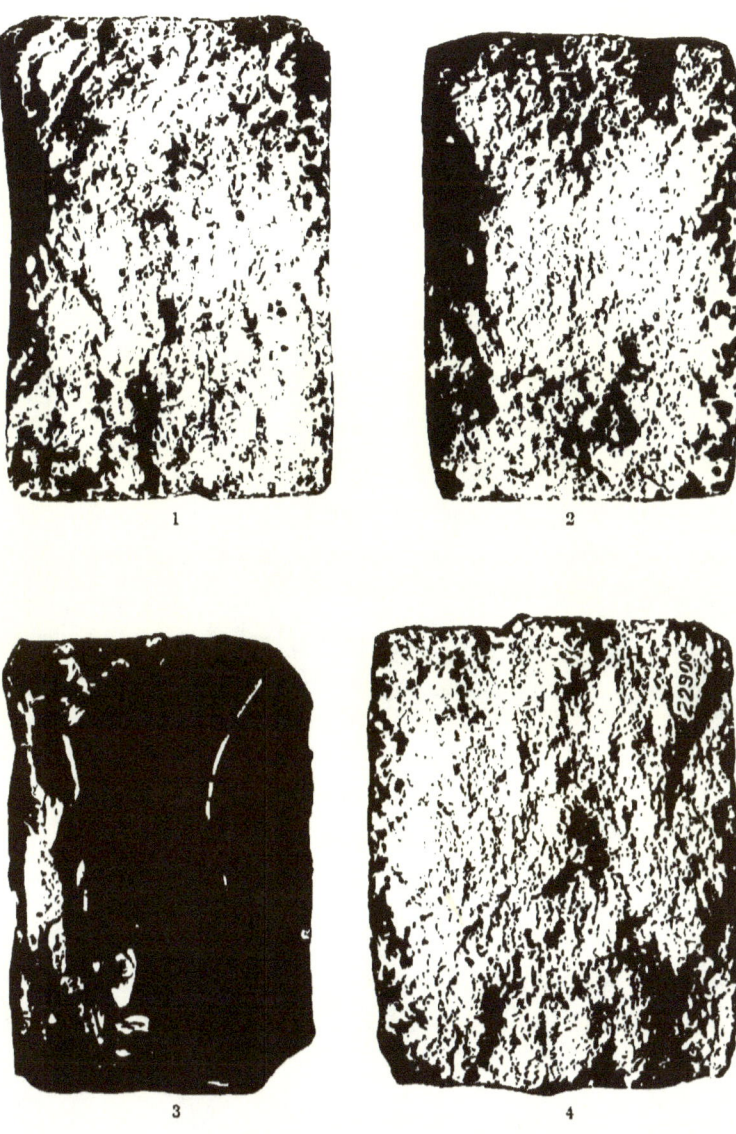

Fig. 1. Liparite, nevadite form.
Fig. 2. Liparite, rhyolite form.
Fig. 3. Liparite, obsidian form.
Fig. 4. Liparite, pumiceous form.

sided, thin platy forms known as tridymite. The accessory minerals are the same as those of the granites and quartz porphyries.

Chemical Composition. — Below is given the composition of: (I) Nevadite, from the northeastern part of Chalk Mountain, Colorado, as given by Cross.[1] (II) That of a rhyolitic form, from the Montezuma Range, Nevada, as given by King,[2] and (III) that of a black obsidian from the Yellowstone National Park, Wyoming, as given by Iddings.[3]

Constituents	I	II	III
Silica (SiO_2)	74.50%	74.02%	74.70%
Alumina (Al_2O_3)	14.72	11.90	13.72
Ferric oxide (Fe_2O_3)	None	1.20	1.01
Ferrous oxide (FeO)	0.56	0.10	0.02
Ferric sulphide (FeS_2)	....	....	0.40
Manganese (MnO)	0.28	....	Trace
Lime (CaO)	0.83	0.30	0.78
Magnesia (MgO)	0.37	....	0.14
Soda (Na_2O)	3.97	2.26	3.90
Potash (K_2O)	4.53	7.70	4.02
Phosphoric anhydride (P_2O_5)	0.01	....	....
Ignition	0.06	1.02	0.02
	100.38%	99.28%	99.91%
Specific gravity	....	2.2	2.3447

Colors. — These are fully as variable as in the quartz porphyries; white, though all shades of gray, green, brown, yellow, pink, and red are common. Black is the more common color for the glassy varieties of obsidian, though they are often beautifully spotted and streaked with red or reddish-brown.

Structure. — The liparites present a great variety of structural features, varying from holocrystalline, through porphyritic and felsitic, to clear, glassy forms. These varieties can be best understood by reference to Plates 5 and 7, prepared from photographs. Fig. 1, Pl. 7, is that of the coarsely crystalline variety nevadite from Chalk Mountain, Colorado; Fig. 2 is

[1] Geology and Mining Industry of Leadville, Monograph XII, U. S. Geol. Survey, p. 349.
[2] Geological Exploration 40th Parallel, Vol. I, p. 652.
[3] Ann. Rep. U. S. Geol. Survey, 1885-86, p. 282.

that of a common felsitic and porphyritic type; Fig. 3 is that of the clear, glassy form, obsidian; Fig. 4 shows also an obsidian, but with a pumiceous structure; Fig. 1 on Pl. 6 shows the hollow spherulites or *lithophysæ*, which have been studied and described by Professor J. P. Iddings, of the United States Geological Survey.[1] Such forms are regarded by Mr. Iddings as resulting "from the action of absorbed vapors upon the molten glass from which they were liberated during the process of crystallization consequent upon cooling." A pronounced flow structure is quite characteristic of the rocks of this group, as indicated by the name *rhyolite*. The microscopic structure of an obsidian is shown in Fig. 4, Pl. 5. Transitions from compact obsidian into pumiceous forms, due to expansion of included moisture, are common.

Classification and Nomenclature. — The following varieties are now generally recognized, the distinctions being based mainly on structural features, as with the quartz porphyries. We thus have the granitic-appearing variety *nevadite*, the less markedly granular and porphyritic variety *rhyolite*, and the glassy forms *hyaloliparite, hyaline rhyolite*, or *obsidian* as it is variously called. Hydrous varieties of the glassy rock with a dull pitch-like lustre are sometimes called *rhyolite pitchstone*.

The name *rhyolite*, from the Greek word ῥέω, to flow, it may be stated, was applied by Richtofen as early as 1860 to this class of rocks as occurring on the southern slopes of the Carpathians. Subsequently Roth applied the name *Liparite* to similar rocks occurring on the Lipari Islands. The first name, owing to its priority, is the more generally used for the group, though Professor Rosenbusch in his latest work has adopted the latter. The name *Nevadite* is from the state of Nevada, and was also proposed by Richtofen. The name *Obsidian* as applied to the glassy variety is stated to have been given in honor of Obsidius, its discoverer, who brought fragments of the rock from Ethiopia to Rome. The name *pantellerite* has been given by Rosenbusch to a liparite in which the porphyritic constituent is an orthoclase.

Rocks of these types occur, in the United States, only in the regions west of the front range of the Rocky Mountains. *Apo-rhyolite* is the name proposed by Dr. Williams for the

[1] Obsidian Cliff, Yellowstone National Park, Ann. Rep. U. S. Geol. Survey, 1885-86.

devitrified and otherwise altered pre-Cambrian rhyolite found at South Mountain in Pennsylvania.

2. THE SYENITE-TRACHYTE GROUP

This group stands next to that of the granites in point of acidity, from which it differs mainly in the lack of free silica (quartz) as an essential constituent. On chemical grounds this and the next group to be described belong to the intermediate series, standing midway between the acid granites and the basic basalts. As with the last, we have plutonic and effusive forms. These may be described as below : —

(1) THE SYENITES

The name *Syenite*, from Syene, a town of Egypt. The word was first used by Pliny to designate the coarse red granite from quarries at Syene, and used by the Egyptians in their obelisks and pyramids. Afterwards (in 1787) Werner introduced the word into geological nomenclature to designate a class of granular rocks consisting of feldspar and hornblende, either with or without quartz. Later, when a more precise classification became necessary, the German geologists reserved the name *syenite* to designate only the quartzless varieties of these rocks, while the quartz-bearing varieties were referred to the hornblendic granites. This is the classification now followed by all the leading petrologists and is therefore adopted here. Much confusion has arisen from the fact that the French geologist Rozière insisted upon designating the quartz-bearing rock as syenite, a practice which has been followed to a considerable extent both in this country and England.

Mineral Composition. — The syenites differ from the granites only in the absence of the mineral quartz, consisting essentially of orthoclase feldspar in company with biotite, or one or more minerals of the amphibole or pyroxene group. A soda-lime feldspar is nearly always present and frequently microcline; other common accessories are apatite, zircon, and the iron ores: more rarely sodalite.

Chemical Composition. — In column I below is given the composition of a hornblende syenite from near Dresden, Saxony, in II that of a mica syenite (minette) from the Odenwald, and in III and IV that of augite-sodalite syenites from Montana.

Constituents	I	II	III	IV
Silica (SiO_2)	60.02%	57.37%	54.15%	56.45%
Alumina (Al_2O_3)	16.66	13.84	18.02	20.08
Ferric iron (Fe_2O_3)	} 7.21	{ 2.44	} 6.79	{ 1.31
Ferrous iron (FeO)		3.44		4.39
Magnesia (MgO)	2.51	6.05	1.90	0.63
Lime (CaO)	3.59	5.53	3.72	2.14
Soda (Na_2O)	2.41	1.53	5.47	5.61
Potash (K_2O)	6.50	4.47	8.44	7.13
Ignition (H_2O)	1.10	3.17	....	1.77
Chlorine (Cl)	....	....	0.42	0.43
Phosphoric acid (P_2O_5)	....	....	....	0.13
	100.00%	97.84%	99.81	100.07%

Structure. — The structure of the syenites is wholly analogous to that of the granites, and need not be further described here. In process of crystallization the apatite, zircon, and iron ores were the first to separate out from the molten magma, and hence are found in more or less perfect forms enclosed by the feldspars and later-formed minerals. These were followed in order by the mica, hornblende, or augite, and lastly the feldspars, the soda-lime feldspars, when such occur, forming subsequent to the orthoclase.

Color. — The prevailing colors are various shades of gray, through pink to reddish.

Classification and Nomenclature. — According as one or the other of the accessory minerals of the bisilicate group predominates we have (1) *hornblende syenite*, (2) *mica syenite*, or minette, and (3) *augite syenite*.

Other varietal names have from time to time been given by various authors. The name *minette*, first introduced into geological nomenclature by Voltz in 1828 (Teall), is applied to a fine-grained mica orthoclase rock, occurring only in the form of dikes and further differing from the typical syenites in having a porphyritic rather than granitic structure. *Vogesite* is the name applied to a similar rock in which hornblende or augite prevails in place of mica. These rocks are placed by Professor Rosenbusch in his latest work in the group of syenitic lamprophyrs. *Monzonite* is a varietal name for the augite syenite of Monzoni in the Tyrol.

The mode of occurrence of the syenites is similar to that of the granites, though they are much more limited in their distribution. In the United States they have thus far been described but sparingly. Marblehead Neck, Massachusetts; Jackson, New Hampshire, are well-known localities; a beautiful hornblende syenite is found among the glacial drift boulders about Portland, Maine, but its exact source is not known. The hornblende syenite described by Hawes as occurring at Red Hill, Moultonborough, New Hampshire, has been shown by Professor W. S. Bayley[1] to carry elæolite, and to belong to the group of elæolite syenites. Hornblende syenites occur in the Vosges Mountains of Germany and in Saxony; mica syenites or minettes in the Odenwald, Germany, Baden, Saxony, and in the Fichtelgebirge. A mica-augite syenite carrying sodalite occurs as a Cretaceous eruptive in Jefferson County, Montana,[2] and a similar rock has been described by Lindgren from the Highwood Mountains in the same state.[3]

(2) THE ORTHOCLASE OR QUARTZ-FREE PORPHYRIES

Mineral Composition. — The essential constituents are the same as those of syenite. They consist therefore of a compact porphyry ground-mass with porphyritic feldspar (orthoclase) and accessory plagioclase, quartz, mica, hornblende, or minerals of the pyroxene group. More rarely occur zircon, apatite, magnetite, etc., as in the syenites.

Chemical Composition. — Being poor in quartz, these rocks are a trifle more basic than the quartz porphyries which they otherwise resemble. The following is the composition of an orthoclase porphyry from Predazzo as given by Kalkowski:[4] Silica, 64.45 %; alumina, 16.31 %; ferrous oxide, 6.49 %; magnesia, 0.30 %; lime, 1.10 %; soda, 5.00 %; potash, 5.45 %; water, 0.85 %.

Structure. — Excepting that orthoclase is the porphyritic constituent, they are structurally identical with the quartz porphyries, and need not be further described here.

Colors. — These are the same as the quartz porphyries already described.

[1] Bull. Geol. Soc. of America, Vol. III, 1892.
[2] Proc. U. S. Nat. Museum, Vol. XVII, 1894.
[3] Proc. Cali. Acad. of Sciences, Vol. III, 2d series, p. 47.
[4] Elemente der Lithologie, p. 86.

76   ROCKS FORMED THROUGH IGNEOUS AGENCIES

Classification and Nomenclature. — The orthoclase or quartz-free porphyries bear the same relation to the syenites as do the quartz porphyries to granite, and the rocks are frequently designated as syenite porphyries. Like the quartz porphyries, they occur in intrusive sheets, dikes, and lava flows associated with the Palæozoic formations. Owing to the frequent absence of accessory minerals of the ferro-magnesia group, the rocks cannot in all cases be classified as are the syenites, and distinctive names based upon other features are often applied. The term *orthophyr* is applied to the normal orthoclase porphyries, and these are subdivided when possible into biotite, hornblende, or augite orthophyr according as either one of these minerals is the predominating accessory. The term *rhombporphyry* has been used to designate an orthoclase porphyry found in southern Norway, and in which the porphyritic constituent appears in characteristic rhombic outlines, and which is further distinguished by a complete absence of quartz and rarity of horn-blende. The name *keratophyr* has been given by Gumbel to a quartzose or quartz-free porphyry containing a sodium-rich alkaline feldspar. So far as can be at present judged, rocks of this type are much more restricted in their occurrence than are the quartz porphyries already described.

(3) THE TRACHYTES

Trachyte, from the Greek word τραχυς, rough, in allusion to the characteristic roughness of the rock. The term was first used by Haüy to designate the well-known volcanic rocks of the Drachenfels on the Rhine.

Mineral Composition. — Under the name of *trachyte* are comprehended those massive Tertiary and post-Tertiary lavas, consisting essentially of sanidin with hornblende augite or black mica, and which may be regarded as the younger equivalents of of the quartz-free porphyries. The common accessory minerals are plagioclase, tridymite, apatite, sphene, and magnetite, more rarely olivine, sodalite, humite, hauyne, and melilite.

Chemical Composition. — The following analyses show the range in chemical composition of these rocks, I being that of the trachyte of Game Ridge, Colorado, and II that of a La Guardia stone.

Constituents	I	II
Silica (SiO_2)	66.03 %	56.09 %
Alumina (Al_2O_3)	18.49	26.09
Ferric oxide (Fe_2O_3)	2.18	....
Manganese oxide (MnO)	Trace	Trace
Lime (CaO)	0.90	3.41
Magnesia (MgO)	0.30	2.70
Potash (K_2O)	5.86	6.49
Soda (Na_2O)	5.22	3.38
Ignition (H_2O)	0.85	1.05
Phosphoric acid (P_2O_5)	0.04	....
Total	100.24 %	100.74 %

Structure. — In structure the trachytes are rarely granular, but possess a fine, scaly or microfelsitic ground-mass, rendered porphyritic through the development of scattering crystals of sanidin, hornblende, augite of black mica. The texture is porous, and the rock possesses a characteristic roughness to the touch; hence the derivation of the name as given above. Perlitic structure is common in the glassy forms. The microscopic structure of the trachyte of Monte Vetta is shown in Fig. 5, Pl. 5.

Colors. — The prevailing colors are grayish, yellowish, or reddish.

Classification and Nomenclature. — They are divided into *hornblende, augite,* or *mica trachytes,* according as any one of these minerals predominates. The name *sanidin-oligoclase trachyte* is sometimes given to trachytes in which both these feldspars appear as prominent constituents. The presence of quartz gives rise to the variety *quartz trachytes.* (See under rhyolite.) The glassy form of trachyte is commonly known under the name of *trachyte pitchstone,* or if with a perlitic structure simply as *perlite.* In his most recent work Professor Rosenbusch has included the glassy forms under the name of *hyalotrachyte.*

3. THE FOYAITE-PHONOLITE GROUP

This group differs from the last mainly in the partial replacement of the potash feldspars by the closely related mineral elæolite or nepheline. It includes therefore those plutonic and effusive rocks commonly known under the name of *elæolite* or *nepheline syenites* and the *phonolites*. In their silica and potash percentages it will be observed they differ not greatly from the syenites proper, but are much more rich in soda and correspondingly poor in lime. They may be described in detail as follows: —

(1) THE NEPHELINE (ELÆOLITE) SYENITES: FOYAITS

Nepheline from the Greek νεφελη, a cloud, since the mineral becomes cloudy on immersion in acid. Elæolite from ελαίον, oil, in allusion to the greasy lustre. Syenite from Syene in Egypt.

Mineral Composition. — The essential constituents of this group are nepheline (elæolite) and orthoclase, with nearly always a pyroxenic or amphibolic mineral and a plagioclase feldspar. The common accessory minerals are sphene, sodalite, cancrinite, zircon, apatite, black mica, and the iron ores ilmenite and magnetite, with occasional leucite, eucolite, melinophane, and also tourmalines, perowskite, and olivine. Calcite, epidote, chlorite, analcite, and sundry minerals of the zeolite group occur as secondary products.

Professor W. S. Bayley has computed [1] the relative proportions of the various constituents in the elæolite syenite of Litchfield, Maine, as follows: Elæolite, 17 %; potash feldspar, 27 %; albite, 47 %; cancrinite, 2 %; and black mica (lepidomelane), 7 %.

Chemical Composition. — The composition of the elæolite syenite from several well-known localities is given below: —

CONSTITUENTS	ALGBAVE, PORTUGAL	HOT SPRINGS, ARKANSAS	LITCHFIELD, MAINE	BEEMERVILLE, NEW JERSEY
Silica (SiO_2)	54.61 %	59.70 %	60.39 %	50.36 %
Alumina (Al_2O_3)	22.07	18.85	22.51	19.84
Ferric oxide (Fe_2O_3) . . .	2.33	4.85	.42	} 0.94
Ferrous oxide (FeO) . . .	2.50	....	2.26	
Magnesia (MgO)	0.88	0.08	0.13	....
Manganese oxide (MnO) . .	....	....	0.08	0.411
Lime (CaO)	2.51	1.34	0.32	3.43
Soda (Na_2O)	7.58	6.29	8.44	7.64
Potash (K_2O)	5.46	5.97	4.77	7.17
Titanium oxide (TiO_2) . .	0.00	....	....	....
Phosphoric anhydride (P_2O_5)	0.15	....	....	....
Water (H_2O)	1.13	1.88	.57	3.512 (loss)

[1] Bull. Geol. Soc. of America, Vol. III, 1892, p. 231.

The essential points to be noted are the larger percentages of the alkalies over those yielded by syenites of the ordinary type, or the granites.

Color. — The colors are light to dark gray, and sometimes reddish.

Structure. — The syenites, like the granites, are massive holocrystalline granular rocks, and as a rule sufficiently coarse in texture to allow a partial determination of the constituent parts by the unaided eye. In the Litchfield (Maine) syenite the elæolite often occurs in crystals upwards of 5 centimetres in length, and zircons 2 centimetres in length are not rare. Neither of the essential constituents occur in the form of perfect crystals, while the apatite, zircon, black mica, and pyroxenic constituents often present very perfect forms. The cancrinite occurs both as secondary after the elæolite and as a primary constituent in the form of long needle-like yellow crystals with a hexagonal outline. This last form is especially characteristic of the Litchfield rock. The sodalite occurs both as crystals and in irregular massive forms, coating the walls of crevices.

Classification and Nomenclature. — Several varietal names have been given to the rocks of this group as described by various authors. *Miascite* was the name given by G. Rose to the syenite occurring at Miask in the Urals; *Ditroite* to that occurring at Ditro in Transylvania, and *Foyaite*, by Blum, to that from Mount Foya, in the province of Algrave in Portugal. The name *zircon syenite*, or *Laurvikite*, has been given to the variety from Laurvig in southern Norway, which is rich in zircons. *Tinguaite* is the name proposed for a varietal form from Serra de Tingua, province of Rio Janeiro, Brazil.

American petrographers have not been at all delinquent in the matter of names, and have added to an already overburdened nomenclature such terms as *Litchfieldite*, *Ouachitite*, *Pulaskite*, and *Fourchite* to varieties from Litchfield (Maine) and the Hot Springs region of Arkansas. *Liebnerite* is the name given to an elæolite syenite porphyry occurring in the Tyrol.

Rocks of this group, although wide-spread in their distribution, are nevertheless not abundant. The more important localities thus far described have already been noted; there remains to be mentioned only the locality at Red Hill, Moultonborough, New Hampshire, the rock of which was first described as an ordinary syenite, and that of Hastings County, Ontario.

(2) THE PHONOLITES

Phonolite, from the Greek word φωνή, sound, and λίθος, stone, in allusion to the clear ringing or clinking sound which slabs of the stone emit when struck with a hammer; formerly called *clinkstone* for the same reason.

Mineral Composition. — The phonolites consist essentially of sanidin and nepheline or leucite, together with one or more minerals of the augite-hornblende group, and generally hauyne or nosean. The common accessories are plagioclase, apatite, sphene, mica, and magnetite; more rarely occur tridymite, melanite, zircon, and olivine. The rock undergoes ready alteration, and calcite, chlorite, limonite, and various minerals of the zeolite group occur as secondary products.

Chemical Composition. — The average of six analyses given by Zirkel[1] is as follows: Silica, 58.02 %; alumina, 20.03 %; iron oxides, 6.18 %; manganese oxide, 0.58 %; lime, 1.89 %; magnesia, 0.80 %; potash, 6.18 %; soda, 6.35 %; water, 1.88 %; specific gravity, 2.58.

Structure. — The phonolites present but little variety in structure, being usually porphyritic, seldom evenly granular. The porphyritic structure is due to the development of large crystals of sanidin, nepheline, leucite, or hauyne, and more rarely hornblende, augite, or sphene, in the fine-grained and compact ground-mass, which is usually microcrystalline, rarely glassy or amorphous.

Colors. — The prevailing colors are dark gray or greenish.

Classification and Nomenclature. — Three varieties are recognized by Professor Rosenbusch, the distinction being founded upon the variation in proportional amounts of the three minerals, sanidin, nepheline, or leucite. We thus have (1) *nepheline phonolite*, consisting essentially of nepheline and sanidin, and which may therefore be regarded as the volcanic equivalent of the nepheline syenite; (2) *leucite phonolite*, consisting essentially of leucite and sanidin; and (3) *leucitophyr*, which consists essentially of both nepheline and leucite in connection with sanidin, and nearly always melanite.

[1] Lehrbuch der Petrographie, II, p. 193.

So far as now known, these rocks are of comparatively rare occurrence in the United States, having been described as occurring only in the Black Hills of South Dakota and the Cripple Creek district of Colorado.

4. THE DIORITE-ANDESITE GROUP

We come now to groups of rocks which show a still greater falling off in their total amount of silica, as indicated by analyses, and a like diminution in the amount of potash. The cause of this falling off is due to the absence as an essential constituent of quartz and potash feldspars, the latter being replaced by soda-lime varieties, and which in their turn cause a corresponding increase in the elements sodium and calcium. The group includes the plutonic type *diorite*, and the effusive types *hornblende porphyrite*, and *andesite*. These may be described as below:—

(1) THE DIORITES (GREENSTONES IN PART)

Diorite, from the Greek word διοριζεῖν, to distinguish. Term first used by the mineralogist Haüy.

Mineral Composition. — The essential constituents of diorite are plagioclase feldspar, either labradorite or oligoclase, and hornblende or black mica. The common accessories are magnetite, titanic iron, orthoclase, apatite, epidote, quartz, augite, black mica, and pyrite, more rarely garnets. Calcite and chlorite occur as alteration products.

Structure. — Diorites are holocrystalline granular rocks, and as a rule, massive, though schistose forms occur. The individual crystals composing the rock are sometimes grouped in globular aggregates, thus forming the so-called *orbicular diorite, kugel diorite*, or *napoleonite* from Corsica. (Fig. 1, Pl. 8.) The texture is, as a rule, fine, compact, and homogeneous, and its true nature discernible only with the aid of a microscope; more rarely porphyritic forms occur as in the camptonites.

Colors. — The colors vary from green and dark gray to almost black.

Chemical Composition. — The following table shows the wide range in chemical composition found in rocks commonly grouped under this head.

Classification. — Accordingly as they vary in mineral composition the diorites are classified as (1) *diorite*, in which horn-

CONSTITUENTS	I	II	III	IV	V	VI
Silica (SiO_2)	67.54 %	61.75 %	56.71 %	50.47 %	43.50 %	39.32 %
Titanic oxide (TiO_2)	....	....	....	....	....	1.70
Alumina (Al_2O_3)	17.02	18.88	18.36	18.73	17.02	14.48
Ferric iron (Fe_2O_3)	2.97	0.52	....	4.19	13.08	2.01
Ferrous iron (FeO)	0.04	3.52	0.45	4.92	....	8.73
Manganese oxide (MnO)	....	....	....	....	....	0.71
Lime (CaO)	2.94	3.54	6.11	8.82	8.15	8.30
Magnesia (MgO)	1.51	1.00	3.92	3.48	0.84	11.11
Potash (K_2O)	2.28	1.24	2.38	3.50	2.84	0.87
Soda (Na_2O)	4.02	3.07	3.52	4.02	2.84	3.70
Phosphoric acid (P_2O_5)						0.01
Carbonic acid (CO_2)	0.55	4.46	....	0.58	....	5.25
Water (H_2O)					4.35	2.57

I. Quartz-mica diorite: Electric Peak, Yellowstone Park (J. P. Iddings). II. Diorite: Penmaen-Mawr, Wales (J. A. Phillips). III. Diorite: Comstock Lode, Nevada (40th Parallel Survey). IV. Augite diorite: Custer County, Colorado (Whitman Cross). V. Porphyritic diorite (camptonite): Fairhaven, Vermont (J. F. Kemp). VI. Porphyritic diorite: Lewiston, Maine (G. P. Merrill).

blende alone is the predominating accessory; (2) *mica diorite*, in which black mica replaces the hornblende, and (3) *augite diorite*, in which the hornblende is partially replaced by augite. The presence of quartz gives rise to the varieties, *quartz, quartz augite*, and *quartz-mica diorites*. The name *tonalite* has been given by Vom Rath to a quartz diorite containing the feldspar andesine and very rich in black mica. *Kersantite* is a dioritic rock occurring, so far as known, only in dikes, and consisting essentially of black mica and plagioclase, with accessory apatite and augite, or more rarely hornblende, quartz, and orthoclase. It differs from the true mica diorite in being, as a rule, of a porphyritic rather than granitic structure. Professor Rosenbusch, in his latest work, has placed the kersantites, together with the porphyritic diorites (camptonites), under the head of dioritic lamprophyrs in the class of dike rocks or "gangesteine." The name, it should be stated, is from Kersanton, a small hamlet in the Brest Roads, department of Finistère, France.

The diorites were formerly, before their exact mineralogical nature was well understood, included with the diabases and melaphyrs under the general name *greenstone* (Ger. *Grünstein*).

PLATE 8

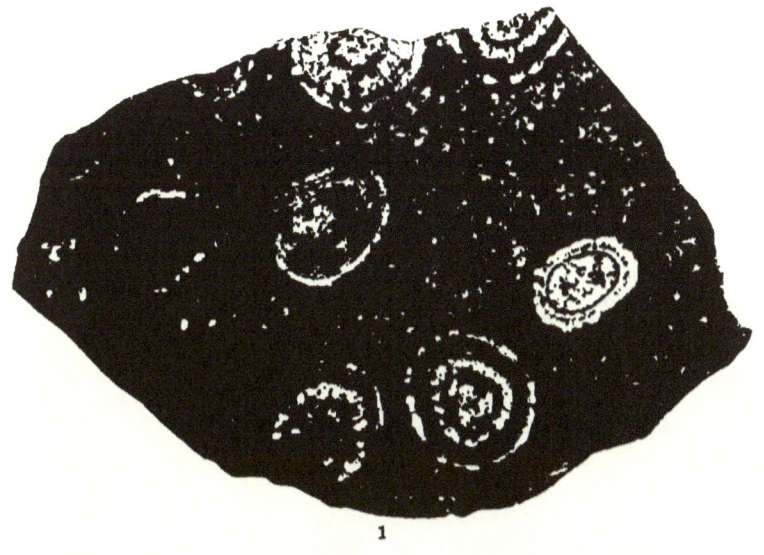

Fig. 1. Orbicular diorite.   Fig. 2. Granite spheroid.

They are rocks of wide geographic distribution, but apparently less abundant in the United States than are the diabases. The lamprophyr varieties are still less abundant, so far as now known.

(2) THE PORPHYRITES

Mineral and Chemical Composition. — The essential constituents of the porphyrites are the same as of the diorites, from which they differ mainly in structure.

Structure. — The porphyrites, as a rule, show a felsitic or glassy ground-mass, as do the quartz porphyries, in which are embedded quite perfectly developed porphyritic plagioclases, with or without hornblende or black mica. At times, as in the well-known "porfido rosso antico," or antique porphyries of Egypt, the ground-mass is microcrystalline, forming thus connecting links between the true diorites and diorite porphyrites. Indeed, the rocks of the group may be said to bear the same relation to the diorites in the plagioclase series as do the quartz porphyries to the granites in the orthoclase series, or better yet, they may be compared with the hornblende andesites, of which they are apparently the Palæozoic equivalents.

Colors. — The prevailing colors are dark brown, gray, or greenish.

Classification. — According to the character of prevailing accessory mineral, we have *hornblende porphyrite*, or *diorite porphyrite*, as it is sometimes called, and *mica porphyrite*. When, as is frequently the case, neither of the above minerals are developed in recognizable quantities, the rock is designated as simply *porphyrite*. The porphyrites are wide-spread rocks, very characteristic of the later Palæozoic formations, occurring as contemporaneous lava flows, intrusive sheets, dikes, and bosses.

(3) THE ANDESITES

The name *Andesite* was first used by L. V. Buch in 1835, to designate a type of volcanic rocks found in the Andes Mountains, South America.

Mineral Composition. — The essential constituents are soda-lime feldspar, together with black mica, hornblende, augite, or a rhombic pyroxene, and in smaller, usually microscopic proportions, magnetite, ilmenite, hematite, and apatite. Common accessories are olivine, sphene, garnets, quartz, tridymite, anorthite, sanidin, and pyrite.

Chemical Composition. — The composition of the andesites varies very considerably, the quartz-bearing members naturally showing much the higher percentage of silica. The following table shows the composition of a few typical forms: —

Constituents	I	II	III	IV	V	VI
Silica (SiO_2)	60.32 %	69.51 %	61.12 %	56.07 %	56.19 %	58.33 %
Alumina (Al_2O_3)	14.33	15.75	11.61	19.06	16.21	18.17
Ferric oxide (Fe_2O_3)	5.53	3.34	11.64	5.39	4.92	....
Ferrous oxide (FeO)	0.25	....	....	0.92	4.43	6.03
Magnesia (MgO)	2.45	2.09	0.61	2.12	4.60	2.40
Lime (CaO)	4.64	1.71	4.33	7.70	7.00	6.19
Soda (Na_2O)	3.90	3.89	3.85	4.52	2.96	3.20
Potash (K_2O)	1.61	3.34	3.52	1.24	2.37	3.02
Water (H_2O)	1.13	....	4.35	0.99	1.03	0.76
	100.16 %	99.63 %	101.03 %	98.01 %	99.62	98.10 %

I. Dacite from Kis Sebes, Transylvania. II. Dacite from Lassens Peak, California. III. Hornblende andesite from hill north of Gold Peak, Nevada. IV. Hornblende andesite from Bogoslof Island, Alaska. V. Hypersthene andesite, Buffalo Peaks, Colorado. VI. Augite andesite from north of American Flat, Washoe, Nevada.

Structure. — To the unaided eye the andesites present as a rule a compact, often rough and porous ground-mass carrying porphyritic feldspars and small scales of mica, hornblende, or whatever may be the prevailing accessory; pumiceous forms are not uncommon. Under the microscope the ground-mass is found to vary from clear glassy through microlitic forms to almost holocrystalline. The minerals of the ground-mass are feldspars in elongated microlites, specks of iron ore, apatite in very perfect forms, and one or more of the accessory ferro-magnesian minerals.

Colors. — The prevailing colors are some shade of gray, greenish or reddish.

Classification and Nomenclature. — Specific names are given dependent upon the character of the prevailing accessory. We thus have: —

Andesites with quartz = *Quartz andesites* or *dacites.*
Andesites in which hornblende prevails = *Hornblende andesites.*

Andesites in which augite prevails = *Augite andesites.*
Andesites in which hypersthene prevails = *Hypersthene andesites.*
Andesites in which mica prevails = *Mica andesites.*

The glassy varieties are often known as *hyaline andesites.* The name *propylite* was given by Richthofen to a group of andesitic rocks prevalent in Hungary, Transylvania, and the western United States, but these rocks have since been shown by Dr. Wadsworth[1] and others to be but altered andesites, and the name has fallen largely into disuse.

5. THE GABBRO-BASALT GROUP

We have here a large and variable group of rocks which on structural and mineralogical grounds might well be subdivided. Thus the gabbros, norites, and hypersthene andesites might well be considered as a group by themselves, while the diabases, augite porphyrites, melaphyrs, and basalts could form a second. Owing, however, to the similarity of the magmas from which they have been derived, it is believed the wants of the student will be best subserved by grouping them all together as above. They may be described in detail as below : —

(1) THE GABBROS

Gabbro, an old Italian name originally applied to serpentinous rocks containing diallage.

Mineral Composition. — The gabbros consist essentially of a basic soda-lime feldspar, either labradorite, bytownite, or anorthite, and diallage or a closely related monoclinic pyroxene, a rhombic pyroxene (enstatite or hypersthene), and more rarely olivine. Apatite and the iron ores are almost universally present, and often picotite, chromite, pyrrhotite, more rarely common pyrites, and a green spinel. Secondary brown mica and hornblende are common. Quartz occurs but rarely.

Chemical Composition. — As with other groups, the percentage amounts of the various constituents obtained by analyses is dependent upon the relative proportion of the constituent minerals. In the tables given below, analyses like I and III, showing very little iron and magnesia, but rich in lime and soda and alumina, are of rocks in which the pyroxenic con-

[1] Proc. Boston Society of Natural History, Vol. XXI, 1881, p. 260.

stituents are almost wholly lacking, and which consist essentially of lime feldspars only.

Constituents	I	II	III	IV	V	VI
Silica (SiO_2)	59.55 %	54.72 %	53.43 %	49.15 %	46.85 %	45.66 %
Alumina (Al_2O_3)	25.62	17.79	28.01	21.90	19.72	16.44
Ferric iron (Fe_2O_3)	0.75	2.08	0.75	6.60	3.22	0.66
Ferrous iron (FeO)	....	6.03	....	4.54	7.99	13.00
Lime (CaO)	7.73	6.84	11.24	8.22	13.10	7.23
Magnesia (MgO)	Trace	5.85	0.63	3.03	7.75	11.57
Potash (K_2O)	0.96	3.01	0.96	1.61	0.09	0.41
Soda (Na_2O)	5.09	3.02	4.85	3.83	1.56	2.13
Ignition and loss	0.45	....	....	1.92	0.56	0.07

I. Anorthosite: Chateau Richer, Canada (T. S. Hunt). II. Gabbro: near Cornell Dam, Croton River, New York (J. F. Kemp). III. Anorthosite: Labrador (A. Wickman). IV. Gabbro: near Duluth, Minnesota (Streng). V. Gabbro: near Baltimore, Maryland (G. H. Williams). VI. Gabbro: Northwest Minnesota (W. S. Bayley).

Structure. — The gabbro structure is quite variable. Like the other plutonic rocks mentioned, they are crystalline granular, the essential constituents rarely showing perfect crystal outlines. As a rule the pyroxenic constituent occurs in broad and very irregularly outlined plates, filling the interstices of the feldspars, which are themselves in short and stout forms quite at variance with the elongated, lath-shaped forms seen in diabases. This rule is, however, in some cases reversed, and the feldspars occur in broad, irregular forms surrounding the more perfectly formed pyroxenes. Transitions into diabase structure are not uncommon. In rare instances the pyroxenic constituents occur in concretionary aggregates as in the peculiar kugel gabbro or potato rock from Smaalanene, in Norway. Through a molecular change of the pyroxenic constituent, the gabbros pass into diorites, as do also the diabases.

Colors. — The prevailing colors are gray to nearly black; sometimes greenish through decomposition.

Classification. — The rocks of this group are divided into (1) the *true gabbros* — that is, plagioclase-diallage rocks — and (2) *norites*, or plagioclase-bronzite and hypersthene rocks. Both varieties are further subdivided according to the presence or absence of olivine. We then have: —

True gabbro   = Plagioclase + diallage.
Olivine gabbro = Plagioclase + diallage and olivine.
Norite   = Plagioclase + hypersthene or bronzite.
Olivine norite = Plagioclase + hypersthene and olivine.

Nearly all gabbros contain more or less rhombic pyroxene, and hence pass by gradual transitions into the norites. Through a diminution in the proportion of feldspar they pass into the peroditites, and a like diminution in the proportion of pyroxene gives rise to the so-called *forellenstein*. *Hyperite* is the name given, by Tornebohm, to a rock intermediate between normal gabbro and norite. Anorthosite, as above indicated, is the name given to the granular varieties poor or quite lacking in pyroxenes.

(2) THE DIABASES

Diabase, from the Greek word διαβασις, a passing over; so called by Brongniart because the rock passes by insensible gradations into diorite.

Chemical Composition. — The table below shows the average range in composition of (I and II) the plutonic diabase and (III, IV, V, and VI) the effusive forms melaphyr and basalt.

Constituents	I	II	III	IV	V	VI
Silica (SiO_2)	53.13 %	45.46 %	56.52 %	51.02 %	57.25 %	46.90 %
Alumina (Al_2O_3) . .	13.74	19.94	13.53	18.86	16.45	10.17
Ferric iron (Fe_2O_3) . .	1.08	} 15.36	12.56	{ 6.57	1.07	1.22
Ferrous iron (FeO) . .	9.10			4.68	1.77	5.17
Lime (CaO)	9.47	8.32	5.31	7.36	7.65	6.20
Magnesia (MgO) . . .	8.58	2.95	2.70	5.57	6.74	20.98
Potash (K_2O) . . .	1.03	3.21	3.59	2.10	1.57	2.04
Soda (Na_2O)	2.30	2.12	3.71	2.54	3.00	1.16
Ignition	0.90	0.30	0.81	2.86	0.45	5.42
Specific gravity . . .	2.96	2.945	....	....	....	2.80

I. Diabase: Jersey City, New Jersey (G. W. Hawes). II. Diabase: Palmer Hill, Au Sable Forks, New York (J. F. Kemp). III. Melaphyr: Hockenberg, Silesia. IV. Melaphyr, Falgendorf, Bohemia (quoted from Zirkel's Lehrbuch der Petrographie). V. Quartz basalt: Snag Lake, California (J. S. Diller). VI. Basalt (absarokite): near Bozeman, Montana (G. P. Merrill).

Mineral Composition. — The essential constituents of diabase are plagioclase feldspar and augite, with nearly always magnetite and apatite in microscopic proportions. The common accessories are hornblende, black mica, olivine, enstatite, hypersthene, orthoclase, quartz, and titanic iron. Calcite, chlorite, hornblende, and serpentine are common as products of alteration. Through a molecular change known as uralitization the augite not infrequently becomes converted into hornblende, as already described (p. 40), and the rock thus passes over into diorite. The plagioclase may be labradorite, oligoclase, or anorthite.

Structure. — In structure the diabases are holocrystalline. Rarely do the constituents possess perfect crystal outlines, but

Fig. 4. — Microstructure of diabase.

are more or less imperfect and distorted, owing to mutual interference in process of formation, the *granular hypidiomorphic* structure of Professor Rosenbusch. The augite in the typical forms occurs in broad and sharply angular plates enclosing the elongated or lath-shaped crystal of plagioclase, giving rise to a structure known as ophitic. (See Fig. 4.) The rocks are, as a rule, compact, fine, and homogeneous, though sometimes porphyritic and rarely amygdaloidal.

Colors. — The colors are sombre, varying from greenish through dark gray to nearly black, the green color being due to a disseminated chloritic or serpentinous product resulting from the alteration of the augite or olivine.

Classification. — Two principal varieties are recognized, the distinction being based upon the presence or absence of the mineral olivine. We thus have: (1) *diabase* proper and (2) *olivine diabase*.

Many varietal names have been given from time to time by different authors. Gumbel gave the name of *leucophyr* to a very chloritic, diabase-like rock consisting of pale green augite and a saussurite-like plagioclase. The same authority gave the name *epidiorite* to an altered diabase rock occurring in small dikes between the Cambrian and Silurian formations in the Fichtelgebirge, and in which the augite had become changed to hornblende. He also designated by the term *proterobase* a Silurian diabase consisting of a green or brown, somewhat fibrous, hornblende, reddish augite, two varieties of plagioclase, chlorite, ilmenite, a little magnetite, and usually a magnesian mica. The name *ophite* has been used by Pallarson to designate an augite plagioclase eruptive rock, rich in hornblende and epidote, and occurring in the Pyrenees. The researches of M. Levy Kuhn [1] and others have, however, shown that both hornblende and epidote are secondary, resulting from the augitic alteration, and that the rock must be regarded as belonging to the diabase.

The Swedish geologist, Törnebohm, gave the name *sahlite diabase* to a class of diabasic rocks containing the pyroxene sahlite, and which occurred in dikes cutting the granite, gneiss, and Cambrian sandstones in the province of Smaaland, and in other localities. The name *teschenite* was for many years applied to a class of rocks occurring in Moravia, and which, until the recent researches of Rohrbach, were supposed to contain nepheline, but which are now regarded as merely varietal forms of diabase. *Variolite* is a compact, often spherulitic, variety occurring in some instances as marginal facies of ordinary diabase. The name *eukrite* or *eucrite* was first used by G. Rose to designate a rock consisting of white anorthite and grayish green augite occurring in the form of a dike cutting the Carboniferous limestone of Carlingford district, Ireland. These rocks were included by Professor Zirkel under the head of "anorthitgesteine." The name is now little used, and rocks of this type are here included with the diabases.

The diabases are among the most abundant and wide-spread of our so-called *trap* rocks, occurring in the form of dikes, intrusive sheets, and bosses. They are especially characteristic of the Triassic formations of the eastern United States. It should be noted, however, that many of these Triassic *traps* have been shown to be true lava flows, and that on both lithological and geological grounds such might with propriety be classed with the basalts.

[1] Untersuchungen über pyrenaeische Ophite, Inaugural Dissertation Universitat, Leipzig, 1881.

(3) THE MELAPHYRS AND AUGITE PORPHYRITES

The term *melaphyr* is used to designate a volcanic rock occurring in the form of intrusive sheets and lava flows, and consisting essentially of a plagioclase feldspar, augite, and olivine, with free iron oxides and an amorphous of porphyry base. The augite porphyrites differ in containing no olivine. The rocks of this group are therefore the porphyritic, effusive, forms of the olivine-bearing and olivine-free diabases and gabbros.

Structure. — As above noted, they are porphyritic rocks with, in their typical forms, an amorphous base, are often amygdaloidal, and with a marked flow structure.

Colors. — In colors they vary through gray or brown to nearly black; often greenish through chloritic and epidotic decomposition.

Classification and Nomenclature. — According as olivine is present or absent, they are divided primarily into melaphyrs and augite porphyrites, the first bearing the same relation to the olivine diabases as do the quartz porphyries to the granites, or the hornblende porphyrites to the diorites, and the second a similar relation to the olivine-free diabases. The augite porphyrites are further divided upon structural grounds into (1) *diabase porphyrite*, which includes the varieties with holocrystalline diabase granular ground-mass of augite, iron ores, and feldspars, in which are embedded porphyritic lime-soda feldspars, — mainly labradorite, — idiomorphic augites, and at times accessory hornblende and black mica; (2) *spilite*, which includes the non-porphyritic compact, sometimes amygdaloidal and decomposed forms such as are known to German petrographers as *dichte diabase, diabase mandelstein* (amygdaloid), *kalk-diabase, variolite,* etc.; (3) the true *augite porphyrite,* including the normal porphyritic forms with the amorphous base, and (4) the glassy variety *augite vitrophyrite.*

(4) THE BASALTS

Basalt, a very old term used by Pliny and Strabo to designate certain blacks rocks from Egypt, and which were employed in the arts in early times.[1]

[1] Teall, British Petrography, p. 136.

Mineral Composition. — The essential minerals are augite and plagioclase feldspar with olivine in the normal forms; accessory iron ores (magnetite and ilmenite), together with apatite, are always present, and more rarely a rhombic pyroxene, hornblende, black mica, quartz, perowskite, hauyne and nepheline, and minerals of the spinel group. Metallic iron has been found as a constituent of certain basaltic rocks on Disco Island, Greenland.

Chemical Composition. — The composition is quite variable, as shown by analyses in columns V and VI on p. 87. The following shows the common extremes of variation: Silica, 45 % to 55 %; alumina, 10 % to 18 %; lime, 7 % to 14 %; magnesia, 3 % to 10 %; oxide of iron and manganese, 9 % to 16 %; potash, 0.058 %; soda, 2 % to 5 %; loss by ignition, 1 % to 5 %; specific gravity, 2.85 to 3.10.

Structure. — Basalts vary all the way from clear glassy to holocrystalline forms. The common type is a compact and, to the unaided eye, homogeneous rock, with a splintery or conchoidal fracture, and showing only porphyritic olivines in such size as to be recognizable. Under the microscope they show a ground-mass of small feldspar and augite microlites, with perhaps a sprinkling of porphyritic forms of feldspar, augite, and olivine, and a varying amount of interstitial brownish glass; the glass may be wholly or in part replaced by devitrification products, as minute hairs, needles, and granules. A marked flow structure is often developed, the feldspars of the ground-mass having flowed around the olivine belonging to the earlier period of consolidation, giving rise to an appearance that may be compared to logs in a mill stream, the olivines representing small islands. Pumiceous and amygdaloidal forms are common.

Colors. — The prevailing colors are dark, some shade of gray to perfectly black. Red and brown colors are also common. Mineralogically it will be observed the basalts resemble the olivine diabases and melaphyrs, of which they may be regarded as the younger equivalents. Indeed, in very many cases it has been found impossible to ascertain from the study of the specimen alone to which of the three groups it should be referred, so closely at times do they resemble one another.

Classification and Nomenclature. — In classifying, the variations in crystalline structure are the controlling factors. As, however, these characteristics are such as may vary almost indefinitely in different portions of the same flow, the rule has not been rigidly adhered to here. We thus have: —

(1) *Dolerite*, including the coarse-grained almost holocrystalline variety; (2) *anamesite*, including the very compact fine-grained variety, the various constituents of which are not distinguishable by the unaided eye; (3) *basalt* proper, which includes the compact homogeneous, often porphyritic, variety, carrying a larger proportion of interstitial glass or devitrification products than either of the above varieties, and (4) *tachylite, hyalomelan,* or *hyalobasalt,* which includes the vitreous or glassy varieties, the mass having cooled too rapidly to allow it to assume a crystalline structure. These varieties, therefore, bear the same relation to normal basalt as do the obsidians to the liparites. Other varieties, though less common, are recognizable and characterized by the presence or absence of some predominating accessory mineral. We have thus *quartz, hornblende,* and *hypersthene basalt*, etc. An olivine-free variety is also recognized.

The basalts are among the most abundant and wide-spread of the younger eruptive rocks. In the United States they are found mainly in the regions west of the Mississippi River. They are eminently volcanic rocks, and occur in the form of lava streams and sheets, often of great extent, and sometimes showing a characteristic columnar structure. According to Richthofen, the basalts are the latest products of volcanic activity. A quartz-bearing basalt has been described by Mr. J. S. Diller as occurring at Snag Lake, near Lassens Peak, California, and which is regarded by him as a product of the latest volcanic eruption within the limits of the state. This lava field covers an area of only some three square miles, and trunks of trees killed at the time of the eruption are still standing.[1]

Under the name of *melilite basalt* is included a group of rocks in which the mineral melilite is the characterizing constituent, with accessory augite, olivine, nepheline, biotite, magnetite, perowskite, and spinel. The normal structure is holocrystalline porphyritic, in which the olivine, augite, mica, or occasionally the melilite, appear as porphyritic constituents. These are rocks of very limited distribution, and at present known in North America only near Montreal, Canada. Professor Rosen-

[1] Bull. No. 79, U. S. Geol. Survey, 1891.

busch, in his latest work, separates this entirely from the basalts, and considers it in a group by itself under the name of Melilite Rocks.

6. THE THERALITE-BASANITE GROUP

This is a small, and so far as now known, comparatively insignificant group of rocks, representatives of which are confined to limited and widely separated areas. They are described as below : —

(1) THE THERALITES

This name, derived from the Greek word $\theta\eta\rho\alpha\nu$, to seek eagerly, is given by Professor Rosenbusch to a class of intrusive rocks consisting essentially of plagioclase feldspar and nepheline, and which are apparently the plutonic equivalents of the tephrites and basanites.

The group is founded by Professor Rosenbusch upon certain rocks occurring in dikes and laccolites in the Cretaceous sandstones of the Crazy Mountains of Montana, and described by Professor J. E. Wolff,[1] of Harvard University.

Mineral Composition. — The essential constituents as above noted are nepheline and plagioclase with accessory augite, olivine, sodalite, biotite, magnetite, apatite and secondary hornblende, and zeolitic minerals.

Chemical Composition. — The chemical composition of a sample from near Martinsdale, as given by Professor Wolff, is as follows: Silica, 43.175 % ; alumina, 15.236 % ; ferrous oxide, 7.607 % ; ferric oxide, 2.668 % ; lime, 10.633 % ; magnesia, 5.810 % ; potash, 4.070 % ; soda, 5.68 % ; water, 3.571 % ; sulphuric anhydride, 0.94 %.

Structure. — The rocks are holocrystalline granular throughout.

Colors. — These are dark gray to nearly black.

The theralites, so far as known, have an extremely limited distribution, and in the United States have thus far been reported only from Gordon's Butte and Upper Shields River basin in the Crazy Mountains of Montana.

---

[1] Notes on the Petrography of the Crazy Mountains and other localities in Montana, by J. E. Wolff. Neues Jahrb. fur. Min., etc., 1885, I, p. 69.

(2) THE TEPHRITES AND BASANITES

Mineral Composition. — The essential constituent of the rocks of this group as given by Rosenbusch are a lime-soda feldspar and nepheline or leucite, either alone or accompanied by augite. Olivine is essential in basanite. Apatite, the iron ores, and rarely zircon occur in both varieties. Common accessories are sanidin, hornblende, biotite, hauyne, melanite, perowskite, and a mineral of the spinel group.

Chemical Composition. — The following is the composition of (I) a nepheline tephrite from Antao, Pico da Cruz, Azores, and (II) a nepheline basanite from San Antonio, Cape Verde Islands, as given by Roth.[1]

Constituents	I	II
Silica (SiO_2)	47.44%	43.09%
Alumina (Al_2O_3)	23.71	17.45
Iron sesquioxide (Fe_2O_3)	6.83	18.99
Iron protoxide (FeO)	3.53	....
Magnesia (MgO)	1.95	4.63
Lime (CaO)	0.47	9.76
Soda (Na_2O)	6.40	5.02
Potash (K_2O)	3.34	1.81
Water (H_2O)	1.73	0.83

Structure. — The rocks of this group are as a rule porphyritic with a holocrystalline ground-mass, though sometimes there is present a small amount of amorphous interstitial matter or base; at times amygdaloidal.

Colors. — The colors are dark, some shade of gray or brownish.

Classification and Nomenclature. — According to their varying mineral composition Rosenbusch divides them into : —

Leucite tephrite. = Leucite, augite, plagioclase rocks.
Leucite basanite = Leucite, augite, plagioclase and olivine rocks.
Nepheline tephrite = Nepheline, plagioclase rocks.
Nepheline basanite = Nepheline, plagioclase and olivine rocks.

The group, it will be observed, stands intermediate between the true basalts and the nephelinites to be noted later. Their distribution, so far as now known, is quite limited.

[1] Abhandlungen der König. Akad. der Wissenschaften zu Berlin, 1884, p. 64.

7. THE PERIDOTITE-LIMBURGITE GROUP

This and the following groups include eruptive rocks in which neither quartz nor feldspars of any kind longer appear as essential constituents, and which are therefore very low in silica, causing them to be classed as ultrabasic. Although in most cases comparatively insignificant as rock masses, they are peculiarly interesting as mineral aggregates, and even more on account of the character of their alteration products. The peridotites are further of interest in presenting the nearest homologues to meteorites of any of our terrestrial rocks. The group includes the plutonic *peridotites* (serpentine in part), and effusive *picrite porphyrites* and *limburgites*. In detail these are as below : —

(1) THE PERIDOTITES

Peridotite, so called because the mineral peridot (olivine) is the chief constituent.

Mineral Composition. — The essential constituent is olivine associated nearly always with chromite or picotite and the iron ores. The common accessories are one or more of the ferromagnesian silicate minerals augite, hornblende, enstatite, and black mica; feldspar is also present in certain varieties and more rarely apatite, garnet, sillimanite, perowskite, and pyrite.

Constituents	I	II	III	IV	V	VI
Silica (SiO_2)	41.58 %	43.84 %	39.103 %	42.94 %	38.01 %	45.68 %
Alumina (Al_2O_3.)	0.14	1.14	4.94	10.87	5.32	0.28
Magnesia (MgO)	49.28	44.33	29.176	16.32	23.29	34.76
Lime (CaO)	0.11	1.71	3.951	9.07	4.11	2.15
Iron sesquioxide(Fe_2O_3)	....	8.76	4.315	3.47	0.70	9.12
Iron protoxide (FeO)	7.49	....	11.441	10.14	4.92	....
Chrome oxide (Cr_2O_3)	....	0.42	0.430	....	....	0.26
Manganese (MnO)	....	0.12	0.276	Trace	....	....
Potash (K_2O)	....	....	Trace	0.15	0.22	....
Soda (Na_2O)	....	....	....	0.90	4.15	....
Nickel oxide (NiO)	0.34	....	....	....	....	....
Water and ignition	1.72	1.06	5.069	6.09	10.60	1.21
Specific gravity	....	3.287	2.93	2.88	2.83	3.209

I. Dunite: Macon County, North Carolina. II. Saxonite: St. Paul's Rocks, Atlantic Ocean. III. Picrite: Nassau, Germany. IV. Hornblende picrite: Ty Cross, Anglesia. V. Picrite: Little Deer Isle, Maine. VI. Lherzolite: Monte Rossi, Piedmont.

Chemical Composition. — The chemical composition varies somewhat with the character and abundance of the prevailing accessory. The preceding table shows the composition of several typical varieties.

Structure. — The structure as displayed in the different varieties is somewhat variable. In the dunite it is as a rule even crystalline granular, none of the olivines showing perfect crystal outlines. In the picrites the augite or hornblende often occurs in the form of broad plates occupying the interstices of the olivines and wholly or partially enclosing them, as in the hornblende picrite of Stony Point, New York. The saxonites and lherzolites often show a marked porphyritic structure produced by the development of large pyroxene crystals in the fine and evenly granular ground-mass of olivines. (See Fig. 5, as drawn by Dr. G. H. Williams.) The rocks belong to the class designated as hypidiomorphic granular by Professor Rosenbusch; that is, rocks composed only in part of minerals showing crystal faces peculiar to their species.

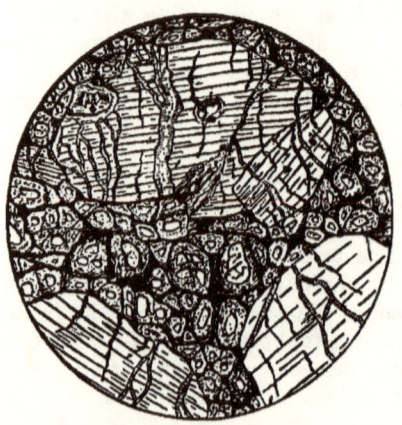

FIG. 5. — Microstructure of porphyritic lherzolite, partly altered into serpentine.

Colors. — The prevailing colors are green, greenish gray, yellowish green, dark green to black.

Nomenclature and Classification. — Mineralogically and geologically it will be observed the peridotites bear a close resemblance to the olivine diabases and gabbros, from which they differ only in the absence of feldspars. Indeed, Professor Judd has shown that the gabbros and diabase both, in places, pass by insensible gradations into peridotites through a gradual diminution in the amount of their feldspathic constituents. Dr. Wadsworth would extend the term *peridotite* to include rocks of the same composition, but of meteoric as well as terrestrial origin, the condition of the included iron, whether metallic or as an oxide, being considered by him as non-essential, since native iron is also found occasionally in terrestrial rocks, as the Greenland basalts and some diabases.

In classifying the peridotites the varietal distinctions are based upon the prevailing accessory mineral. We thus have:—

Dunite, consisting essentially of olivine only.
Saxonite, consisting essentially of olivine and enstatite.
Picrite, consisting essentially of olivine and augite.
Hornblende picrite, consisting essentially of olivine and hornblende.
Wehrlite (or eulysite), consisting essentially of olivine and diallage.
Lherzolite, consisting essentially of olivine, enstatite, and augite.

The name *Dunite* was first used by Hochstetter and applied to the olivine rock of Mount Dun, New Zealand. *Saxonite* was given by Wadsworth, rocks of this type being prevalent in Saxony. The same rock has since been named *Harzburgite* by Rosenbusch. The name *Lherzolite* is from Lake Lherz in the Pyrenees.

The peridotites are, as a rule, highly altered rocks, the older forms showing a more or less complete transformation of their original constituents into a variety of secondary minerals, the olivine going over into serpentine or talc and the augite or hornblende into chlorite. The most common result of this alteration is the rock serpentine, the transformation taking place through the hydration of the olivine and the liberation of free iron oxides and chalcedony. (See Fig. 5.) Recent investigations have shown that a large share of the serpentinous rocks were thus originated. The chemistry of the process has been already discussed under the head of olivine, p. 24.

Since in this process of hydration the combined iron becomes converted into the sesquioxide form, and the calcium of the lime-magnesian silicates separates out in large part as free calcite, or as mixed carbonates of lime and magnesia, so these serpentinous rocks are rarely uniform in color or composition. The prevailing color is some shade of green, though not infrequently brown, yellow, red, or nearly black. Through the presence of still unaltered grains of pyroxene, many varieties are porphyritic. The rock is almost universally badly jointed, an evident necessary accompaniment to the alteration, and into these joints have filtered the lime or magnesia carbonate solutions, where, depositing their load, they have formed the numerous white, yellow, and greenish veins with which the stone is traversed. Many varieties indeed, like the *rosso de Levante*, *verde di Pegli*, and *verde di Genora* of Italy, are but breccias of serpentinous fragments cemented by calcareous and ferruginous cements.[1]

It is, perhaps, as yet too early to state definitely that all peridotites are eruptive. In many instances their eruptive nature is beyond dispute. Others are found in connection with the crystalline schists, so situated as to suggest that they may themselves be metamorphic.

(2) THE PICRITE PORPHYRITES

Under this head is placed a small group of rocks so far as now known very limited in their distribution, and which are regarded as the effusive forms of the plutonic picrites, as bearing the same relation to these rocks as do the melaphyrs to the olivine diabases. The essential constituents are therefore olivine and augite with accessory apatite, iron ores, and other minerals mentioned as occurring in the true picrites. Structurally they differ from these rocks in presenting an amorphous base rather than being crystalline throughout. Rocks of this type are supposed to have had an important bearing on the origin of the diamond, the diamond-bearing rocks of South Africa being picrite porphyrite (kimberlite) cutting highly carbonaceous shales. An examination of the Kentucky peridotite locality, where the same rock occurs under quite similar conditions, failed to show that similar results have been there produced, a fact which is supposed to be due in part to the small amount of carbonaceous matter in the surrounding shales.

The group is very limited, and is represented in the United States only in Elliott County, Kentucky; Pike County, Arkansas; Syracuse, Onondaga County, New York.

(3) THE LIMBURGITES

This is a small group of lavas described by Rosenbusch in 1872 as occurring at Limburg, or the Kaiserstuhl in the Rhine. The essential constituents are augite and olivine with the usual iron ores. Structurally the rock is so far as known never holocrystalline, but glassy and porphyritic. The composition of the

[1] See the Stones for Building and Decoration, Wiley & Sons, New York.

Prussian limburgite is given as below. So far as known, the group has no representatives in the United States.

Constituents	Per cent
Silica (SiO_2)	42.24
Alumina (Al_2O_3)	18.66
Iron sesquioxide (Fe_2O_3)	7.45
Magnesia (MgO)	12.27
Lime (CaO)	11.76
Soda (Na_2O)	4.02
Potash (K_2O)	1.08
Water (H_2O)	3.71
	99.19

8. THE PYROXENITE-AUGITITE GROUP

Here are included a small group of eruptive rocks differing from the last mainly in the absence of olivine as an essential constituent. They are represented, so far as now known, only by the plutonic pyroxenites and effusive augitites.

(1) THE PYROXENITES

Pyroxenite, a term applied by Dr. Hunt to certain rocks consisting essentially of minerals of the pyroxene group, and which occurred both as intrusive and as beds or nests intercalated with stratified rocks. The author here follows the nomenclature and classification adopted by Dr. G. H. Williams.[1]

Mineral Composition. — The essential constituents are one or more minerals of the pyroxene group, either orthorhombic or monoclinic. Accessory minerals are not abundant and limited mainly to the iron ores and minerals of the hornblende or mica groups.

Chemical Composition. — The following analyses serve to show the variations which are due mainly to the varying character of the pyroxenic constituents: —

[1] American Geologist, Vol. VI, July, 1890, pp. 35-49.

Constituents	I	II	III
Silica (SiO$_2$)	50.80 %	53.98 %	55.14 %
Alumina (Al$_2$O$_3$)	3.40	1.32	0.00
Chrome oxide (Cr$_2$O$_3$)	0.32	0.53	0.25
Ferric oxide (Fe$_2$O$_3$)	1.39	1.41	3.48
Ferrous oxide (FeO)	8.11	3.90	4.73
Manganese (MnO)	0.17	0.21	0.03
Lime (CaO)	12.31	15.47	8.39
Magnesia (MgO)	22.77	22.50	26.06
Soda (Na$_2$O)	Trace	....	0.30
Potash (K$_2$O)	Trace	....	....
Water (H$_2$O)	0.52	0.83	0.38
Chlorine (Cl)	0.24	....	0.23
	100.03 %	100.24 %	100.25 %

I. Hypersthene-diallage rock: Johnny Cake Road, Baltimore County, Maryland. II. Hypersthene-diallage rock: Hebbville post-office, Baltimore County, Maryland. III. Bronzite-diopside rock from near Webster, North Carolina.

Structure. — The pyroxenites are holocrystalline granular rocks, at times evenly granular and saccharoidal, or again porphyritic, as in the websterite from North Carolina. The microscopic structure of this rock is shown in Fig. 6 from the original drawing by Dr. Williams.

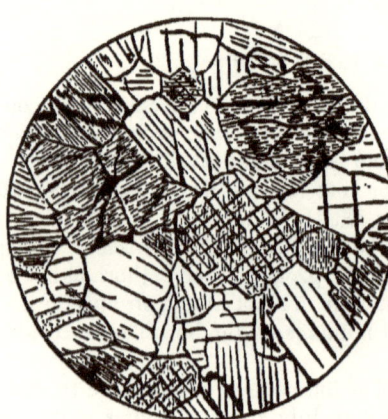

Fig. 6. — Microstructure of websterite, Webster, North Carolina.

Colors. — The colors are, as a rule, greenish or bronze.

Classification and Nomenclature. — The pyroxenites, it will be observed, differ from the peridotites only in the lack of olivine. Following Dr. Williams's nomenclature, we have the varieties *diallagite*, *bronzitite*, and *hypersthenite*, according as the mineral diallage, bronzite, or hypersthene forms the essential constituent. *Websterite* is the name given to the enstatite-diopside variety, such as occurs near Webster, North Carolina, and *hornblendite* to the hornblende-augite variety. The pyroxenites rank, in geological importance, next to the peridotites. Through processes of hydration and other chemical changes, these rocks pass into amphibolic and steatitic masses to which the name *soapstone* or *potstone* is not infrequently applied. These are dark gray or greenish rocks, soft enough to be readily cut with a knife and with a pronounced soapy or greasy feeling; hence the name *soapstone*. The name *potstone* was given on account of their having been utilized for making rude pots, for which their softness and fireproof properties render them well qualified. Although it is commonly stated in the text-books that soapstone is a compact form of steatite or talc, few are even approximately pure forms of this mineral, but all contain varying proportions of chlorite, mica, and tremolite, together with perhaps unaltered residuals of pyroxene, granules of iron ore, iron pyrites, quartz, and, in seams and veins, calcite and magnesian carbonates. The variation in chemical composition is shown in the following analyses, I being that of a compact, homogeneous-appearing, quite massive variety from Alberene, in Albemarle County, Virginia, and II one from Francestown, New Hampshire.

Constituents	I	II
Silica (SiO_2)	39.00%	42.43%
Alumina (Al_2O_3)	12.84	6.08
Ferric and ferrous iron (Fe_2O_3) and (FeO)	12.00	13.07
Lime (CaO)	5.08	3.27
Magnesia (MgO)	22.76	25.71
Potash (K_2O)	0.19	0.32
Soda (Na_2O)	0.11	0.16
Ignition	6.56	8.45
	100.40%	99.49%

(2) AUGITITE

The effusive form, augitite, differs from the pyroxenite proper mainly on structural grounds. In common with many lavas it has a glassy base, in which are embedded the crystals of augite and iron ores. The composition of an augitite from the Cape Verde Islands, as given by Roth, is as below: —

102 ROCKS FORMED THROUGH IGNEOUS AGENCIES

Constituents	Per cent
Silica (SiO$_2$)	41.83
Alumina (Al$_2$O$_3$)	18.60
Iron sesquioxide (Fe$_2$O$_3$)	16.11
Magnesia (MgO)	4.08
Lime (CaO)	11.83
Soda (Na$_2$O)	4.70
Potash (K$_2$O)	2.47
Water (H$_2$O)	0.91
	101.43

9. THE LEUCITE-NEPHELINE ROCKS

Under this head are grouped two small but interesting groups of effusive rocks, having, so far as known, no exact equivalent among the plutonics, and characterized by the presence of leucite or nepheline as essential constituents and which here seem to play the rôle of feldspars. In detail they are as below: —

(1) THE LEUCITE ROCKS

Mineral Composition. — The essential constituent is leucite and a basic augite. A variety of accessories occur, including biotite, hornblende, iron ores, apatite, olivine, plagioclase, nepheline, melilite, and more rarely garnets, hauyne, sphene, chromite, and perowskite. Feldspar as an essential fails entirely.

Chemical Composition. — The average chemical composition as given by Blaas[1] is as follows: Silica, 48.9%; alumina, 19.5%; iron oxides, 9.2%; lime, 8.9%; magnesia, 1.9%; potash, 6.5%; soda, 4.4%.

Structure. — The rocks of this group are, as a rule, fine grained and only slightly vesicular, presenting to the unaided eye little to distinguish them from the finer-grained varieties of ordinary basalt.

Colors. — The prevailing colors are some shades of gray, though sometimes yellowish or brownish.

Classification and Nomenclature. — The varietal distinctions are based upon the presence or absence of the mineral olivine

---

[1] Katechismus der Petrographie, p. 117.

and upon structural grounds and various minor characteristics. We have the olivine-free variety *leucitite* and the olivine-holding variety *leucite basalt*.

These rocks have also a very limited distribution, and, so far as known, are found within the limits of the United States only at the Leucite Hills, Wyoming.

(2) THE NEPHELINE ROCKS

Mineral Composition. — These rocks consist essentially of nepheline with a basaltic augite and accessory sanidin, plagioclase, mica, olivine, leucite, minerals of the sodalite group, magnetite, apatite, perowskite, and melanite.

Chemical Composition. — Below is given the composition of (I) a nephelinite from the Cape Verde Islands, and (II) a nepheline basalt from the Vogelsberg, Prussia.[1]

Constituents	I	II
Silica (SiO_2)	46.95 %	42.37 %
Alumina (Al_2O_3)	21.50	8.88
Iron sesquioxide (Fe_2O_3)	8.09	11.20
Iron protoxide (FeO)	....	7.80
Magnesia (MgO)	2.49	13.01
Lime (CaO)	7.97	10.93
Soda (Na_2O)	8.93	4.51
Potash (K_2O)	2.04	1.21
Water (H_2O)	2.09	0.34
Specific gravity	....	3.103

Colors. — The prevailing colors are various shades of gray to nearly black.

Structure. — Structurally they are porphyritic, with a holocrystalline or in part amorphous base, usually fine grained and compact, at times amygdaloidal.

Classification and Nomenclature. — These rocks differ from the basalts, which they otherwise greatly resemble, in that they bear the mineral nepheline in place of feldspar. Based upon the presence or absence of olivine, we have, first, *nepheline basalt*,

[1] Roth, Abhandl. der König. Preus. Akad. der Wiss. zu Berlin, 1884.

and second, *nephelinite*. The name *nepheline dolerite* has been given in some cases to the coarser, holocrystalline, olivine-bearing varieties.

Like the leucite rocks, the members of this group are somewhat limited in their distribution.

II. AQUEOUS ROCKS

1. ROCKS FORMED THROUGH CHEMICAL AGENCIES

This comparatively small, though by no means unimportant, group of rocks comprises those substances which, having once been in a condition of aqueous solution, have been deposited as rock masses either by cooling, evaporation, by a diminution of pressure, or by direct chemical precipitation. It also includes the simpler forms of those produced by chemical changes in pre-existing rocks. Water, when pure or charged with more or less acid or alkaline material, and particularly when acting under great pressure, is an almost universal solvent. Thus, heated alkaline waters, permeating the rocks of the earth's crust at great depths below the surface, are enabled to dissolve from them various mineral matters with which they come in contact. On coming to the surface or flowing into crevices, the pressure is diminished, or evaporation takes place, and the water, no longer able to carry its load, deposits it wholly or in part as vein material or a surface coating. In other cases alkaline or acid water, bearing mineral matters, may, in course of their percolations, be brought in contact with neutralizing solutions, and these dissolved materials be thus deposited by direct precipitation. In these various ways were formed the rocks here described. It will be observed that the various members of the group are composed mainly of minerals of a single species only.

This group cannot, however, be separated by any sharp lines from that which is to follow, inasmuch as many rocks are not the product of a single agency, acting alone, but are rather the result of two or more combined processes. This is especially the case with the limestones. It is safe to assume that few of these are due wholly to accumulations of calcareous, organic remains, but are, in part at least, chemical precipitates, as is well illustrated by the oölitic varieties.

According to their chemical nature, the group is divided into (1) Oxides, (2) Carbonates, (3) Silicates, (4) Sulphates, (5) Phosphates, (6) Chlorides, and (7) the Hydrocarbon Compounds.

(1) OXIDES

Here are included those rocks consisting essentially of oxygen combined with a base, though usually other constituents are present as impurities.

Hematite. — Anhydrous sesquioxide of iron. Fe_2O_3 = oxygen, 30 %; iron, 70 %. In nature nearly always more or less impure through the mechanical admixture of argillaceous silicates or calcareous matter, manganese oxides, sulphur, phosphates, etc. Several forms are recognized, the distinction being based mainly upon physical properties. Specular hematite is a micaceous or foliated variety with a black, metallic, often splendent lustre; this variety is mainly a metamorphic form, and properly should be classed with the metamorphic rocks. Compact, columnar, fibrous, and earthy forms also occur, the latter often known as ochre, as are similar forms of limonite. Although classified here under the head of aqueous rocks, it does not follow that the hematites have all originated in precisely the same manner. To a limited extent the specular variety is found about volcanic craters and fumaroles, where it was originally deposited by a process of sublimation. Through a process of oxidation, beds of magnetic iron become locally altered into hematite, giving rise to pseudomorphous granular, octahedral, and dodecahedral forms, to which the name *martite* is given. Many extensive beds undoubtedly arise from the dehydration by dynamic agencies — the folding and metamorphosing of the enclosing rocks — of beds of limonite. Others, like the fossil and oölitic ores of the Clinton formations, arise in part from a process of chemical precipitation and subsequent segregation, the ore being originally disseminated throughout a ferruginous limestone, and having accumulated as an insoluble residue as the lime carbonate was carried away through the action of carbonated waters. The extensive hematite deposits of the Lake Superior region of Michigan are regarded as oxidation products from pre-existing carbonates (siderite), the oxide having been precipitated from solution in synclinal troughs, and subsequently crystallized by metamorphism.[1] The ores of the Mesabi

[1] Van Hise Monograph XIX, U. S. Geol. Survey, 1892.

PLATE 9

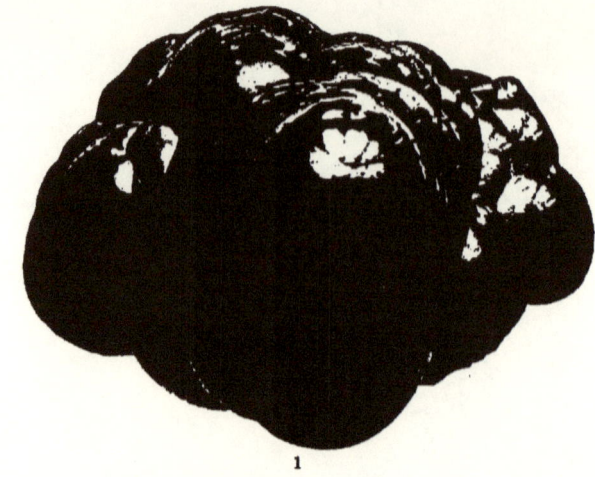

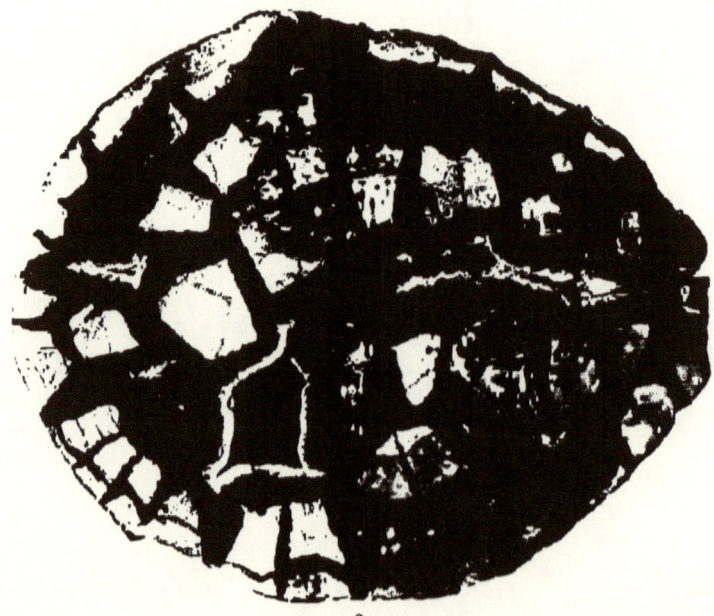

Fig. 1. Botryoidal hematite. Fig. 2. Clay-iron stone septarian nodule.

range, on the other hand, are regarded by at least one writer as having originated through a somewhat complicated process of oxidation and metasomatosis, whereby a pre-existing glauconitic rock (a ferruginous silicate) became converted into an admixture of free iron oxide and silica, the one or the other, according to the intermittent character of the permeating solutions, being leached out and redeposited at no great distance in a fair condition of purity.[1] A discussion of this subject belongs more properly to economic geology, and need not be dwelt upon further here.

Limonite (Brown Hematite). — Iron sesquioxide plus water. $H_6Fe_2O_6 + Fe_2O_3$. An earthy or compact dark brown, black, or ochreous-yellow rock, containing, when pure, about two-thirds its weight of pure iron. It occurs in beds, veins, and concretionary forms, associated with rocks of all ages, and forms a valuable ore of iron. (See Fig. 1, Pl. 9.) On the bottoms of lakes, bogs, and marshes it often forms in extensive deposits, where it is known as bog-iron ore. The formation of these deposits is described as follows: Iron is widely diffused in rocks of all ages, chiefly in the form of (1) the protoxide, which is readily soluble in waters impregnated with carbonic or other feeble acids, or (2) the peroxide, which is insoluble in the same liquids. Water percolating through the soils becomes impregnated with these acids from the decomposing organic matter, and then dissolves the iron protoxide with which it comes in contact. On coming to the surface and being exposed to the air, as in a stagnant lake or marsh, this dissolved oxide absorbs more oxygen, becoming converted into the insoluble sesquioxide, which floats temporarily on the surface as an oil-like, iridescent scum. Finally this sinks to the bottom, where it gradually becomes aggregated as a massive iron ore. This same ore may also form through the oxidation of pyrite, or beds of ferrous carbonate. At the Ktaadn Iron Works, in Piscataquis County, Maine, the ferrous salt as it oxidizes is brought to the surface by water and deposited as a coating over the leaves and twigs scattered about, forming thus beautifully perfect casts, or fossils.

Pyrolusite, Psilomelane, and Wad. — These are names given to the anhydrous and more or less hydrated forms of manganese

---

[1] J. E. Spurr, Bull. No. 10, Geol. and Nat. Hist. Survey of Minnesota, 1894.

oxides, and which, though wide in their distribution, are found in such abundance as to constitute rock masses in comparative rarity. The origin of such deposits is at times somewhat obscure. In all cases they are doubtless secondary. The original source of the material appears to have been the manganiferous silicates of Archæan and more recent eruptive rocks, whence it was derived by leaching, being transported in the form of soluble salts and finally precipitated as oxide or carbonate, the latter being subsequently converted into oxide. The deposits which are of sufficient extent to be of commercial value occur as a rule in residual clays, as interbedded strata in shales and sandstones, or as occupying superficial seams and joints, and in the form of pockets and nests. True fissure veins of manganese oxide are not known. It is often associated with the form of limonite known as bog-iron ore, and, apparently, has been deposited contemporaneously.

Bauxite (so called from Beaux, near Arles, France) is the name given to a somewhat indefinite mixture of alumina and iron oxides, and occurring in the form of compact concretionary grains of a dull red, brown, or nearly white color, and also in compact and earthy forms. The mode of occurrence of the mineral is somewhat variable. At Beaux and several other localities it occurs in pockets in limestone, and also in beds alternating with limestones, sandstones, and clays belonging to the Cretaceous period. In the Puy-de-Dôme the beds rest directly upon gneiss, and are overlaid by basalt. At Oberhessen, Germany, the mineral occurs in rounded masses embedded in clay, as is also the case at Vogelsberg. In America, bauxite has been found in Alabama, Georgia, and Arkansas. In Alabama and Georgia it occurs in beds of irregular extent, associated with limestones of Upper Cambrian age (the Knox dolomite); in Arkansas the deposits are Tertiary.

The origin of the bauxite is somewhat obscure. It has been argued that the beds at Beaux, and those of Var, are deposits from mineral springs. Those of the Puy-de-Dôme, the Westerwald, Vogelsberg, and of Ireland, on the other hand, are regarded as derived from basalt by a metasomatic process. The Alabama and Georgia deposits, like those of Beaux, are regarded as of chemical origin.[1]

[1] See *résumé* of the subject, by R. L. Packard, in Mineral Resources of the United States for 1891.

According to C. Willard Hayes,[1] the prevailing rocks of this region are dolomites underlaid by aluminous shales. It is assumed that heated waters, in their passage upward from greater depths, have oxidized the iron sulphides of the shale, giving rise to sulphates of iron, of alumina, and the double sulphates of alumina and potash. As the ascending water, carrying these salts in solution, passes through the dolomite, it becomes charged with calcium carbonate, which causes the precipitation of the aluminum salts in the concretionary, pisolitic form so characteristic.

Beauxite has, of late, come to be of considerable economic value as an ore of aluminum, and as a source of alum, in place of clay.

The material from various sources varies greatly in chemical composition, as shown by the following analyses:—

Constituents	I	II	III	IV	V
Silica (SiO_2)	2.8 %	1.10 %	21.08 %	2.80 %	10.38 %
Alumina (Al_2O_3)	57.6	50.92	48.92	52.21	55.04
Iron sesquioxide (Fe_2O_3)	25.3	15.70	2.14	13.50	1.95
Water (H_2O)	10.08	27.75	23.41	27.72	27.02
Titanium oxide (TiO_2)	3.1	3.20	2.52	3.52	3.50

I. Beaux, France. II. Vogelsberg, Germany. III. Jacksonville, Alabama. IV. Floyd County, Georgia. V. Pulaski County, Arkansas.

Silica. — Silica, as has been already noted under the head of rock-forming minerals, is one of the most abundant constituents of the earth's crust. In its various forms, which are sufficiently extensive to constitute rock masses, it is always of chemical origin, that is, results by deposition from solution, by precipitation, or evaporation, as noted above. Varietal names are given to the deposits, dependent upon their structure, method of formation, color, and degree of purity. *Siliceous sinter*, *geyserite*, or *fiorite* is the name given to the nearly white, often soft and friable, hydrated varieties formed on the evaporation of the siliceous waters of hot springs and geysers, or through the eliminating action of algous vegetation, as described by W. H. Weed in the reports of the United States

[1] Trans. Am. Inst. of Mining Engineers, February, 1894.

Geological Survey.[1] The material is, in reality, an impure form of opal. Throughout the geyser regions of the Yellowstone Park, Iceland, and New Zealand, the sinter has been deposited as a comparatively thin crust over the surface, or in the form of cones about the throats of the geysers. The varieties of silica known as opal are hydrous forms occurring in veins and pockets, in a variety of rocks. Not infrequently it forms the replacing material in silicified or "petrified" woods. In the old lake beds of the Madison valley, Montana, may not infrequently be found large logs composed wholly of this material, no sign of organic matter remaining, but yet with the woody structure beautifully preserved.

The origin of these silicified logs, so far as it has been traced, appears to have been somewhat as follows: The water which permeated the lake beds in which these logs lay, was more or less alkaline, and carried small amounts of silica in solution. As the logs slowly decayed, there were given off minute quantities of organic acids which, neutralizing the alkaline water, caused a gradual precipitation of the silica, building up thus an exact cast of the decaying structure. *Chalcedony* is the translucent, massive, cryptocrystalline variety of silica occurring mainly in cavities in older rocks, where it has been deposited by infiltration. It is a common secondary product formed during the decomposition of many rocks, and, like opal, not infrequently forms the petrifying medium of fossil woods and other organisms. Not infrequently, also, it occurs in continuous layers of several inches in thickness, interstratified with limestone, as may be seen in the walls of the Wyandotte caves in southern Indiana, or, more rarely, in beds from 2 to 8 feet thick, interstratified with coal and fire-clay, as at the well-known "Flint Ridge" of Licking County, Ohio. Such deposits are considered to be due to accumulations of the siliceous tests of diatoms. *Flint* is a variety of chalcedony formed by segregation in chalky limestone, and is composed, in part, of the broken and partially dissolved spicules of sponges, and the siliceous casts of infusoria. The source of the silica is, doubtless, the sponge spicules above noted and diatomaceous remains. Chert is an impure flint containing not infrequently fossil nummulitic remains, and with sometimes a pronounced

[1] 0th Ann. Rep. U. S. Geol. Survey, 1887-88. See also Bischof's Chemical and Physical Geology, Vol. I, pp. 184-200.

oölitic structure. It occurs in rounded, nodular, concretionary masses interbedded with limestones, particularly Palæozoic varieties, and doubtless originated as did the flints in the chalky limestones. *Jasper* is a dull or bright red, or yellow variety of chalcedony containing alumina, and owing its color to iron oxides. It is sometimes used in jewellery.

The name *novaculite* is frequently given to very fine-grained and compact quartz rocks, such as are suitable for hones. As commonly used, the name is made to include rocks of widely different origin, some of which are evidently chemical precipitates, while others are indurated clastic or schistose rocks. The well-known novaculites of Arkansas are clear white masses of chalcedonic silica, containing scattering quartz granules, minute grains of garnet, and numerous small rhomboidal cavities which seemingly were once occupied by crystals of calcite or dolomite. Opinions differ as to the origin of this rock. Owen[1] regarded it as a sandstone metamorphosed by percolating hot water. Branner[2] looked upon it as a metamorphosed chert; Griswold,[3] as a chemical deposit in the form of a siliceous slime on a sea-bottom, while Rutley[4] argues that it is but a siliceous replacement of beds of dolomite or dolomitic limestone. It seems probable that the views of Branner or Rutley are the most nearly correct.

Quartz is a massive form of crystalline silica occurring in veins, disseminated granules, and pockets in rocks of all kinds and all ages. It is one of the most wide-spread and commonest of minerals, and is frequently quarried and crushed for abrasive purposes or use in pottery manufacture. It is not infrequently of a pink or rose color from metallic oxides. It is a common gangue of ores of the precious metals, particularly of gold. *Lydian stone* is an exceedingly hard impure quartz rock, of a black color and splintery fracture. It was formerly much used in testing the purity of precious metals.

(2) CARBONATES

Water carrying small amounts of carbonic acid readily dissolves the calcium carbonate of rocks with which it comes in

[1] 2d Rep. Geological Reconnaissance of Arkansas, 1860.
[2] Ann. Rep. Geol. Survey of Arkansas, Vol. I, 1886, p. 49.
[3] Ann. Rep. Geol. Survey of Arkansas, Vol. III, 1890.
[4] Quarterly Journal Geological Society of London, August, 1894.

contact; on evaporation and through loss of a portion of the carbonic acid, this is again deposited. In this way are formed numerous and at times extensive deposits, to which are given varietal names dependent upon their structure and the special conditions under which they originated. *Calc sinter* or *tufa* is a loose friable deposit made by springs and streams either by evaporation or through intervention of algous vegetation. Such are often beautifully arborescent and of a snow-white color, as seen at the Mammoth Hot Springs of the Yellowstone National Park. Somewhat similar deposits are formed by springs in Virginia, California, Mexico, New Zealand. Others, like those from Niagara Falls, New York, and Soda Springs, Idaho, were formed by the deposition of the lime on leaves and twigs, forming beautifully perfect casts of these objects.

Tufa deposits of peculiar imitative shapes have been described by Mr. I. C. Russell of the United States Geological Survey, as formed by the evaporation of the waters of Pyramid Lake, Nevada. Oölitic and pisolitic limestones are so called on account of their rounded, fish-egg-like structure, the word *oölite* being from the Greek word ωον, *an egg*. (See Pl. 12.) These are in part chemical and in part mechanical deposits. The water in the lakes and seas in which they were formed became so saturated that the lime was deposited in concentric coatings about the grains of calcareous sand on the bottom, and finally the little granules thus formed became cemented into firm rock by the further deposition of lime in the interstices. This structure will be best understood by reference to Fig. 7. Rocks of this nature are now forming along the beaches of Pyramid Lake. Concerning the occurrence of these Mr. Russell writes: —

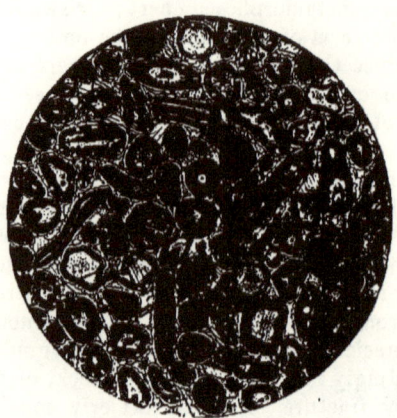

FIG. 7. — Microstructure of oölitic limestone.

"Among The Needles the rocky capes are connected by cres-

PLATE 10.—View in a limestone cavern showing stalactitic and stalagmitic masses.

cent-shaped beaches of clean, creamy sands, over which the summer surf breaks with soft murmurs. These sands are oölitic in structure, and are formed of concentric layers of carbonate of lime which is being deposited near where the warm springs rise in the shallow margin of the lake. In places these grains have increased by continual accretion until they are a quarter of an inch or more in diameter, and form gravel, or pisolite, as it would be termed by mineralogists. In a few localities this material has been cemented into a solid rock, and forms an oölitic limestone sufficiently compact to receive a polish. No more attractive place can be found for the bather than these secluded coves, with their beaches of pearl-like pebbles, or the rocky capes, washed by pellucid waters, that offer tempting leaps to the bold diver."

Such forms as these may or may not show a nucleus. It seems safe to assume that such a nucleus, at first, in all cases existed, though it may be in microscopic dimensions only.

Travertine is a compact and usually crystalline deposit formed, like the tufas, by waters of springs and streams. The travertines are often beautifully veined and colored by metallic oxides and form some of the finest marbles. Such are the so-called "onyx marbles" of Mexico and Arizona.[1]

Stalactite and *stalagmite* are the names given to the deposits formed from the roofs and on the floors of caves; water, percolating through the limestone roof, by virtue of the carbonic acid it contains, dissolves out a small amount of the lime, which, on evaporation, is again deposited either as pendent cones from the ceiling, or as massive and pillar-like forms upon the floor. The pendants are known as stalactites; the corresponding growths upon the floor as stalagmites. Stalactite and stalagmite sometimes meet, forming thus continuous pillars, or columns extending from floor to ceiling. The lime of these deposits, it may be said, is as a rule in the form of calcite, though sometimes, as in the old portions of the Wyandotte caves in Indiana, it is aragonite. The so-called "oriental alabaster" of the ancients is a stalagmitic deposit derived in part from crevices and pockets in the Eocene limestones of the Nile valley.

Magnesite, a carbonate of magnesia, occurs frequently as a

[1] The Onyx Marbles, Ann. Rep. U. S. National Museum for 1893. Also Stones for Building and Decoration, Wiley & Sons, New York, 2d ed., p. 120.

secondary mineral in the form of veins in serpentinous rocks, but rarely itself forms rock masses of any importance. *Rhodochrosite*, a carbonate of manganese, sometimes occurs in rock masses, but is found most commonly in the form of veins associated with ores of silver, lead, or copper.

Another carbonate, less common than that of lime, but which sometimes occurs in such quantities as to constitute true rock masses, is *siderite*, or carbonate of iron. A common form of this is dull brownish or nearly black in color, very compact and impure, containing varying amounts of calcareous, clayey, and organic matter. In this condition it is found in stratified beds and in the shape of rounded and oval nodules, or concretions, which are called *clay-ironstone nodules*, *septaria*, and *sphærosiderite*. (See Fig. 2, Pl. 9.) These septarian nodules are often beautifully veined with calcite, and when cut and polished form not undesirable objects of ornamentation. Other forms of siderite are massive, coarsely crystalline, and of a nearly white or yellowish color, becoming brownish on exposure. Pure siderite yields about 48 % metallic iron, and is of value as an ore.

(3) SILICATES

Silica, combined with magnesia and water, gives rise to an interesting group of serpentinous and talcose substances, which are often sufficiently abundant to constitute rock masses. Pure serpentine consists of about equal parts of silica and magnesia, with from 12 to 13 % of water. It is a compact, amorphous, or colloidal rock, soft enough to be cut with a knife, with a slight greasy feeling and lustre, and of a color varying from dull greenish and almost black, through all shades of yellow, brownish, and red. It also occurs in fibrous and silky forms, filling narrow veins in the massive rocks, and is known as *amianthus*, or *chrysotile*. These fibres, when sufficiently long, are used for the manufacture of fireproof material, and the mineral is commercially confounded with asbestos, a fibrous variety of amphibole. It is very doubtful if serpentine is ever an original rock; it is rather an alteration product of other and less stable magnesian minerals. Here will be considered only those which have originated by a series of chemical changes known as *metasomatosis*, a process of indefinite substitution and replacement, in simple mineral aggregates occurring associated with the older metamorphic rocks. Such are the serpentines derived from non-aluminous pyroxenes, like those of Montville, New Jersey, and Moriah, New York, and those from Easton, Pennsylvania, derived from a massive tremolite rock. The analyses given below will serve to illustrate the chemical changes which occur in this process of metasomatosis, I being that of a nearly white pyroxene, and II that of the serpentine derived therefrom.

CONSTITUENTS	I	II
Silica (SiO_2)	54.215 %	42.38 %
Magnesia (MgO)	19.82	42.14
Lime (CaO)	24.71	0.00
Alumina (Al_2O_3)	0.59	0.07
Ferric oxide (Fe_2O_3)	0.20	0.07
Ferrous oxide (FeO)	0.27	0.17
Ignition (H_2O)	0.14	14.20
	99.945 %	99.85 %

The pyroxene, it should be observed, occurs in nodular masses in a crystalline granular dolomite. Various stages of the process are shown in Fig. 8, in which the white and gray central portions are nucleal masses of unchanged pyroxenes, surrounded by the darker crusts of secondary serpentine.[1] Serpentine as an alteration product of the mineral chondrodite is also known to occur, though this form is less common. At Brewster, New York, are extensive deposits of this nature. (See further on, p. 158.)

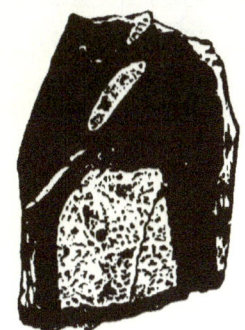

FIG. 8. — Pyroxene partially altered to serpentine.

Several varieties of serpentine are popularly recognized. *Precious* or *noble* serpentine is simply a very pure compact variety of a deep oil-yellow or green color. *Amianthus*, or *chrysotile*, as noted above, is the name given to the fibrous variety. *Williamsite* is a deep bright green, translucent, and somewhat scaly variety, occurring asso-

[1] See On the Serpentine of Montville, New Jersey, Proc. U. S. National Museum, Vol. XI, 1888, p. 105.

ciated with the chrome iron deposits in Fulton township, Lancaster County, Pennsylvania. *Deweylite* is a hard, translucent variety occurring in veins in altered dunite beds. *Bowenite* is a pale green variety forming veins in limestone at Smithfield, Rhode Island. *Picrolite, marmolite,* and *retinolite* are varieties of minor importance. Serpentine alone, or associated with calcite and dolomite, forms a beautiful marble, to which the names *verd antique, ophite,* and *ophiolite* are given. The so-called Eozoon Canadense, a supposed fossil rhizopod, is a mixture of serpentine and calcite or dolomite. The name *serpentine* is from the Latin *serpentinus*, a serpent, in allusion to its green color and often mottled appearance.

Those serpentines which were derived from basic eruptives, or complex metamorphic rocks are described with those rocks with which, in their unaltered state, they would naturally be grouped.

The mineral *steatite*, or *talc*, when pure, differs from serpentine in containing 63.5 % of silica, 31.7 % of magnesia, and 4.8 % of water. Its common form is that of white or greenish inelastic scales, forming an essential constituent of the talcose schists. As is the case with serpentine, it sometimes results from the alteration of eruptive magnesian rocks, such as the pyroxenites, and rarely occurs as a direct result of precipitation. It will be described more fully under the head of schists and pyroxenites. *Rensselaerite* is a closely related rock of a white or gray color, found in St. Lawrence County, New York. Its composition is essentially that of talc.

Pyrophyllite, or *agalmatolite*, is a hydrous silicate of alumina, somewhat harder than talc, which it otherwise resembles, and which is used in making slate pencils and small images. It occurs in a schistose form in the Deep River region of North Carolina.

Kaolin, also a hydrous silicate of alumina, is a chemical product in that it is a residue left by the chemical decomposition of the feldspars. These minerals, as explained elsewhere, consist of silicates of alumina and lime, with more or less of the alkalies potash and soda, and iron oxides. In the process of decomposition new compounds are formed, the more soluble of which are leached out, leaving the less soluble silicates, including kaolin, behind in a condition of more or less purity. The mineral is of great value for fictile purposes, and is described more fully under the head of argillaceous fragmental rocks.

(4) SULPHATES

Gypsum. — The rock gypsum is chemically a hydrous sulphate of lime, that is to say, consists of sulphur, lime, and water, and in the proportion of 32.6 parts of lime and 20.9 parts of water, combined with 46.5 parts of sulphur trioxide. When crystallized, the mineral is nearly colorless and transparent, and splits readily into thin, inelastic sheets. The compact massive varieties are white, gray to black, and sometimes pink from various impurities. The most characteristic feature is its softness, which is such that it can be readily cut with a knife or even by the thumbnail.

Four varieties of gypsum are recognized: (1) The common massive form, dull in color and often more or less impure; (2) the pure white, fine-grained variety, *alabaster;* (3) the fibrous variety, *satin spar;* and (4) the broadly foliated, transparent variety, *selenite*, so called from the Greek word σελενέ, the moon, in allusion to its soft and pleasing lustre.

The following is an analysis of a commercial gypsum from Ottawa County, Ohio, as given by Professor Orton:[1] —

Lime (CaO)	32.52%
Sulphuric acid (SO_3)	46.56
Water (H_2O)	20.14
Magnesia (MgO)	0.56
Alumina (Al_2O_3)	0.16
Insoluble residue	0.68
	99.02%

Gypsum occurs mainly associated with stratified rocks, and is regarded as a chemical deposit resulting from the evaporation of waters of inland seas and lakes; it may also originate through the decomposition of sulphides and the action of the resultant sulphuric acid upon limestone; through the mutual decomposition of the carbonate of lime (limestone) and the sulphates of iron, copper, and other metals; through the hydration of anhydrite; and through the action of sulphurous vapors and solutions from volcanoes upon the rocks with which they come in contact. According to Dana,[2] the gypsum deposits in western

---
[1] Geology of Ohio, 1888, Vol. VI, p. 700.
[2] Manual of Geology, p. 234.

New York do not form continuous layers in the strata, but lie in embedded, sometimes nodular, masses in limestones. In all such cases this authority says the gypsum is the product of the action of sulphuric acid from springs upon the limestone. "The sulphuric acid, acting on the carbonate of lime, drives off its carbonic acid and makes sulphate of lime or gypsum; and this is the true theory of its formation in New York." W. C. Clarke, however, regards it as a product of deposition from solution in sea-water.[1]

The gypsum deposits of northern Ohio form apparently continuous beds over thousands of square miles, and are regarded by Professors Newberry and Orton as deposits from the evaporation of landlocked seas at the same time as was the rock-salt which overlies it.

Geological Age and Mode of Occurrence. — As may be readily inferred from what has gone before, beds of gypsum have formed at many periods of the earth's history, and are still forming wherever proper conditions exist.

In New York there are extensive deposits belonging to the Salina period of the Upper Silurian. In Ohio, gypsum associated with limestones and shales of Lower Helderberg age occur over areas comprising thousands of square miles. The following section of beds in Ottawa County, this state, will serve to show the conditions under which the rock may occur: —

Drift clays	12 to 14 feet
Gray rock carrying impure gypsum	5 to 14 feet
Blue shale	½ to 14 feet
Boulder bed carrying gypsum embedded in shaly limestone	5 to 14 feet
Blue limestone	1 to 14 feet
Main gypsum bed	7 to 14 feet
Gray limestone	1 to 14 feet
Gypsum	3 to 5 feet

Anhydrite is an anhydrous variety of calcium sulphate somewhat less common than gypsum. *Barite*, or *heavy spar*, the sulphate of barium, also occurs in nature, but less abundantly than the calcium sulphates. It is found commonly in connection with metallic ores (silver, lead, and zinc), or as a secondary mineral associated with limestone, sometimes in distinct veins, or, as in southwest Virginia, filling irregular fractures in certain beds of the Cambrian limestones, or in

[1] Bull. New York State Museum, Vol. III, No. 1, 1893.

part replacing the limestone itself. It is easily distinguished from coarsely crystalline calcite, for which it might possibly be mistaken, by its weight, the specific gravity being about 4.5 as against 2.7 for the latter.

(5) PHOSPHATES

The mineral apatite, a phosphate of lime, as already noted, is a common accessory, in the form of small crystals, in crystalline rocks of all ages, both metamorphic and eruptive. In rare instances, as among certain Laurentian rocks of Canada, it occurs in coarsely granular aggregates of a green or pinkish color and of such dimensions as to constitute true rock masses. Here we have to do, however, more with the amorphous, fibrous, or concretionary forms to which the name *phosphorite* is commonly applied. These occur nearly if not quite altogether as secondary products, due to the leaching out of phosphatic material from older rocks, and its redeposition in clefts and cavities at lower levels. It is thus that the phosphorites of Estremadura, Spain, are accounted for. From these very pure, semi-crystalline masses, to the amorphous nodular and earthy forms, such as are found in the eastern Carolinas and in Florida, there are no well-defined lines of demarcation. All have resulted apparently either from the leaching out of the phosphate as above, or from the dissolving and carrying away of the lime carbonate in a phosphatic limestone, leaving the phosphatic material to accumulate as a residual product. Some of the latter products, like the phosphatic sandstones of the Carolinas, might with equal propriety be classed with the fragmental rocks, as are the residual clays. (See p. 151.)

(6) CHLORIDES

Sodium chloride, or common salt, is one of the most widespread constituents of the earth's crust, and from the standpoint of human comfort a most important constituent as well. The theoretically pure mineral consists of 66.6 parts of sodium and 39.4 parts of chlorine, though in nature it is almost universally contaminated with chlorides, sulphates, and carbonates of potassium, calcium, and magnesium, together with oxides of iron and aluminum. A large number of analyses of rock-salts from world-wide sources show them to range from 94 to 99% sodium chloride. The pure mineral is white in color, but shows often yellow, red, or purplish hues due to iron oxides or organic matter. When crystallizing freely from solution, it ordinarily assumes the form of a cube, the faces being frequently cavernous or hopper-shaped; rarely it occurs in octahedrons, and occasionally in fibrous forms. Sodium chloride in solution is an almost universal constituent of carbonated waters, though often in but the merest traces. Its prevailing solid form is that of coarsely granular aggregates constituting the so-called *rock-salt*, the beds of which are often of such thickness and extent as to constitute true rock masses and entitle them to consideration here. These rock masses are invariably products of deposition from solution, a deposition brought about through the evaporation of saline waters in enclosed lakes or seas. They are not limited to any particular geological period, but are to be found wherever suitable conditions have existed for their formation and preservation. Some of the more important beds now known belong to either the Upper Silurian, Carboniferous, Triassic, or Tertiary ages, and vary in thickness from a mere film to upwards of 1200 feet. In the United States, beds of rock-salt are known to occur in the states of New York, Pennsylvania, Ohio, Virginia, West Virginia, Michigan, Kansas, Kentucky, Texas, Wyoming, California, and Nevada. Canada, England, the Carpathian Mountains, the Austrian and Bavarian Alps, West Germany, the Vosges, the Jura, Spain, the Pyrenees and Celtiberian mountains, all contain important beds. With the rock-salt are not infrequently associated other salts, as above noted. In the celebrated Stassfurth deposits, sixteen different compounds in the shape of chlorides and sulphates of sodium, potassium, magnesium, calcium, and iron have been determined, many of them in sufficient quantity to be of commercial value.

(7) THE HYDROCARBON COMPOUNDS

Under this head are included a series of hydrocarbon compounds varying in physical properties from solid to gaseous, and in color from coal-black through brown, greenish, red, and yellow to colorless. Unlike the other members of the hydrocarbon series yet to be described, they are not the residual products of plant decomposition *in situ*, but are rather distillation products from deeply buried organic matter of both animal and vegetable origin. The different members of the series differ so widely in their properties and uses that each must be discussed independently. The grouping of the various compounds as given below is open to many objections from a strictly scientific standpoint, but, all things considered, it seems best suited for our present purposes.[1]

	Gaseous	Marsh gas (natural gas)
	Fluidal	Petroleum (naphtha)
	Viscous and semi-solid	Pittasphalt (maltha) / Mineral tar
Bituminous	Elastic	Asphalt (bitumen) / Elaterite / Wurtzilite
	Solid	Albertite / Grahamite / Uintaite
Resinous		Succinite / Copalite / Ambrite
Cerous (waxy)		Ozokerite / Hatchettite

Marsh Gas (Natural Gas). — This is a colorless and odorless gas arising from the decomposition of organic matter protected from the oxidizing influence of atmospheric air. By itself it burns quietly with a slightly luminous flame, but when mixed with air forms a dangerous explosive. It is this gas which forms the dreaded fire-damp of the miners.

Under this head may properly be considered the so-called *natural gas*, which has of late years become of so much importance from an economic standpoint. This is, however, by no means a simple compound, but an admixture of several gases, samples from different wells showing considerable variation in composition, as well as those from the same well collected at different periods. This last is shown by the six analyses following, and which may serve well to illustrate the average composition, though in some instances the percentage of marsh gas has been found greater.

[1] W. P. Blake, Trans. Am. Inst. of Mining Engineers, Vol. XVIII, 1890, p. 582.

Constituents	I	II	III	IV	V	VI
Mash gas	57.85 %	75.16 %	72.18 %	65.25 %	60.70 %	49.58 %
Hydrogen.	9.64	14.45	20.02	26.16	29.03	35.92
Ethylic hydride. . . .	5.20	4.80	3.60	5.50	7.92	12.30
Olifiant gas	0.80	0.60	0.70	0.80	0.98	0.60
Oxygen	2.10	1.20	1.10	0.80	0.78	0.80
Carbonic oxide	1.00	0.30	1.00	0.80	0.58	0.40
Carbonic acid	0.00	0.30	0.80	0.60	0.00	0.40
Nitrogen	23.41	2.89	0.00	0.00	0.00	0.00
	100.00 %	99.70 %	99.40 %	99.01 %	99.99 %	100.00 %

Natural gas in quantities sufficient to be of economic importance is necessarily limited to rocks of no particular horizon. The tendency of recent studies seems to be to show that it results, as above stated, from the deeply buried organic matter, of both plant and animal origin. It is not, however, indigenous to the rocks in which it is now found, but occurs in an overlying, more or less porous, sand or lime rock into which it has been forced by hydrostatic pressure. The first necessary condition for the presence of gas in any locality may, indeed, be said to depend upon the existence of such a porous rock as will serve as a reservoir to hold it, and also the presence of an impervious overlying stratum to prevent its escape. In Pennsylvania the reservoir rock is a sandstone of Carboniferous or Devonian age; in Ohio and Indiana, a cavernous dolomitic limestone of Silurian (Trenton) age.

Natural gas, as may readily be understood, is still in process of formation, though at a rate vastly slower than it is being utilized, or *wasted*, in many regions. It is a necessary consequence that the available supply must sooner or later become exhausted. Indeed this contingency has already made itself apparent in many fields, necessitating continuous activity in prospecting, and in more than one instance all known sources of supply are already exhausted. Few more marked illustrations of man's unreasonable squandering of nature's resources have ever been offered than that relating to the utilization of natural gas.

Petroleum. — This is the name given to a complex hydrocarbon compound, liquid at ordinary temperatures, though varying greatly in viscosity, of a black, brown, greenish, or more rarely, red or yellow color, and of extremely disagreeable odor. Its specific gravity varies from 0.6 to 0.9. Through becoming more and more viscous, the material passes into the solid and semi-solid forms, asphalt and maltha. Chemically it is considered as a mixture of the various hydrocarbons included in the marsh gas, ethyline, and paraffin series.

An ultimate analysis of several samples, as given by the reports of the 10th Census of the United States (1880), showed the following percentages of the three essential constituents:—

Localities	Hydrogen	Carbon	Nitrogen
West Virginia	13.359 %	85.200 %	0.540 %
Mecca, Ohio	13.071	86.310	0.230
California	11.810	86.034	1.100

As with marsh gas, petroleum is considered as a product of organic decomposition, which has been for the most part forced up from the rocks in which it originated into overlying strata. It is therefore limited to no particular geological horizon, but is found in rocks of all ages, from the Cambrian to the most recent, its existence in quantities sufficient for economic purposes being dependent upon local conditions for its generation and subsequent preservation. Inasmuch as its accumulation in large quantities necessitates a rock of porous nature to act as a reservoir, the petroleum-bearing rocks are mostly sandstones, though not uniformly so. Petroleums are found in California and Texas, in Tertiary sands; in Colorado, in the Cretaceous; in West Virginia, both above and below the Crinoidal (Carboniferous) limestones; in Pennsylvania, in the Mountain sands (Lower Carboniferous) and the Venango sands (Devonian); in Canada, in the Corniferous (Lower Devonian) limestone; in Kentucky, in the Hudson River shales (Lower Silurian); and in Ohio, in the Trenton limestone, also of Lower Silurian age.

In some instances petroleum oozes naturally from the ground, forming at times a thin layer on the surface of pools of water, whence in times past it has been gathered and used for chemical and medicinal purposes. The so-called "Seneca oil" thus used some fifty or sixty years ago was obtained from a spring in Cuba, Alleghany County, in New York. The immense supply now demanded for commercial purposes is, however, obtained altogether from artificial wells of varying depths, and which are in some cases self-flowing, while in others the oil is raised by means of pumps. Wells of from 500 to 1500 feet in depth are of common occurrence, while those upwards of 2000 feet are not rare. The principal sources of petroleum, in the United States, are in New York, Pennsylvania, and Ohio, with smaller fields in West Virginia, Kentucky, Tennessee, Indiana, Texas, Colorado, and California. The chief foreign source is the Baku region, on the Caspian Sea, and Galicia, in Austria.

The quantity of petroleum and semi-solid bituminous compounds contained in the rocks of certain areas is sometimes enormous. Dr. Hunt estimated that the dolomite underlying the city of Chicago and vicinity contains for each square mile over 7,000,000 barrels. A like computation by Professor Orton [1] led to the conclusions given in the following quotation relative to the water-lime stratum of Ohio, which is almost universally petroliferous: —

"Estimating its petroleum contents at one-tenth of one per cent, and the thickness of the stratum at 500 feet, both of which estimates are probably within the limits, we find the petroleum contained in it to be more than 2,500,000 barrels to the square mile. The total production of the great oil field of Pennsylvania and New York to January, 1885, is 261,000,000 barrels. It would require only three ordinary townships, or a little more than 100 square miles, to duplicate this enormous stock from the water-lime alone. But if the rate of one-tenth of one per cent should be maintained through a descent of 1500 feet at any point in the state, each square mile would, in that case, yield 75,000,000 barrels, or nearly one-third of the total product of the entire Pennsylvania and New York oil fields. These figures pass at once beyond clear comprehension, but they serve to give some idea of the vast stock of petroleum contained in the earth's crust. If petroleum is generally distributed through a considerable series of rocks in any appreciable percentage, it is easy to see that the aggregate amount must be immense. Even one-thousandth of one per cent would yield 750,000 barrels to the square mile in a series of rocks 1500 feet deep, but this amount is nearly equal to the greatest actual production per square mile of any part of the leading Pennsyl-

---

[1] Ann. Rep. U. S. Geol. Survey, 1886-87, Part II, p. 507.

vania fields. It is obvious that the total amount of petroleum in the rocks underlying the surface of Ohio is large beyond computation, but in its diffused and distributed state it is entirely without value. It must be accumulated in rocks that serve as reservoirs before it becomes of economic interest. In respect to the importance of concentration, it agrees with most other forms of mineral wealth."

Asphaltum (Bitumen, or Mineral Pitch). — These are names given to what are rather indefinite admixtures of various hydrocarbons, in part oxygenated, and which, for the most part solid or at least highly viscous at ordinary temperatures, pass by insensible gradations into pittasphalts or mineral tar, and these in turn into the petroleums. They are characterized by a black or brownish black color, pitchy lustre, and bituminous odor. The solid forms melt ordinarily at a temperature of from 90° to 100° F., and burn readily with a bright flame, giving off dense fumes of a tarry odor. The fluidal varieties become solid on exposure to the atmosphere, owing to evaporation of the more volatile portions.

The crude asphalt of Trinidad has the following composition and physical characteristics: [1] —

Specific gravity, 1.28; hardness at 70° F., 2.5 to 3, Dana's scale; color, chocolate-brown. Composition: —

Bitumen	39.83 %
Earthy matter	33.99
Vegetable matter	9.31
Water	16.87
	100.00 %

The mode of occurrence of asphalt deposits varies greatly, owing to the fact that, as with petroleum and natural gas, it has come up through fissures and cracks in the earth's surface, and as a rule no longer occupies its place of origin. On the island of Trinidad is an immense superficial deposit having an area of about 114 acres and a depth varying from 18 to 78 feet. The surface is sufficiently solid over nearly every part for the passage of teams, is of a brownish black color, and nearly level. The deposit has in numerous publications been compared to a lake, and stated to be fluidal and at a high temperature in the centre. This statement is quite erroneous and misleading.

[1] Trans. Am. Inst. Mining Engineers, Vol. XVII, 1880, p. 363.

In Ventura County, California, the material occurs in a fissure vein in siliceous clay of Miocene age, the vein being from 7 to 15 inches thick on the surface, but widening rapidly in descent to a thickness of 5 feet at a depth of 65 feet below the surface. The material of the vein is, however, far from pure asphalt; but rather an asphaltic sand. In western Kentucky the asphalt exudes from the ground in the form of "tar springs," and occurs also disseminated through sandstones and limestones of sub-Carboniferous age. Frequently, as in the dolomite underlying Chicago, Illinois, the bituminous matter is so diffused throughout the rock as to give it, on exposure, a brownish black appearance, and cause it to exhale an odor of petroleum appreciable for some distance. In the Dead Sea, bituminous masses of considerable size have in times past risen like islands to the surface of the water, and furnished thus the material used by the ancients in pitching the walls of buildings and rendering vessels water-tight. The ancient name of this body of water was *Lake Asphaltites*, and from it our word *asphalt* is derived.

The above illustrations are sufficient to indicate the numerous conditions under which the substance occurs. The material is world-wide in its geographic distribution and equally cosmopolitan in its geological range, being found in gneissic rocks of presumably Archæan age in Sweden, and in rocks of all intermediate horizons down to late Tertiary.

Elaterite (Mineral Caoutchouc). — This is the name given to a soft and elastic variety of asphalt much resembling pure indiarubber. It is easily compressible in the fingers, to which it adheres slightly, of a brownish color, and of a specific gravity varying from 0.905 to 1.00. It has been described from mines in Derbyshire and elsewhere in England, but, so far as the writer is aware, is of no commercial value. Its composition so far as determined is, carbon, 85.47 %; hydrogen, 13.28 %.

The name *wurtzilite* has been given by Professor W. P. Blake to a hydrocarbon very similar in appearance to the uintaite (described below), but differing in physical and chemical properties. It is described as a firm black solid, amorphous in structure, brittle when cold, breaking with a conchoidal fracture, but when warm, tough and elastic, its elasticity being best compared with that of mica. If bent too quickly, it snaps like glass. It cuts like horn, has a hardness between 2 and 3, a specific gravity of 1.03, gives a brown streak, and in very thin flakes shows a garnet-red color. It does not fuse or melt in boiling water, but becomes softer and more elastic; in the flame of a candle it melts and takes fire, burning with a bright, luminous flame, giving off gas and a strong bituminous odor. It is not soluble in alcohol, but sparingly so in ether, in both of which respects it differs from elaterite proper.

Albertite. — This is a brilliant jet-black compound, breaking with a lustrous, conchoidal fracture, having a hardness of between 1 and 2 of Dana's scale, a specific gravity of 1.097, a black streak, and showing a brown color on very thin edges. In the flame of a lamp it shows signs of incipient fusion, intumesces somewhat, and emits jets of gas, giving off a bituminous odor; when rubbed it becomes electric. According to Dana, it softens slightly in boiling water, is scarcely at all soluble in alcohol, and only slightly so in ether and in turpentine. The following is the composition as given by Witherill: Carbon, 86.04 %; hydrogen, 8.96 %; oxygen, 1.97 %; nitrogen, 2.93 %; ash, 0.10 %. The mineral occurs in fissures in rocks of sub-Carboniferous age, at the Albert Mines, in Hillsborough County, Nova Scotia; hence the name.

Formerly it was used for the distillation of oils for illuminating purposes. Since the discovery of petroleum its use has been discontinued.

Grahamite. — This variety resembles the last in its general appearance and its conduct toward solvents, and it is a question if it is not identical therewith. It was described by Dr. Wurtz from Ritchie County, in West Virginia, where it occurred in a vein some four feet in width in Carboniferous sandstones.

Uintaite (Gilsonite). — This is a black, brilliant, and lustrous variety giving a dark-brown streak, breaking with a beautiful conchoidal fracture, and having a hardness of 2 to 2.5 and a specific gravity of 1.065 to 1.07. It fuses readily in the flame of a candle, is plastic but not sticky while warm, and unless highly heated will not adhere to cold paper. Its deportment is stated to be much like that of sealing wax or shellac. Like albertite and grahamite, it dissolves slightly in turpentine and is not soluble in alcohol. It is a good non-conductor of electricity, but like albertite becomes electric by friction. Its composition as given is, carbon, 80.88 %; hydrogen, 9.76 %;

nitrogen, 3.30 %; oxygen, 6.05 %; and ash, 0.01 %. The mineral as first described occurred in a vertical vein from 3 to 5 feet in thickness, cutting through nearly horizontal sandstones some 3 miles east of Fort Duchesne, on the reservation of the Uinta Indians.

Succinite (Amber). — The mineral commonly known as amber is a fossil resin, consisting of some 78.94 parts of carbon, 10.53 parts of oxygen, and 10.53 parts of hydrogen, together with usually from two to four-tenths of a per cent of sulphur. It is not a simple resin, but a compound of four or more hydrocarbons. According to Berzelius, as quoted by Dana, it "consists mainly of (85 % to 90 %) two other resins in soluble alcohol and ether, and an oil, and $2\frac{1}{2}$ % to 6 % of succinic acid."

The mineral, as found, is of a yellow, brownish, or reddish color, frequently clouded, translucent, or even transparent, tasteless, becomes negatively electrified by friction, has a hardness of 2 to 2.5, a specific gravity, when free from enclosures, of 1.096, a conchoidal fracture, and melts at 250° to 300° Fahr. without previous swelling, but boils quietly, giving off dense white fumes with an aromatic odor and very irritating effect on the respiratory organs.

Amber, or closely related compounds, has been found in varying amounts at numerous widely separated localities, but always under conditions closely resembling one another. The better known localities are the Prussian coast of the Baltic; on the coast of Norfolk, Essex, and Suffolk, England; the coasts of Sweden, Denmark, and the Russian Baltic provinces; in Galicia, Westphalia; Poland; Moravia; in Norway; Switzerland; France; Upper Burma; Sicily; Mexico; the United States at Martha's Vineyard, and near Trenton and Camden, New Jersey, and at Cedar Lake in Northwest Canada.

The amber of commerce comes now, as for the past 2000 years, mainly from the Baltic, where it occurs in a stratum of blue earth of from 4 to 20 feet in thickness underlying the brown coal formation.

Ozokerite (Mineral Wax; Native Paraffin). — This is a wax-like hydrocarbon, usually with a foliated structure, soft and easily indented with the thumb nail; of a yellow, yellow brown, or sometimes greenish color, translucent when pure, with a greasy feeling, and fusing at 56° to 63° F.; specific gravity, 0.955. It is essentially a natural paraffin. The name is derived from two Greek words, signifying *to smell*, and *wax*. Below is given the composition of (I) samples from Utah, and (II) from Boryslaw, in Galicia.

Constituents	I	II
Carbon	85.47 %	85.78 %
Hydrogen	14.57	14.29
	100.04 %	100.07 %

The substance is completely soluble in boiling ether, carbon disulphide, or benzine, and partially so in alcohol.

Ozokerite occurs in the United States, in Emery and Uinta counties, Utah, where in the form of small veins in Tertiary rocks it extends over a wide area. It is also found in Galicia, Austria, in Miocene deposits, in Roumania, Hungary, Russia, and other Asiatic and European sources. As a rule the deposits are in beds of Tertiary or Cretaceous age. The Galician deposits are the most noted of the above. According to Boverton Redwood,[1] the material occurs here in the form of veins from the thickness of a few millimetres to some feet, and is accompanied by petroleum and gaseous hydrocarbons.

The names *scheerite*, *hatchettite*, *fichtellite*, and *konlite* are applied to simple hydrocarbons allied to ozokerite found in beds of peat and coal, but so far as the writer is aware never in such abundance as to be of commercial value.

The name *retinite* includes a considerable series of fossil resins allied to amber, differing mainly in containing no succinic acid. They occur in beds of brown coal of Tertiary and Cretaceous age. The so-called *copalite*, a hard brittle, clear yellow, or brownish variety used in making varnishes, belongs here.

2. ROCKS FORMED AS SEDIMENTARY DEPOSITS AND FRAGMENTAL IN STRUCTURE: CLASTIC

The rocks of this group differ from those just described in that they are composed mainly of fragmental materials derived from the breaking down of older rocks, or are but the more or

[1] Jour. Soc. of Chem. Industry, February, 1892.

less consolidated accumulations of organic and inorganic débris from plant and animal life. The group shows transitional forms into the last, as will be illustrated by certain of the limestones and the quarzites. They are water deposits, and, as a rule, are eminently stratified or bedded, although this structure is not always apparent in the hand specimen.

As will be readily comprehended when one considers from what a multitude of materials the fragmental rocks have been derived, the amount of assorting, admixture with other substances, solution, and transportation by streams these materials have undergone, they cannot be classified by any hard and fast lines, but one variety may grade into another, both in texture and structure as well as in chemical composition, almost indefinitely. Indeed, many of them can scarcely be considered as more than indurated muds, and only very general names can be given them.

Accordingly as these rocks consist of mechanically formed inorganic particles of varying composition and texture, or of the more or less fragmental débris from plant and animal life, they are here divided into two main groups, each of which is subdivided as below: —

I. Rocks formed by mechanical agencies, and mainly of inorganic materials.

(1) The Arenaceous group — Psammites: Sand, gravel, sandstone, conglomerate, and breccia.

(2) The Argillaceous group — Pelites: Kaolin, clay, wacke, shale, clayey marl, argillite.

(3) The Calcareous group: — Arenaceous and brecciated limestones. The rocks of this group are often in part organic, and in part chemical deposits. Only those are considered here in which the fragmental nature is the most pronounced characteristic.

(4) The Volcanic group: — Fragmental rocks composed mainly of ejected volcanic material: Tuffs, lapilli, sand and ashes, pumice-dust, trass, peperino, pozzuolano, etc.

II. Rocks formed largely or only in part by mechanical agencies and composed mainly of the débris from plant and animal life — Organagenous.

(1) The Siliceous group — Infusorial earth.

(2) The Calcareous group — Fossiliferous and oölitic limestone, marl, shell-sand, shell-rock.

PLATE 11

FIGS. 1 and 2. Shell limestones.　　FIG. 3. Crinoidal limestone.

(3) The Carbonaceous group — Peat, lignite, coals, oil shale, etc.

(4) The Phosphatic group — Phosphatic sandstone, guano, coprolite nodules.

(1) ROCKS COMPOSED MAINLY OF INORGANIC MATERIAL

(1) **The Arenaceous Group: Psammites.** — Arenaceous, from the Latin *arenaceous*, sandy or sand-like; psammite from the Greek ψαμμίτης, sandy.

These rocks are composed mainly of the siliceous materials derived from the disintegration of older crystalline rocks and which have been rearranged in beds of varying thickness through the mechanical agency of water. They are, in short, more or less consolidated beds of sand and gravel. In composition and texture, they vary almost indefinitely. Many of them having suffered little during the process of disintegration and transportation, are composed of essentially the same materials as the rocks from which they were derived. Others, in which the fragmental materials suffered more prior to their final consolidation, have had the softer and more soluble minerals removed, leaving the sand composed mainly of the hard, almost indestructible mineral quartz.

In structure, the sandstones also vary greatly,

FIG. 9. — Microstructure of sandstone, Portland, Connecticut.

in some the grains being rounded, while in others they are sharply angular. Figure 9 shows the microscopic structure of a brown Triassic sandstone from Portland, Connecticut.

The material by which the individual grains of a sandstone are bound together is as a rule of a calcareous, ferruginous, or siliceous nature; sometimes argillaceous. The substance has been deposited between the granules by percolating water or during the process of sedimentation, and forms a natural cement. It sometimes happens that the siliceous cement is deposited about the rounded grains of quartz in the form of a new crystalline growth, converting the stone into quartzite; such are in this work classed with the crystalline rocks.

Upon the character of this cementing material and the closeness with which the grains are bound together, is very largely dependent the power of the stone to resist disintegration under the trying action of percolating carbonated waters and the mechanical action of heat and frost. The calcareous, and to a less extent the ferruginous cements are liable to removal in solution, allowing the rock to fall away to sand, or at least allowing it to absorb water, which, on freezing, brings about the disintegration. The argillaceous cementing material, while in itself inert, also permits a high degree of absorption, with like results. Those sandstones cemented by silica, and which therefore partake of the nature of quartzite (see p. 169), are by far the more refractory.

The following analyses will serve to indicate the considerable range in composition of rocks of this class:—

CONSTITUENTS	I	II	III	IV
Silica (SiO_2)	69.94%	84.40%	95.24%	90.86%
Alumina (Al_2O_3)	13.15	7.49	0.56	4.76
Iron oxides (Fe_2O_3) and (FeO)	2.48	3.87	1.28	1.58
Manganese (MnO)	0.70	....	....	....
Lime (CaO)	3.09	0.74	1.40	0.15
Magnesia (MgO)	Trace	2.11	1.23	0.59
Potash (K_2O)	3.30	0.24	....	1.06
Soda (Na_2O)	5.43	0.56	....	0.45
Loss	1.01	....	0.56	....
Totals	99.10%	99.41%	99.27%	99.45%

I. Brown Triassic sandstone: Portland, Connecticut. II. Gray sub-Carboniferous sandstone: Berea, Ohio. III. Red Carboniferous sandstone: Anan, Scotland. IV. Cambrian sandstone: Siskowit Bay, Wisconsin.

The table given on p. 166 will serve to show the close chemical relationship existing between many rocks of this group, and their metamorphic equivalents.

The colors of sandstone are dependent upon a variety of circumstances. The red, brown, and yellowish colors are due to iron oxides in the cementing constituent. Some of the dark colors are due to carbonaceous matter.

Many varieties of sandstone are popularly recognized. *Calcareous, ferruginous, siliceous,* or *argillaceous* sandstones are those in which the cementing materials are of a calcareous, ferruginous, siliceous, or argillaceous nature. The name *arkose* is given to a coarse feldspathic sandstone derived from granitic rocks, with a minimum amount of loss of original material. *Conglomerate* or *puddingstone* is merely a coarse sandstone; it differs from ordinary sandstone only as gravel differs from sand. *Breccia* is a fragmental rock differing from conglomerate in that the individual particles are sharply angular instead of rounded. The term is made to include also certain volcanic rocks with a brecciated structure. (See Pl. 4.)

Greywacke or *Grauwacke* is an old German name for brecciated fragmental rocks made up of argillaceous particles. The name is now little used. Other names, as *flagstone, freestone,* and *brownstone,* are applied to such as are used for flagging or other structural purposes. *Itacolumite* is a feldspathic sandstone, or perhaps more properly quartzite, in which the feldspathic material plays the rôle of a binding constituent to the quartz granules. The so-called *flexible* sandstone is an itacolumite from which the feldspathic portions have been removed by decomposition leaving the interlocking quartz grains with a small amount of play between them. The rock is in no sense elastic, but merely loose jointed.

The name *greensand, greensand marl,* and *glauconitic sand* are given to a prevailing dull green, loosely coherent, clayey or arenaceous deposit which owes its peculiarities to the presence of the hydrous silicate of iron and potassium *glauconite*, but which is variously contaminated with minute particles of quartz and siliceous minerals such as feldspar, hornblende, augite, garnet, epidote, tourmaline, zircon, and the iron ores, clay, rock fragments, and particles of shells.

Beds of glauconitic sand are most abundant among terranes of Cretaceous age, but are by no means limited to them, as has been already intimated on p. 31. They are aqueous deposits, formed during processes of slow sedimentation along coasts receiving débris from the continental slopes and of a nature such as is derived from the breaking down of granitic and other feldspathic rocks. The depth at which such deposits form is naturally quite variable, but conditions most favorable to their accumulation seem to lie just beyond the reach of wave agitation and under a depth of 900 fathoms.

The following table of analyses of glauconitic marls is from the Report of the Geological Survey of New Jersey, for 1893.

Constituents	I	II	III	IV	VI	X	XI	XII	XIII
	%	%	%	%	%	%	%	%	%
Phosphoric acid ..	1.15	0.58	1.51	1.14	0.84	0.19	0.50	6.87	3.73
Sulphuric acid ...	1.28	....	2.40	0.14	0.12	0.41	0.34	3.12	2.44
Silica and sand...	34.50	45.50	55.69	38.70	52.07	51.15	47.50	44.08	40.08
Potash.........	1.54	3.79	5.27	3.65	6.46	7.08	5.29	3.97	4.98
Lime..........	2.52	1.51	0.65	9.07	1.01	0.49	0.56	4.97	4.14
Magnesia	2.15	2.20	0.79	1.50	1.53	2.02	2.70	2.97	0.47
Alumina	6.00	5.80	6.61	10.20	6.96	8.23	8.60	6.04	?
Oxide of iron....	31.50	24.50	21.63	18.63	21.55	23.13	20.52	18.97	28.71
Water.........	18.80	15.40	8.85	10.00	9.31	6.67	13.57	8.63	5.54
Carbonic acid ...	....	....	....	6.14	....	....	....	....	....
Carbonate of lime .	....	....	....	....	....	....	....	.,,..	....
	99.43	99.18	102.40	99.16	99.85	99.37	99.58	99.32	99.09

I. Clay marl, from near Mattawan. II. Clay marl, from Matchaponix Creek, three miles south of Spottswood. III. Lower marl, from Navesink Highlands. IV. Lower marl, from north shore of Navesink River, at Red Bank. VI. Lower marl, from northwest slope of Mount Pleasant Hills. X. Middle marl, from near Eatontown. XI. Middle marl, from southeast of Freehold. XII. Upper marl, from Poplar. XIII. Upper marl, from Shark River.

The most extensive and best known deposits in the United States are included in what are known as the Upper, Middle, and Lower marl beds of the Cretaceous formations in southeastern New Jersey, and which has been very thoroughly described in the various reports of the State Survey.[1] The marl is somewhat variable in different localities, but may in a general way be described as a dull green, arenaceous deposit of such consistency as to be easily removed by the shovel alone, or pick and shovel. The beds vary from 30 to 60 feet in thickness, but the glauconitic layers are not uniformly distributed through it. Through weathering, the ferruginous constituents become more highly oxidized, and the color changed from dull green to red and yellow.

[1] The reader is especially referred to Professor W. B. Clarke's paper on "The Cretaceous and Tertiary Formations of New Jersey," in the Ann. Rep. State Geologist of New Jersey for 1892.

Rocks belonging to the arenaceous group are world wide in their distribution, covering not infrequently thousands of square miles of territory to depths, it may be, of thousands of feet. They are, in some of their varieties, among the most common and wide-spread of materials. Being themselves the products of disintegration and decomposition of pre-existing rocks, and having become consolidated under conditions not greatly different from those now existing at or near the surface of the earth, the rocks of this group are as a whole in a state of comparatively stable chemical equilibrium. Unless including calcareous matter or readily oxidizing ferruginous compounds, such are subject to disintegration more through physical than chemical agencies, as will be noted later.

(2) **The Argillaceous Group: Pelites.** — The rocks of this group are composed of more or less hydrated aluminous silicates admixed in almost indefinite proportions with siliceous sand, various silicate minerals in a more or less fragmental and decomposed condition, and calcareous and carbonaceous matter. In their least consolidated form they are best represented by the common plastic clays used for brick and pottery manufacture. Such, although alike in their general physical or even ultimate chemical nature, have widely diverse origins. In fact, the term *clay*, like *silt*, indicates physical condition rather than chemical or mineralogical composition, and it may perhaps be defined as an indefinite admixture of more or less hydrated aluminous silicates, free silica, iron oxides, carbonates of lime, and various silicate minerals which in a more or less decomposed and fragmental condition have survived the destructive agencies to which they have been subjected. About the only feature characteristic of all clays, is that of plasticity, when wet, and this is dependent, apparently, wholly upon texture and structure, *i.e.* upon the size and shape of the individual particles. Pure quartz, chalcedony, flint, feldspar, or other silicates, will, when reduced to an impalpable powder, possess the plasticity and even odor usually ascribed to clay, and in the pages following, the term is used only with reference to degree of comminution, regardless of mineral nature or chemical composition. It includes residual products of any or all forms of rock degeneration, and which may or may not have been reassorted through the agency of water. (See further under The Regolith, Part V.) The oft-repeated statement that kaolin

forms the basis of clays, or that clay is impure kaolin, is therefore to a certain extent misleading, and if accepted at all it must be with the reservations made by Johnson and Blake,[1] who limit the term *kaolin* itself to the impure material, quite distinct from true *kaolinite*, which is a definite chemical compound corresponding to the formula $H_4Al_2Si_2O_9$.

Throughout the glaciated region of the northeastern United States the clays are mostly glacial or water deposits from the floods of the Champlain epoch. The latter are often beautifully and evenly stratified, as shown in the illustration on Pl. 24. The plastic clays and siliceous sands about Woodbridge, New Jersey, are regarded as derived from the Azoic rocks and deposited by sea-water in enclosed basins. The exact source of the material is not always apparent; the porcelain clays of Lawrence County, Indiana, on the other hand, are residual deposits resulting from the decomposition of impure Carboniferous (Archimides) limestones, the lime carbonate being removed in solution, while the less soluble clay remains. Kaolin, as already noted, is a residual deposit from the decay of feldspathic and other aluminous rocks, while the ordinary brick and tile clays of the Southern states, as well as the clayey soils, are residual aluminous deposits resulting from the decay and leaching out of soluble constituents from a variety of rocks, both sedimentary and eruptive. (See chapter on rock weathering.)

As showing the comparative compositions of kaolins and clays, the following table is given: —

Constituents	I	II	III	IV	V	VI
SiO_2 (combined)	46.4 %	39.00%	34.70 %	28.30 %	42.71 %	} 60.97 %
SiO_2 (free)	....	....	12.20	27.80	0.70	
Al_2O_3	39.7	36.00	31.34	27.42	39.24	26.38
H_2O (combined)	13.9	14.00	12.00	6.60	13.32	} 8.93
H_2O at 212°	....	9.50	8.00	2.90	1.58	
CaO and MgO	....	0.63	0.10	0.18	0.20	} 1.90
Alkalies	....	0.54	0.05	2.71	0.80	
Fe_2O_3	....	....	0.16	2.08	0.40	1.40
	99.00 %	99.67%	99.45 %	98.59 %	99.10 %	99.04%

I. Kaolin. II. Indianite, a white clay residual from St. Lawrence County, Indiana. III. Potter's clay, from Pope County, Illinois. IV. Brick clay from New Jersey. V. Fire clay from New Jersey. VI. Fire clay from Illinois.

[1] Am. Jour. of Science, 1867, p. 351.

Amongst the older formations the clays have undergone induration, giving rise to what are known as *argillites*, or if fissile, *slates* or *clay slates*, such as are used for roofing and similar purposes, the fissile property having been imparted by pressure or shearing. Such forms pass by imperceptible gradations into argillaceous schists which are classed with the metamorphic rocks. (See p. 170.) The argillites are, as a rule, among the most indestructible of rocks, since they are themselves composed of the least destructible débris of pre-existing rocks. Their ultimate chemical composition is much like that of the clays, and scarce any two samples will show similar results when submitted to analysis. The table given below shows the composition of some schistose argillites used for roofing purposes from (I) Harford County, Maryland, (II) Lancaster County, Pennsylvania, and (III) Llangynog, North Wales.

Constituents	I	II	III
Silica (SiO_2)	58.37 %	60.32 %	60.150 %
Sulphuric acid (H_2SO_4)	0.22	....	....
Alumina (Al_2O_3)	21.985	23.10	24.20
Iron oxides (FeO) and (Fe_2O_3)	10.661	7.05	7.05
Lime (CaO)	0.30	....	....
Magnesia (MgO)	1.203	0.87	
Soda (Na_2O)	1.933	0.49	4.278
Potash (K_2O)	....	3.83	
Water (H_2O)	4.03	4.08	3.72
	98.099 %	99.74 %	99.998 %

Shale is a somewhat loosely defined term, indicating structural rather than chemical or mineralogical composition. The word is perhaps best used in its adjective sense, as a *shaly sandstone*, or *shaly limestone*. By many authors it is used with reference more particularly to thinly stratified or laminated, clayey rocks. Many shales are but the finer, more fissile portions of sandstone beds; such may represent the off-shore or deep-water portions of arenaceous sediments, which, beginning with gravels near the shore-line, become gradually finer as the distance from the shore increases, passing through coarse to finer sands and finally to sandy clays and silts as the water, through the lessening of its carrying power, lays down its load. Or they may represent later stages in the cycle of sedimentation; the finer silts brought down after erosion have so far reduced the level of the land as to greatly diminish the currents and consequent carrying power of the seaward-flowing streams. Such beds, on consolidation, yield then what are commonly known, in the order of their formation, as conglomerates, sandstones, shales and argillites, or clay slates, the shales occupying, both in texture and composition, a position intermediate between the argillites and sandstones.

The following table will serve to show the varying character of the rocks included under this name. Those such as given in columns I and II carry their sulphur in combination with iron, as iron pyrites (FeS_2). This, on decomposing, through the action of meteoric waters, yields iron sesquioxides and sulphuric acid, the latter combining with a portion of the alumina in the rock to form sulphate of aluminum, or common alum. Hence they have been called *alum shales*.

Constituents	I	II	III
Silica (SiO_2)	50.13%	72.40%	66.06%
Alumina (Al_2O_3)	10.73	16.45	15.626
Iron sesquioxide (Fe_2O_3)	5.78	1.05	8.38
Lime (CaO)	0.40	0.17	0.493
Magnesia (MgO)	1.00	1.48	0.677
Potash (K_2O)	....	5.08	3.205
Soda (Na_2O)	....	0.53	0.628
Sulphur (S)	4.02	1.21	....
Carbon (C)	22.83	Undet.	3.787
Water (H_2O) [1]	2.21	Undet.	....
Phosphoric acid (P_2O_5)	....	....	0.154

I. An alum shale from Garnsdorf, near Saalsfeld. II. An alum shale from Bornholm. III. A "marly shale" from Breckenridge County, Kentucky.

The name *till* or boulder clay is given to a sandy clay of glacial origin and consisting of the usual indefinite mixture. Professor W. O. Crosby, who has studied the composition of the normal till of the Boston Basin, reports it as composed, exclusive of the larger pebbles, of "about 25%, or one-fourth, of coarse material which may be classed as gravel; about 20%,

[1] Ignition.

or one-fifth, of sand; 40 to 45 % of extremely fine sand, or rock flour, and less than 12 % of clay."[1]

Laterite is a red, ferruginous residual clay found in tropic and semitropic regions. (See p. 310.) *Catlinite*, or Indian pipe-stone, is an indurated clay rock formerly used by the Dakota Indians for pipe material. The name *porcellainite* has been given to a compact porcelain-like rock consisting of clay indurated by igneous agencies. The name *wacke* is sometimes used to designate an earthy or compact, dark-colored clayey material resulting from the decomposition *in situ* of basaltic rocks. *Adobe* is the name given to a calcareous clay of a general gray-brown or yellowish color, very fine grained and porous, and which is widely distributed throughout the more arid regions of the West. It is described in greater detail under the head of soils (p. 333). *Loess* is a somewhat similar material forming the surface soil over wide areas in the Mississippi valley, and at times sufficiently plastic for brick making. (See also p. 327.)

(3) **The Calcareous Group.** — Here are brought together a small series of fragmental rocks composed mainly of calcareous material, but of which the organic nature, if such it had, is not apparent. These rocks form at times beautifully brecciated marbles. Their structure may be best comprehended by remembering that the original beds, whether crystalline or amorphous, whether fossiliferous or originating as chemical precipitates, have by geological agencies been crushed and shattered into a million fragments, and then, by infiltration of lime and iron-bearing solutions, been slowly cemented once more into solid rock. The composition is essentially the same as the ordinary sedimentary limestones and need not be further dwelt upon here. It may be stated, however, that owing to the softness and ready solubility of their materials limestones do not, on breaking down, except under rare instances, give rise to extensive beds of arenaceous rocks, as do the siliceous varieties. One of the best known rocks of this group is the breccia marble near Point of Rocks in Maryland, which has been used in the United States Capitol building at Washington.

(4) **The Volcanic Group: Tuffs.** — Under this head are included a great variety of fragmental rocks, composed of the more or less finely comminuted materials ejected from vol-

[1] Proc. Boston Society of Natural History, Vol. XXV, 1890, p. 123.

canoes as ashes, dust, sand, and lapilli. Some of them are made up of minute shreds of pumiceous glass. These occur, in many instances, interbedded with lava flows of the same lithological nature, and which are a product of the same periods of volcanic activity, the eruption of molten lava being accompanied by intervals of explosive action, during which only fragmental material was ejected. To such materials the name *pyro*clastic (Greek πυρός, fire) is appropriately given.

The lithological character of the materials varies almost indefinitely, and only very general names are given them in the majority of cases. The name *tuff* or *tuffa* is given to the entire group of volcanic materials formed as above, and also by some authorities to fragmental rocks resulting from the breaking down and reconsolidation of older volcanic lavas. It would seem advisable to designate these last, as has F. Löwinson-Lessing,[1] as *pseudotuffs* or *tuffoids*.

The names *volcanic ashes*, *sand*, and *dust* are applied to the finer of these volcanic materials, and *lapilli* or *rapilli* to the coarser fragments.

The dusts and sands are not infrequently composed of minute shreds of volcanic glass, which were blown from the volcanic vents and carried unknown distances, to be ultimately deposited as stratified beds in comparatively shallow water. Such are described more in detail under the head of Æolian rocks (p. 153). The term *trass* is used to designate a compact or earthy fragmental rock composed of pumice dust, in which are embedded fragments of trachytic and basaltic rocks, carbonized wood, etc., and which occupies some of the valleys of the Eifel. *Peperino* is a tufaceous rock composed of fragments of basalt, leucite, lava, and limestone, with abundant crystals of augite, mica, leucite, and magnetite. It occurs among the Alban Hills, near Rome, Italy. *Palagonite tuff* is composed of dust and fragments of basaltic lava, with pieces of a pale yellow, green, reddish, or brownish glass called *palagonite*. The general name of *volcanic mud* is given to the finely comminuted volcanic material which in a more or less pasty or liquid condition is thrown from volcanic vents during the incipient stages of eruption.

The tuffs are as a rule more or less distinctly stratified, of very uneven texture, and with rarely a pisolitic structure. They are found associated with volcanic rocks of all ages, and

[1] Tschermaks Min. u. Petrog. Mittheilungen, Vol. IX, 1889, p. 530.

at times so highly metamorphosed as to render the original nature of some doubt. Certain English authorities have contended that a part of the so-called argillites and fire clays were of finely comminuted volcanic materials.

The composition of the tuffs naturally varies with that of the character of the lava from which they were derived. Being in a more or less finely comminuted condition, often porous and readily permeated by water or rootlets, they undergo decomposition, forming soils the character of which is dependent to some extent upon their lithological nature. The following table shows the varying composition of rocks of this class: —

Kinds and Localities	Silica (SiO_2)	Alumina (Al_2O_3)	Ferric Oxide (Fe_2O_3)	Lime (CaO)	Magnesia (MgO)	Potash (K_2O)	Soda (Na_2O)	Water (H_2O)	Miscellaneous	Totals
	%	%	%	%	%	%	%	%	%	%
Pozzuolana, Naples, Italy	50.14	21.28	4.76	1.90	....	4.37	6.23	....	....	100.24
Tuff, Crater of Monte Nuova, Italy	56.31	15.23	7.11	1.74	1.30	6.54	4.84	0.12	Chlorine 0.27	100.22
Trass, Andernach, Prussia	54.00	16.50	6.10	4.00	0.70	10.00		7.00	....	99.00
Tuff, Lacher See, Prussia	60.49	19.95	9.37	3.12	1.43	3.40		1.33	....	99.09

(2) ROCKS COMPOSED MAINLY OF DÉBRIS FROM PLANT AND ANIMAL LIFE

(1) **The Siliceous Group: Infusorial or Diatomaceous Earth.** — This is a fine white or pulverulent rock, composed mainly of the minute shells, or tests, of diatoms, and often so soft and friable as to crumble readily between the thumb and fingers. It occurs in beds which, when compared with other rocks of the earth's crust, are of comparatively insignificant proportions, but which are nevertheless of considerable geological importance. Though deposits of this material are still forming, and have been formed in times past at various periods of the earth's history, they appear most abundantly associated with rocks belonging to the Tertiary formations.

The beds are wide-spread, and some of them of economic importance as a source of tripoli, absorbents for nitro-glycerine

AQUEOUS ROCKS

compounds, non-conducting materials, etc. A deposit in Bilu, Bohemia, is some 14 feet in thickness, and is estimated by Ehrenberg to contain 40,000,000 shells to every cubic inch. Beds occur in the United States at South Beddington, Maine;

FIG. 10.— Section through lake basin showing the formation of infusorial earth. *a*, bed rock; *bb*, floating peat; *cc*, decayed peat; *d*, infusorial earth.

Lake Umbagog, New Hampshire; in Morris County, New Jersey; near Richmond, Virginia; in Calvert and Charles counties, Maryland; in New Mexico; Graham County, Arizona; near Reno, Nevada, and in various parts of California and Oregon.

The New Jersey deposit covers about 3 acres, and varies from 1 to 3 feet in thickness; the Richmond bed extends from Herring Bay, on the Chesapeake, to Petersburgh, Virginia, and is in some places 30 feet in thickness; the New Mexico deposit is some 6 feet in thickness and has been traced some 1500 feet. Professor Leconte states that near Monterey, in California, is a bed some 50 feet in thickness, while the geologists of the Fortieth Parallel Survey report beds not less than 300 feet in thickness, of a pure white, pale buff, or canary-yellow color, as occurring near Hunter's Station, west of Reno, Nevada.

The earth is used mainly as a polishing powder, and is sometimes designated as *tripolite*. It has also been used to some extent to mix with nitro-glycerine in the manufacture of dynamite. Chemically the rock is impure opal, as will be seen from the following analyses made on samples from (I) Lake Umbagog, New Hampshire, (II) Morris County, New Jersey, and (III) Paper Creek, Maryland: —

CONSTITUENTS	I	II	III
Silica (SiO_2)	80.53 %	80.60 %	81.53 %
Iron oxides (Fe_2O_3 and FeO)	1.03	....	3.33
Alumina (Al_2O_3)	5.89	3.84	3.43
Lime (CaO)	0.35	0.58	2.61
Water (H_2O)	11.05	14.00	6.04
Organic matter	0.98	....	....
	99.38 %	99.02 %	96.94 %

Number III showed also small amounts of potash and soda.

PLATE 12

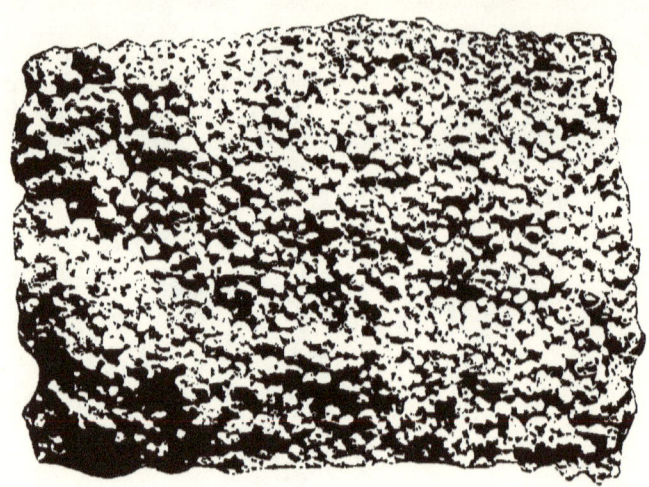

1

2

FIG. 1. Pisolitic limestone.   FIG. 2. Oölitic limestone.

(2) **The Calcareous Group.** — These rocks are made up of the more or less fragmental remains of molluscs, corals, and other marine and fresh-water animals. Many of them are but consolidated beds of calcareous mud, full of more or less fragmentary shells or casts of shells, as shown in Fig. 1, Pl. 11. The name *coquina* (Spanish for *shell*) is given to such as that shown in Fig. 2, Pl. 11, from St. Augustine, Florida. The rock, it will be observed, is composed almost wholly of very perfect shells of a bivalve mollusc, loosely cemented by calcareous materials in a finely divided condition. From such forms as these we have all possible gradations to compact crystalline limestone. Special names are often given these calcareous rocks, designating the character of materials from which they are derived. *Coral* and *shell* limestones, as the names denote, are composed mainly of the débris from these organisms. In like manner such names as *crinoidal*, *fusulina*, etc., are applied.

Lumachelle is the name given to a shell limestone from the Tyrol, in which the shells still retain their pearly lining and original beauty. *Nummulitic* limestone carries fossil nummulites. Rocks of this type were used in the construction of the pyramids of Cheops. *Chalk* is a fine-grained, white, pulverulent rock, composed of finely broken shells of marine molluscs, among which minute foraminifera are abundant. *Shell sand* is a loose aggregate of shell fragments, formed on sea-beaches by the action of the winds and waves. On certain Hawaiian beaches, such sands give out a distinct note, or peculiar crunching sound when walked over, or even when shaken in a closed vessel, and are popularly known as *sounding*, or *singing*, sands. The property is manifested only when the sand is dry and is assumed to be due to the minute air cavities enclosed by the shells. *Oölitic* and *pisolitic* limestones, as previously noted, are made up of rounded concretionary masses of calcium carbonate, and are in part of mechanical origin, and in part chemical deposits (Pl. 12).

The microscopic structure of an oölitic limestone from Princeton, in Caldwell County, Kentucky, is shown in the accompanying figure (p. 144). It will be noticed that the first step in the formation of this stone was the deposition of concentric coatings of lime about a nucleus which is sometimes nearly round, but more frequently quite angular and irregular. After the concretions were completed there were formed in all cases about each one, narrow zones of minute radiating crystals of clear, colorless calcite; then the larger crystals formed in the interstices. The nuclei are composed in some cases of single fragments or, again, of a group of fragments. Certain of the oölites present no distinct concentric structure, but appear as mere rounded masses merging gradually into the crystalline interstitial portions. Recent microscopic studies have tended to show that many of the oölitic limestones owe their structure to the lime-secreting power of microscopic algæ.[1]

Limestones vary almost indefinitely in structure and color. From the soft tufaceous or highly fossiliferous varieties there is a constant gradation to dense compact rocks breaking with a conchoidal or splintery fracture and the true nature of which is sometimes to be ascertained only by chemical tests. There is a like variation in color. White through all shades of gray to black is common, and more rarely occur yellow, brown, pink, or red varieties, the colors depending on organic matter and metallic oxides, mainly ferruginous.

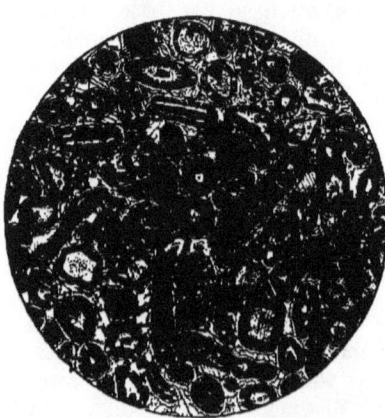

FIG. 11. — Microstructure of oölitic limestone.

Owing to the readiness with which calcium carbonate undergoes crystallization, even at ordinary temperatures, few limestones are wholly amorphous, but grade insensibly into holocrystalline varieties such as are classed with the metamorphic rocks. The name *marble* is given to such limestones as are of sufficiently close texture to take a polish and of such colors as to make them desirable for ornamental work. A large proportion of the marbles belong, however, to the metamorphic group. (See p. 162.) Figure 12 shows the microscopic structure of a dark gray, variegated, highly fossiliferous limestone belonging to the Cincinnati group, near Hamilton, Ohio. It is a natural result of their method of formation that few

[1] American Geologist, Vol. X, No. 5, 1892.

limestones are of pure calcium carbonate. A portion of the calcium is not infrequently replaced by magnesium, giving rise to *magnesian limestones*, or when the proportion of magnesia rises to 45.65 % to *dolomite*. This last can as a rule be distinguished from limestone only by its increased hardness (3.5–4.5) and specific gravity (2.8–2.95). Frequently chemical tests are necessary, limestone effervescing readily when treated with dilute hydrochloric acid, while dolomite is unacted upon.

Mechanically included materials, as sand and clay, are common, giving rise to siliceous and argillaceous varieties. The so-called *hydraulic* limestone is one containing 10 % and upwards of these impurities, and which, when burnt and ground, forms a cement characterized by its property of setting under water. Many limestones, like the dolomitic varieties in Cook County, Illinois, contain so large a proportion of bituminous matter as to give off a distinct odor of petroleum when struck with a hammer, or even to become blackened on the surface by its exudation when exposed to the weather. Others contain phosphatic matter, and pass by insensible gradations through what are known as phosphatic limestones to true phosphates (phosphorites, etc.).

FIG. 12. — Microstructure of fossiliferous limestone.

In chemical composition the limestones vary, like other sedimentary rocks, almost indefinitely, as will naturally be inferred from what is said above. As a general rule, those varieties, which have been formed in deep waters and at a distance from the shores, will be of greatest purity, since less likely to have become contaminated through detrital materials washed in from the land. Even these may, however, be intermingled to a very considerable extent with the fine siliceous and ferruginous matter, such as deep-sea dredgings have shown to be common to our modern sea-bottoms, and which are assumed to be in part at least of volcanic origin. (See under Æolian Rocks, p. 153.) The following table will give some idea of the wide range in chemical composition to be found in rocks of this class: —

CONSTITUENTS	White crystalline marble, West Rutland, Vermont.	White crystalline Dolomite, Westchester, New York.	Magnesian limestone, Minneapolis, Minnesota.	Hydraulic limestone, Lehigh, Pennsylvania.	Oölitic limestone, Bedford, Indiana.
Carbonate of lime ($CaCO_3$)	98.00 %	54.02 %	41.48 %	72.95 %	96.60 %
Carbonate of magnesium ($MgCO_3$)	....	45.04	24.55	3.84	0.13
Oxides of iron (FeO and Fe_2O_3)	....	}0.23	}4.03	1.34	0.98
Oxide of aluminum (Al_2O_3)	....			4.50	....
Silica (SiO_2) and insol. silicates	0.57	....	29.03	14.79	0.50
Potash (K_2O)	....	....	1.22	....	0.31
Soda (Na_2O)	....	....	1.12	....	0.40
Water (H_2O)	....	....	....	....	0.96
Sulphate of lime ($CaSO_3$)	....	....	....	1.75	....
Organic matter	....	....	....	1.46	....
Totals	98.57 %	99.89 %	98.33 %	100.64 %	99.88 %

Researches by the Kentucky Geological Survey have shown that the older limestones are, as a general rule, richer in soda, phosphoric acid, and, when non-magnesian, in lime carbonate, than are the younger more recently formed, and correspondingly poorer in silica and insoluble silicates. This inverse ratio is shown in the table on the opposite page, in which the rocks are arranged by geological horizons, the oldest at the bottom.

The name *shell marl*, or merely *marl*, is given to an illy defined, often arenaceous, soft and earthy rock consisting essentially of shell material in a more or less fragmental condition, and usually intermixed with more or less clayey matter or siliceous sand and silt. Geikie [1] would limit the term to fresh-water accumulations of remains of mollusca, entomostra, and fresh-water algæ, but unfortunately the word has not been so used in much of the literature extant. These marls, being easily decomposed, and on account of their occasional richness in phosphoric acid, or, perhaps, merely on account of the lime they contain, are of value as fertilizers. The following analyses of North Carolina marls, consisting largely of comminuted

[1] Text-book of Geology. 3d ed.

COMPOSITION OF LIMESTONES

GENERAL AVERAGE COMPOSITION OF LIMESTONES OF THE SEVERAL GEOLOGICAL FORMATIONS

LIMESTONES	Specific Gravity	Lime Carbonate	Magnesia Carbonate	Alumina and Iron and Manganese Oxides	Phosphoric Acid (P_2O_5)	Sulphuric Acid (SO_3)	Potash	Soda	Silica and Silicates
	%	%	%	%	%	%	%	%	%
Average composition of 6 Coal-measure limestones	....	72.958	9.608	4.883	0.152	1.003	0.486	0.196	9.657
Average composition of 10 upper sub-Carboniferous limestones	2.604	80.014	3.887	1.202	.123	.571	.181	.132	4.706
Average composition of 3 lower sub-Carboniferous limestones (hydraulic)	2.715	53.800	24.641	3.001	.111	.218	.304	.141	16.900
Average composition of 6 Black Slate limestones	2.728	51.097	26.258	6.578	.230	1.460	.505	.207	12.002
Average composition of 4 Carboniferous limestones: of which 2 are magnesian (hydraulic?) and 2 are non-magnesian	N.E. N.E.	42.048 90.810	17.838 5.090	8.065 .080	.155	N.E. .320	N.E. .160	N.E. .116	25.000 2.630
Average composition of 14 Niagara group (or Upper Silurian) limestones	N.E.	62.292	19.150	4.321	.250	.040	.267	.123	8.100
Average composition of 3 Clinton group limestones	N.E.	49.107	24.233	10.051	.510	.364	.314	.122	12.467
Average composition of 3 upper Hudson group limestones	N.E.	80.250	2.034	3.287	.356	1.003	.418	.139	11.000
Average composition of 7 middle Hudson group ("siliceous mudstone")	N.E. N.E.	1.185 90.312	1.061 1.163	9.216 1.753	.450 .288	.265 .332	.413 .270	.330 .203	84.646 0.070
Average composition of 9 lower Hudson limestones									
Average composition of 7 Trenton limestones (non-magnesian)	2.608	90.970	1.828	2.165	.480	.463	.470	.266	3.704
Average composition of 11 Trenton limestones (magnesian)	2.681	64.323	23.641	3.410	.414	.632	.500	.278	6.078
Average composition of 2 Bird's-eye limestones	....	95.216	2.002	.505	.091	.233	.207	.040	1.880
Average composition of 3 Chazy limestones	....	88.767	25.970	1.000	N.E.	N.E.	N.E.	N.E.	4.346

shells and sometimes coprolite nodules, will serve to show the widely varying character of the materials grouped under this name.[1]

Constituents	I	II	III	IV	V	VI	VII
	%	%	%	%	%	%	%
Silica	6.97	61.61	18.84	58.25	25.28	39.36	5.65
Oxide of iron and alumina	0.86	2.80	2.72	11.28	3.02	3.47	3.30
Lime	47.62	19.60	41.48	13.40	37.52	28.96	48.51
Magnesia	1.03	....	....	....	0.12	0.16	1.96
Potash	0.37	....	0.56	....	0.22	0.75	0.23
Soda	0.15	....	0.09	....	0.25	0.17	0.30
Phosphoric acid	0.10	....	0.18	....	0.40	0.11	trace
Sulphuric acid	0.41	0.06	0.31	....	0.40	0.18	0.31
Carbonic acid	38.15	15.37	32.07	10.59	29.02	22.73	39.80
Organic matter and water	4.25	....	3.42	....	2.98	4.11	0.60

(3) **The Carbonaceous Group: Peat, Lignite, and Coal.** — Here are included a variety of more or less oxygenated hydrocarbons varying widely in physical and chemical properties, but alike in originating from decomposing plant growth protected from the oxidizing influences of the air. Plants, when decomposing upon the surface of the ground, give off their carbon to the atmosphere in the shape of carbonic acid gas (CO_2), leaving only the strictly inorganic or mineral matter behind. When, however, protected from this oxidizing influence by water, or other plant growth, decomposition is greatly retarded, varying portions of the carbonaceous and volatile matters are retained, and the material becomes slowly converted into coal. According to the amount of change that has taken place in the original plant material, the amount of volatile matter still retained by it, its hardness and burning qualities, several varieties are recognized, which, however, pass into each other by insensible gradations.

Peat is the plant matter in its least changed condition. It results from the gradual accumulation in bogs and marshes of growths consisting mainly of sphagnous mosses, a low order of plants having the faculty of continuing in growth upwards as they die off below. In this way the deposits often assume a very considerable thickness. Where sufficiently thick, the

[1] Geology of North Carolina, Vol. I, 1875, p. 105.

lower portions have sometimes been converted into a dense brownish black mass somewhat resembling true coal. The deposits of peat are all comparatively recent and occur only in humid climates. They are developed to an enormous extent in Ireland — about one-seventh of the entire country being covered by them — and average in some cases 25 feet in thickness. They are also abundant in Europe and various parts of North America. In Europe, and especially in Ireland, the material is extensively utilized for fuel, and there would seem no good reason for not so utilizing it in America. An impure variety containing a considerable quantity of siliceous sand, and locally known as " muck," is used as a fertilizer and for " multching " throughout New England. Below are given the results of analyses of (I) peat from the bog of Allan, Ireland, (II) Maine (United States), and (III) Commander Islands in Bering Sea.

Constituents	I	II	III
Carbon	61.04 %	21.00 %	60.48 %
Volatile matter	37.53	72.00	30.53
Ash	1.83	7.00	3.30

Lignite, or brown coal, is the name given to a brownish black variety characterized by a brilliant lustre, conchoidal fracture, and brown streak. Such contain from 55 % to 65 % of carbon, and burn easily, with a smoky flame, but are inferior to the true coals for heating purposes. They are also objectionable on account of the soot they create, and their rapid disintegration and general deterioration when exposed to the air. They occur in beds under conditions similar to the true coals, but are of more recent origin. The lignitic coals of the regions of the United States, west of the Mississippi River, are mainly of Laramie (Upper Cretaceous) age, and, as a rule, show easily recognizable traces of their organic origin, such as compressed and flattened stem and trunks of trees with traces of woody fibre.

Bituminous Coal. — Under this name are included a series of compact and brittle products in which no traces of organic remains are to be seen on casual inspection, but which, under the microscope, often show traces of woody fibre, spores of lycopods, etc. These coals are usually of a brown to black color, with a brown or gray brown streak, breaking with a cubical or conchoidal fracture, and burning readily with a yellow, smoky flame. They contain from 35 % to 70 % of fixed carbon, 18 % to 60 % of volatile matter, and from 2 % to 20 % of water, and only too frequently show traces of sulphur due to included iron pyrites. Several varieties of bituminous coals are recognized, the distinctions being based upon their manner of burning. *Coking* coals are so called from the facility with which they may be made to yield coke; such give a yellow flame in burning, and make a hot fire. They are soft, and break with a cubical fracture. Other varieties of apparently the same composition and general physical properties, cannot, for some unexplained reason, be made to yield coke, and are known as *non-coking coals*. *Cannel* coal has a very compact structure, breaks with a conchoidal fracture, has a dull lustre, ignites easily, and burns with a yellow flame. It does not coke. Its chief characteristic is the large amount of volatile matter given off when heated, whereby it is rendered of particular value for making gas. Before the discovery of petroleum it was used for the distillation of oils. Below is given the composition of (I) a coking coal from the Connelsville Basin of Pennsylvania, and (II) a cannel coal from Kanawha County, West Virginia.[1]

CONSTITUENTS	I	II
Water	1.05 %	....
Volatile matter	29.885	....
Fixed carbon	57.754	58.00 %
Ash	9.895	23.50
Sulphur	1.330	18.50
	100.00 %	100.00 %

Anthracite Coal. — This is a deep black, lustrous, hard and brittle variety, and represents the most highly metamorphosed variety of the coal series. Such have been generally regarded as bituminous coals from which a very large proportion of the

[1] F. P. Dewey, Bull. 42, U. S. National Museum, 1891.

volatile constituents have been driven off by the agencies involved in the production of mountain systems, or by the heat incident to the injection of igneous rocks. Traces of organic nature are almost entirely lacking in the matter of the anthracite itself, though impressions of ferns, lycopods, sigillaria, and other coal-forming plants are frequently associated with the beds in such a manner as to leave little doubt as to their origin. Anthracite is ignited with difficulty and burns with little flame, but makes a hot fire. Below is given the average composition of anthracite from the Kohinoor Colliery, Shenandoah, Pennsylvania.[1]

Water	3.163 %
Volatile matter	3.717
Fixed carbon	81.143
Sulphur	0.899
Ash	11.078
	100.000 %

Like the other coals, anthracite occurs in true beds, but is confined mostly to rocks of the Carboniferous age. Thin seams of anthracite sometimes occur in Devonian and Silurian rocks, but which are too small to be of economic value. Rarely coals of more recent geological horizon have been found locally altered into anthracite by the heat of igneous rocks. Through a still further metamorphism, whereby it loses all its volatile constituents, coal passes over into graphite.

The principal anthracite coal regions of the United States are in eastern Pennsylvania. From here westward throughout the interior states to the front range of the Rocky Mountains the coals are all soft, or bituminous coals. Those of the Rocky Mountain regions proper are largely lignitic, passing into the bituminous varieties.

(4) **Phosphatic Group: Phosphatic Sandstone; Bone Breccia; Guano; Coprolite Nodules.** — This is a group of rocks limited in extent, but nevertheless of considerable economic importance, owing to the high values of certain varieties for fertilizing purposes.

Guano consists mainly of the excrements of sea fowls, and is to be found in beds of any importance only in rainless regions like those of the western coast of South America and southern Africa. The most noted deposits are on small islands off the

[1] Bull. 42, U. S. National Museum, 1891.

coast of Peru. Immense flocks of sea fowls have, in the course of centuries, covered the ground with an accumulation of their droppings to a depth of sometimes 30 to 80 feet, or even more.

An analysis of American guano gave : Combustible organic matter and acids, 11.3 % ; ammonia (carbonate, etc.), 31.7 % ; fixed alkaline salts, sulphates, phosphates, chlorides, etc., 8.1 % ; phosphates of lime and magnesia, 22.5 % ; oxalate of lime, 2.6 % ; sand and earthy matter, 1.6 % ; water, 22.2 % (Geikie).

Coprolite nodules are likewise the excrements of vertebrate animals; those among the Carboniferous shales of the basin of the Firth of Forth are regarded as accumulated excretions of ganoid fishes.

Phosphatic sandstones, as the name denotes, are arenaceous rocks containing more or less phosphatic matter. Inasmuch as the phosphatic material is derived largely by leaching and segregation, these rocks have been already described under the head of chemical deposits (p. 119). In the river beds of the Carolinas are found rounded and nodular masses of this nature, consisting of siliceous and calcareous sand, with embedded bones, teeth of sharks, and other animal remains. Bone breccia consists of fragmental bones of mammals cemented by argillaceous, earthy, or calcareous matter.

III. ÆOLIAN ROCKS

This group comprises a small and comparatively insignificant class of rocks formed from materials drifted by the winds, and more or less compacted into rock masses. They are, as a rule, of a loose and friable texture and of a fragmental nature. Many of the volcanic fragmental rocks (tuffs) are grouped here, their materials having been thrown from the volcanic vents in small fragments and drifted long distances by wind prior to falling upon the surface of the ground or into the water for their final consolidation.

One of the most common results of wind action on the land is the production of sand-dunes — billowy masses of loose sand which, like drifts of snow, though more slowly, gradually change their outlines and creep onward under the restless goading of the wind.

Such, owing to their superficial nature, recent origin, and loose state of consolidation, are considered more in detail in the chapter on The Regolith, p. 299. On undergoing consolidation, these dune sands may give rise to sandstones in many instances indistinguishable from those of aqueous origin, though less regularly bedded. The finely disintegrated shell and coral material thrown up by the waves on the beaches of Bermuda is caught up by the winds and drifted inland, forming hills which, in some instances, are 250 feet in height. Being soluble in the water from rainfalls, these become shortly reconsolidated through the deposition of lime carbonate in the interstices of the fragments, and form thus the drift rock which comprises a large portion of the mass of the islands above tide level.

The finely comminuted materials ejected from volcanic vents may be likewise transported by atmospheric currents and, far from their source, again deposited in beds of no insignificant proportions. These, on induration, give rise to fine-grained tuffs, and, where the final deposition has taken place in water, to distinctly laminated, fine white rocks the lithological nature of which can be made out only by means of the microscope. Such are many of the Pliocene sandstones of Idaho and Montana.[1] The following analyses of samples of tuffs from (I) Marsh Creek Valley, Idaho; and (II) Little Sage Creek, Montana, will serve to show their composition.

CONSTITUENTS	I	II
Ignition (H_2O)	7.00 %	7.02 %
Oxide of iron and aluminium (Fe_2O_3 and Al_2O_3)	16.22	18.24
Silica (SiO_2)	68.92	65.56
Lime (CaO)	1.62	2.58
Magnesia (MgO)	Trace	0.72
Soda (Na_2O)	1.56	2.08
Potash (K_2O)	4.00	3.94
	99.02 %	100.74 %

[1] On the Composition of Certain Pliocene Sandstones from Montana and Idaho, Am. Jour. of Science, Vol. XXVII, 1886, p. 199.

IV. METAMORPHIC ROCKS

Before proceeding to describe in detail the metamorphic rocks, it will be well to devote a brief space to a discussion of the processes by which this metamorphism has been brought about.

The word *metamorphism* as used in geology includes changes in the structure of rocks induced through agencies in part physical, and in part chemical, in their nature. It is, in fact, a very general term, and indicates any transformation taking place in the composition and structural features of rocks of any kind, whether sedimentary or igneous, and from any cause whatever. Rocks laid down in the form of sediments may become so deeply buried as to be subject to intense heat from the earth's interior, as well as to pressure from weight of the overlying material. In this way, a partial or complete fusion of the constituents takes place, which is followed by a crystallization whereby the original fragmental nature may be wholly or in part obscured. This form of change is included under the general name of regional metamorphism. In this manner, it was once assumed, were formed the gneisses, a part of the granites, and the vast series of crystalline schists and calcareous rocks (marbles, etc.). It has, however, been shown that the banded and foliated structure shown by gneisses and schists is not in all cases necessarily an indication of an original bedded structure, but may be due to pressure acting throughout long periods of time, and accompanied by the heat thereby generated. A common and readily understood illustration of this principle of metamorphism by pressure is offered by the roofing slates. These, first laid down as fine silts, rarely show their eminent cleavages whereby they are rendered so useful to man, parallel to their original bedding, but inclined at any and all angles thereto. In such cases the bedding is not infrequently indicated by the dark bands or "ribbons" which are so evident on a split surface.

But it is not alone the fragmental rocks which thus become schistose under pressure. Originally massive, igneous rocks, in regions of profound disturbance have been found converted into schistose aggregates, indistinguishable from rocks ordinarily assumed to be sedimentary. Thus the greenstone schists of the Menominee and Marquette regions of Michigan have been shown by Williams[1] to be highly altered eruptive rocks, mainly gabbros, diabases, and diorites, originally massive, but now foliated, schistose, and variously crumpled through the squeezing and shearing to which they have been subjected since the period of their first extrusion. The changes in these and similar cases, is rarely purely physical, though at times the chemical alterations may be quite inconspicuous. The ultimate composition of the rock may remain essentially the same, while the method of combination of its various elements may have undergone extensive alteration. Quartzes and feldspars may be crushed and distorted, drawn out into lens-shaped and variously elongated forms, while secondary minerals like feldspars, quartz, zoisite, garnet, hornblende, epidote, and the micas may be abundantly generated.

One of the commonest results of pressure effects upon igneous rocks is the conversion of augite or other minerals of the pyroxene group into hornblendes. The coarse hypersthene gabbro occurring about Baltimore is found locally altered into a rock consisting essentially of a schistose aggregate of hornblende and plagioclase feldspars, or what, on mineralogical grounds, might be classed as a diorite.[2] The chemical composition in this case has undergone no appreciable change; there has been simply a molecular rearrangement of the particles. In such cases proof of the character of the change that has taken place is usually found in the fractured and otherwise distorted condition of many of the constituent minerals, as well as intermediate stages of alteration, whereby a residual augite crystal is found enclosed in an envelope of secondary hornblende, as shown in Fig. 1, on p. 40. To the secondary minerals formed in this way the technical name *paramorphic* is applied. To such changes as are above described the name *dynamic metamorphism* is given.

The protrusion of a mass of molten matter into the over-

[1] Bull. 62, U. S. Geol. Survey, 1890.
[2] Bull. 28, U. S. Geol. Survey, 1886.

lying strata may give rise to a series of changes differing from the last in that they are due mainly to heat and to the chemical action of accompanying vapors and solutions. Since these changes are confined to limited areas along the line of the contacts between the two bodies, they are defined as *contact metamorphisms*. As illustrative of such changes, a few cases may be described.

Near Gefrees, in Bavaria, an eruptive biotite granite has been protruded into clay slates and phyllites. At the line of contact both phyllites and slates are converted into a hard, compact blue-black "hornfels" consisting of a crystalline granular aggregate of quartz, deep reddish brown mica (biotite), a little muscovite and andalusite. This zone, some 120 paces in width, is succeeded by a second some 380 paces in width in which the rocks are converted into andalusite mica schists, and this by a third zone some 500 paces wide in which the gradually failing energy was sufficient only to give rise to a spotted mica schist (knoten schiefer), and lastly, a zone some 400 paces wide in which the clay slates has become converted into a chiastolite schist, and the phyllites to a biotite-bearing variety. In all these cases the chemical character of the rock remains essentially the same. Through the metamorphosing action of intruded basic rocks crystalline schists near Peekskill, New York, have near the line of contact become puckered and filled with lens-shaped eyes of quartz containing garnets and other minerals, while crystals of staurolite, sillimanite, cyanite, and garnet appear, the amount of change being directly proportional to the nearness of the line of contact. At the contact the schistose structure is almost completely obliterated and the schists become hard and massive, appearing more or less fused with the eruptive, and consist of a large number of minerals. Briefly expressed, the progressive change, approaching the line of contact, consists in a gradual decrease in the proportional amount of silica and alkalies, with a corresponding increase in iron and alumina, this being accompanied by a disappearance of the quartz and muscovite and the development of biotite, sillimanite, staurolite, cyanite, and garnet, as above mentioned. Where limestones abound, they have become bleached and rendered more closely crystalline, while a variety of metamorphic minerals, as lime-bearing pyroxenes, hornblendes, zoisite, sphene, and scapolite have been developed.

A common form of metamorphism is manifested in the production of a quartzite from siliceous sandstone. This, in its simplest form, is brought about by a secondary deposit of silica about the original rounded granules of sand, whereby the entire mass is converted into an aggregate of quartz crystals, the outlines of which are more or less imperfect through mutual interference in process of growth. The microscopic structure of a quartzite of this nature is shown in Fig. 13. In this case the original rounded granules are readily recognized from the fact that not merely did they frequently contain small cavities and needle-like enclosures, but exteriorly they were covered with a thin pellicle of iron oxide, while the secondary deposit, which now fills all the interspaces, is free from enclosures of all kinds and quite pellucid.

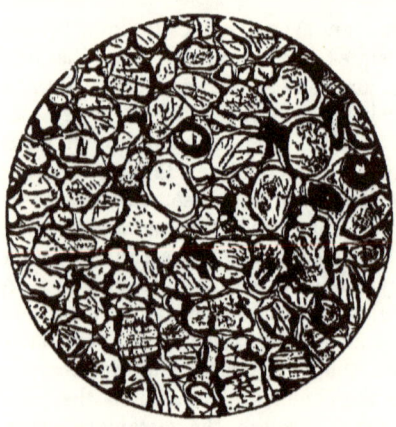

FIG. 13. — Microstructure of quartzite, showing secondary deposit of silica about the original quartz grains.

In many quartzites a shearing force has acted a prominent part, whereby the granules have become elongated and more or less pulverized along their margins by the friction of rubbing one over the other. In such cases mica and other secondary minerals are often developed, and the rock passes over into a mica schist.

Still another form of change, or metamorphism, is that known by the name of *metasomatosis*, a process of indefinite substitution and replacement. Through the chemical action of percolating solutions certain constituents of a rock may be leached out and replaced by others in indefinite proportions. It is by such processes that have originated a large share of the serpentinous rocks, dolomites, etc. The mineral olivine, an anhydrous ferruginous silicate of magnesia, passes over into serpentine by a simple process of hydration, and a more or less complete change of its combined iron from the ferrous to the ferric state; this constituent not infrequently separating out during the process of change, and crystallizing as magnetite, or remaining as an amorphous hematite or limonite. Provided there be no loss in silica, this change in the olivine, according to T. Sterry Hunt, must be accompanied by an increase of volume amounting to some 33 %. Through the hydration of eruptive olivine-bearing rocks, or rocks rich in other magnesian silicate minerals, have originated a large proportion of the so-called serpentines and verd-antique marbles. Many serpentines and serpentinous limestones are derived from metamorphic rocks rich in lime-magnesian pyroxenes or amphiboles, as malacolite and tremolite. To such an origin are to be referred such serpentinous limestones as those of Essex County, New York; Easton, Pennsylvania, and Montville, New Jersey. In the last-named instance the original rock was a coarsely crystalline dolomitic limestone containing numerous nodular masses of white pyroxene (malacolite). Under this metasomatic process they yielded up their calcium, which recrystallized as calcium carbonate or calcite, while the silica and magnesia, combined with some 13 % of water, remained as a beautiful green and yellow serpentine. The transformation was accompanied by a considerable increase in bulk, whereby the exterior of the nodules, pressed against the rough walls of the enclosing rock, became scratched and polished like boulders from the glacial drift, or the entire mass even took on a platy, schistose structure. Figure 8, from a specimen in the National Museum, illustrates a transitional phase of this change, the interior rounded mass of a gray color being of still unaltered pyroxene, while the dark material forming the exterior shell, or traversing the gray in fine thread-like veins, is the secondary serpentine. In a like manner in all probability originated the peculiar structure imitative of animal organisms known as *Eozoon Canadense*.[1]

The conversion of a limestone into a dolomite is believed to have been brought about by a somewhat similar process. Indeed it is doubtful if this last-named rock is ever a product of direct sedimentation or precipitation. Although sea-water contains

[1] See On the Serpentine of Montville, New Jersey, Proc. U. S. National Museum, Vol. XI, 1888; Notes on the Serpentinous Rocks of Essex County, New York, etc., ibid., Vol. XII, 1889; and On the Ophiolite of Thurman, Warren County, New York, Am. Jour. of Science, Vol. XXXVII, 1889.

from three to four times as much magnesia as lime, evidence is wanting to show that the material is ever secreted in appreciable quantities by marine animals, and hence the sedimentary deposits, resulting from the accumulation of the remains of these animals, must be correspondingly lacking in this constituent. It has been argued by Beaumont and others that through a process of partial molecular replacement (metasomatosis) pre-existing limestones were converted into dolomites, the process consisting in the replacement of every other molecule of calcium carbonate by one of the magnesium carbonate. As the dolomite molecule is the more dense of the two, such replacement, in any given limestone bed, must result in a contraction amounting to some $12\frac{1}{2}$ %. Assuming that a dolomitic mass resulting in this way is of the same bulk as the original limestone, this shrinkage must manifest itself in the production of interstitial rifts and cavities, such as do actually occur in many dolomitic limestones, as those of the Ohio Trenton formations. The principal objection to this theory lies in the difficulty of accounting for the large amount of magnesia in solution; whence its source, etc. The same objections apparently apply to the explanation given by M. C. Klement.[1] This writer describes a series of experiments in which solutions of sodium chloride and magnesium sulphate were made to act upon pulverized calcite and aragonite. From the results obtained, he concludes that dolomite is formed by the action of sea-water, concentrated in enclosed basins and heated by the sun, on the aragonite deposited by marine organisms, in such a way that a mixture of carbonates of calcium and of magnesium is first produced, and which is subsequently converted into dolomite.

Still another theory is that which regards the dolomite as a residuary product formed by the leaching out of the lime carbonate from beds of impure, slightly magnesian limestone, leaving behind the less soluble magnesian carbonate. The amount of material lost, and the consequent contraction of the original beds, must necessarily vary with their purity; but in any case where the residual mass has reached the condition of a true dolomite, the proportional loss must have been enormous, since in no cases are unaltered sediments known to contain more than 4 or 5 % of magnesian carbonate. Although on first thought this theory seems the more plausible of the two,

[1] Bull. de la Société Geologique de Belge, Tome IX, 1895.

it is apparently rendered invalid by the presence in these dolomites of very perfect casts of fossils which have undergone no crushing or distortion whatever, and which tend to show that the beds as a whole, so far from having undergone a shrinkage of 95 % and upwards, are of essentially the same bulk as when laid down. The subject is too large for complete discussion here, and the reader is referred to standard works on chemical geology, as well as the current literature.[1]

Still another form of change in the structure and mineral composition of a rock is that brought about through the action of water below the zone of oxidation and of true weathering. It may be best described as a process of hydro-metamorphism, since the influence of water is paramount. It is to this form of metamorphism that is due the production, in part, of secondary epidote, chlorite, sericite, leucoxene, kaolin (?) pyrite, and various zeolitic compounds from pre-existing minerals, but without in any way changing the character as a geological body of the rock mass in which they occur. Such changes are in part metasomatic, and in many instances are rendered more intense by dynamic causes. This form of change has, unfortunately, been too frequently confounded with weathering and decomposition.[2]

Under the head of metamorphic, then, is grouped a large series of rocks which have been changed from their original condition through the dynamical and chemical agencies above described, and which may have been in part of aqueous and in part of eruptive origin. Were it possible, it might have been better to describe each class of these rocks together with the corresponding igneous or aqueous form from which it was derived by this process of change, or metamorphism. In only too many cases, however, the change has been so complete as to quite obliterate all such traces of the original character as would lead to safe and satisfactory conclusions, and consistency demands that all be grouped together.

[1] See The Magnesian Series of the Northwestern States, by C. W. Hall and F. W. Sardeson. Bull. Geol. Soc. of America, Vol. VI, 1895, p. 167.

[2] While it is true that no new compound can be formed without first a breaking up, or decomposition, of those already existing, still, as this decomposition affects only the individual minerals, and not the integrity of the rock mass as a whole, it would seem preferable to include such changes under the name of alteration and metamorphism. Weathering it certainly is not, though 'it is essentially the form of change which Roth (Allegemeine u. Chemische Geologie, Vol. I, pp. 159–412) has designated as complex weathering (*Complicirte Verwitterung*).

Accordingly as they vary in structure, we may divide these metamorphic rocks into two general groups as below: 1. Stratified or bedded; 2. foliated or schistose.

1. STRATIFIED OR BEDDED

(1) THE CRYSTALLINE LIMESTONES AND DOLOMITES

Here are included the metamorphosed form of the sedimentary rocks described on p. 143.

Mineral Composition. — The essential constituent of the crystalline limestones is the mineral calcite. The common accessories are minerals of the mica, amphibole, or proxene group, and frequently sphene, tourmaline, garnets, vesuvianite, apatite, pyrite, graphite, etc.

Chemical Composition. — As may be inferred from the mineral composition, these rocks, when pure, consist only of calcium carbonate. They are, however, rarely if ever found in a state of absolute purity, but show more or less magnesia, alumina, and other constituents of the accessory minerals. The analyses given on pp. 146–47 will serve equally well here, and need not be repeated.

Structure. — The limestones are eminently stratified rocks, though this peculiarity is not always sufficiently marked to be seen in the hand specimen. The purest and finest crystalline varieties often show a granular texture like that of loaf sugar, and hence are spoken of as *saccharoidal limestones*. Statuary marble is a good illustration of this type. Under the microscope the stone is shown to be made up of small grains, which, having mutually interfered in process of growth, do not possess perfect crystal outlines, but are rounded and irregular in outline, as shown in Fig. 14. All grades of textures are common, the coarser forms sometimes showing individual crystals an inch in length. Though in their unchanged conditions highly fossiliferous or tufaceous, these structural features may be wholly or in part obliterated by crystallization.

Colors. — The color of pure limestone is snow-white, as seen in statuary marble. Other common colors are pink or reddish, greenish, blue-gray through all shades of gray to black. The pink and red colors are due to iron oxides, the greenish as a rule to micaceous minerals, the blue-gray and black to carbonaceous matter.

Geological Age and Mode of Occurrence. — The crystalline limestone and dolomites are but the metamorphosed sedimentary deposits such as have already been described on p. 143. They occur associated with rocks of all ages, but only in regions that have been subjected to disturbances such as the folding and faulting incident to mountain-making, or the heat from intruded igneous rocks. From an economic standpoint, the rocks of this group are not infrequently of great economic value for structural and decorative purposes.

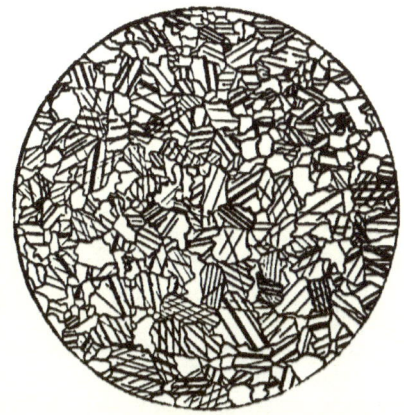

FIG. 14. — Microstructure of crystalline limestone (marble).

Classification and Nomenclature. — It is common to speak of this entire group of rocks as simply limestones, though many varietal names are often rather indefinitely applied. The name *marble* is given to any calcareous or magnesian rock sufficiently beautiful to be utilized in decorative work. Argillaceous and siliceous limestones carry clayey matter and sand. *Dolomite* (so named after the French geologist Dolomieu) consists of 45.50 % carbonate of magnesia and 54.40 % carbonate of lime, as already noted. The names *ophiolite* and *ophicalcite* are popularly applied to stones consisting of a granular aggregate of calcite and serpentine, such as occur in Essex County, New York, and are used as marbles. The so-called *Eozoon Canadenses*, a supposed fossil rhizopod, belongs here. The serpentinous matter in such cases originates from a non-aluminous pyroxene by a process of hydration, as already explained.

2. FOLIATED OR SCHISTOSE

(1) THE GNEISSES

Gneiss, from the German *Gneis*, a term used by the miners of Saxony to designate the country rock in which occur the ore deposits of the Erzgebirge. The word is pronounced as though spelled *nīce*.

Mineral and Chemical Composition. — The composition of the gneisses is essentially the same as that of the granites, from which they differ only in structure and origin. They, however, present a greater variety and abundance of accessory minerals, chief among which may be mentioned (besides those of the mica, hornblende, or pyroxene group) garnet, tourmaline, beryl, sphene, apatite, zircon, cordierite, pyrite, and graphite.

Structure. — Structurally the gneisses are holocrystalline granular rocks, as are the granites, but differ in that the various constituents are arranged in approximately parallel bands or layers, as shown in Pl. 13.

In width and texture these bands vary indefinitely. It is common to find bands of coarsely crystalline quartz several inches in width, alternating with others of feldspar, or feldspar, quartz, and mica, or hornblende. A lenticular structure is common, produced by lens-shaped aggregates of quartz or feldspar, about and around which are bent the hornblendes or mica laminæ. The rocks vary from finely and evenly fissile through all grades of coarseness, and become at times so massive as to be indistinguishable in the hand specimens from granites. The causes of the foliated structure are mentioned below.

Colors. — Like the granites, they are all shades of gray, greenish, pink, or red.

Geological Age and Mode of Occurrence. — The true gneisses are among the oldest crystalline rocks, and have been considered by many geologists as representing "portions of the primeval crust of the globe, traces of the surface that first congealed upon the molten nucleus." By others they are regarded as metamorphosed sedimentary deposits resulting from the breaking down of still older rocks, and may not in themselves, therefore, be confined to any particular geological horizon. They are in large

PLATE 13

1

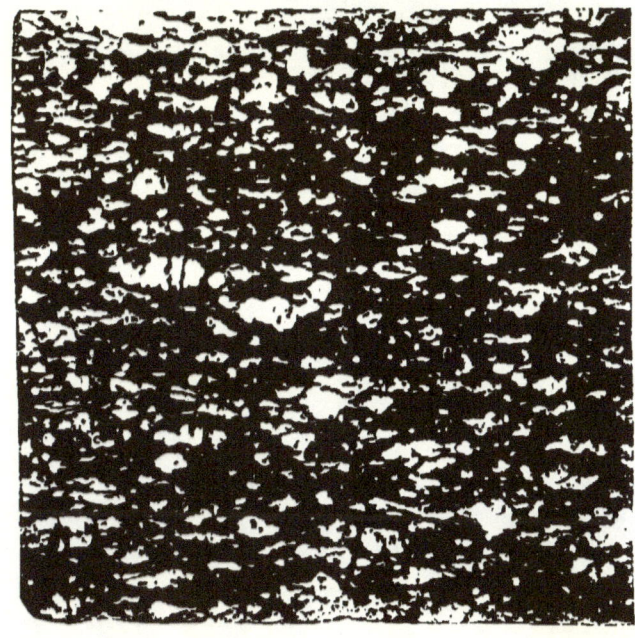

2

part, however, indisputably the oldest known rocks, lying beneath or being cut by all rocks of later formation or injection.

The origin of the gneisses, as above suggested, is in many cases somewhat obscure, the banded or foliated structure being considered by some as representing the original bedding of the sediments, the different bands representing layers of varying composition. This structure is now, however, considered to be due to mechanical causes, and in no way dependent upon original stratification. The name, as commonly used, is made to include rocks of widely different structure, and which are beyond doubt in part sedimentary and in part eruptive, but in all cases altered from their original conditions.

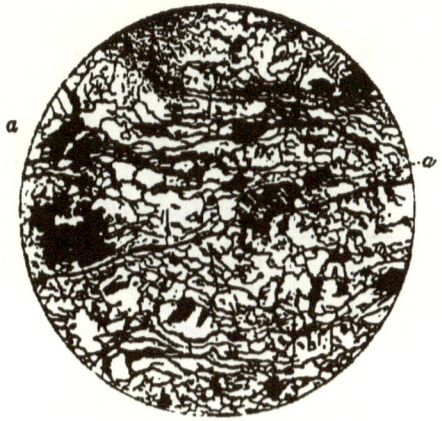

FIG. 15. — Microstructure of gneiss, showing at the points a broken feldspars.

This alteration, it should be stated, has been brought about not by heat and crystallization alone, but in many cases by processes of squeezing, crumpling, and folding so complex as almost to warrant the application of the term *kneading* thereto. It is even possible to conceive that some of them may be original massive or foliated rocks into which eruptive materials have since been injected along lines of foliation or of weakness due to shearing, and the entire mass again submitted to such a kneading as to render it practically impossible to now decide what are portions of the original rock and what of the subsequently injected.

The close chemical relationship which may exist between clastic, metamorphic, and eruptive rocks is shown in the selected series of analyses here given.

Constituents	Granite	Gneiss	Gneiss	Sandstone	Shale	Slate	Disintegrated Granite
	I	II	III	IV	V	VI	VII
	%	%	%	%	%	%	%
Silica (SiO_2)	68.18	61.96	69.24	69.94	61.91	60.32	65.69
Titanium oxide (TiO_2)	....	1.06	....	Not det.	....	....	0.31
Alumina (Al_2O_3)	16.20	19.73	14.85	13.15	21.73	23.10	15.23
Ferric oxide (Fe_2O_3)	....	....	2.62	2.48	4.73	7.05	4.30
Ferrous oxide (FeO)	4.10	4.60	....	....	....	....	....
Ferrous sulphide (FeS_2)	....	4.33	....	....	....	....	....
Manganese oxide (MnO)	....	Trace	0.45	0.70	....	....	Not det.
Lime (CaO)	1.75	0.35	2.10	3.08	0.09	....	2.63
Magnesia (MgO)	0.48	1.81	0.96	Trace	0.50	0.87	2.64
Soda (Na_2O)	2.88	0.79	4.30	5.43	0.25	0.49	2.12
Potash (K_2O)	6.48	2.50	4.33	3.30	3.16	3.83	2.00
Ignition	....	1.82	0.70	1.01	7.43	4.08	4.70
	100.07	99.55	99.56	99.10	99.80	99.74	99.71

I. Granite: Syene, Egypt. II. Gneiss: St. Jean de Matha, Province of Quebec, Canada. III. Gneiss: Trembling Mountain, Province of Quebec, Canada. IV. Sandstone: Portland, Connecticut. V. Shale: England. VI. Slate: Lancaster County, Pennsylvania. VII. Disintegrated granite: District of Columbia.

Figures 1 and 2 on Pl. 13 shows two rather extreme types of these gneissoid rocks. Figure 1 is that of a banded gneiss from Madison County, Montana, and which, so far as we know, may be an altered sedimentary rock. In Fig. 2 of the same plate is shown a foliation rather than a banded rock, and whatever may have been its origin, it undoubtedly owes its foliated structure to dynamic agencies. The effect of the shearing force whereby the foliation was produced is evident in the figure, even to the unaided eye, to the left and just above the centre, where an elongated feldspar is seen broken transversely in four pieces. The same features are brought out even more plainly in Fig. 15, which shows the structure of this same gneiss as seen under the microscope.

As in the present state of our knowledge it is in most cases impossible to separate what may be true metamorphosed sedimentary rocks from those in which the foliated or banded structure is in no way connected with bedding and which may or may not be altered eruptives, all are grouped together here.

Classification and Nomenclature. — The varietal distinctions are based upon the character of the prevailing accessory mineral, as in the granites, forming a parallel series. We thus have *biotite gneiss, muscovite gneiss, biotite-muscovite gneiss, hornblende gneiss*, etc. Rarely the mineral cordierite occurs in sufficient abundance to become a characterizing accessory. Such forms occur in Gilford County, Connecticut, and in Saxony.

The name *granulite* or *leptynite* is applied to a banded quartz-feldspar rock, the constituents of which occur in the form of small grains and show under the microscope a mosaic structure. The Saxon granulites are regarded by Lehman as eruptive rocks altered by pressure. *Halleflinta* is a Swedish name for a rock resembling in most respects the eruptive felsites or quartz porphyries already described. Such, however, show a banded structure and are, as a rule, regarded as metamorphic rocks. *Porphyroid* is also a felsitic rock with a more or less schistose structure, and with porphyritic feldspar or quartzes. Such have been described from the Ardennes, France.

GNEISS

Analogous Massive Type	Of Igneous Origin	Origin Unknown
Granite: Biotite granite . . . Hornblende granite . .	Granite gneiss: Biotite granite gneiss . Hornblende granite gneiss . . . . }	Granitic gneiss: Biotite granitic gneiss. Hornblende granitic gneiss.
Syenite: Hornblende syenite . .	Syenite gneiss: Hornblende syenite gneiss . . . . }	Syenitic gneiss: Hornblende syenitic gneiss.
Mica syenite	Mica syenite gneiss . .	Mica syenitic gneiss.
Pyroxene syenite . .	Pyroxene syenite gneiss	Augite syenitic gneiss.
Diorite: Mica diorite	Diorite gneiss: Mica diorite gneiss . .	Dioritic gneiss: Mica dioritic gneiss.
Gabbro	Gabbro gneiss	Gabbroic gneiss, or gabbric gneiss.
Pyroxenite	Pyroxenite gneiss . . .	Pyroxenitic gneiss.

Inasmuch as the structure characteristic of gneisses is found developed in rocks of diverse types, many petrologists now use the term in an almost wholly structural sense, as in itself noncommittal as to composition or origin, but merely designating a rock of foliated or schistose structure. C. H. Gordon has proposed [1] a scheme of classification of gneissoid rocks as above, and which has much in its favor.

(2) THE CRYSTALLINE SCHISTS

Under this head are grouped a large and extremely variable series of rocks, differing from the gneisses mainly in the lack of feldspar as an essential constituent. They consist, therefore, essentially of granular quartz, with one or more minerals of the mica, chlorite, talc, amphibole, or pyroxene group. In accessory minerals the schists are particularly rich. The more common of these are feldspar, garnet, cyanite, staurolite, tourmaline, epidote, rutile, magnetite, menaccanite, and pyrite. Through an increase in the proportional amount of feldspar the schists pass into the gneisses, and through a decrease in mica, hornblende, or whatever may be the characterizing mineral, into the quartz schists, in which quartz alone is the essential constituent. Occasional forms are met with quite lacking in quartz and other accessory minerals and consisting only of schistose aggregates of minerals of a single species, as is the case with the pyrophyllite schists (or, more properly, schistose pyrophyllites) from North Carolina, talcose schists, and with the more massive "soapstones."

The rocks of this group are characterized as a whole by a pronounced schistose structure, due to the parallel arrangement of the various constituents, this structure being most pronounced in those varieties in which mica is the predominating accessory mineral. They are ordinarily considered as having originated from the crystallization of sediments, and in many cases the microscope still reveals existing "traces of the original grains of quartz sand and other sedimentary particles of which the rocks at first consisted." Like the gneisses, they are in part, however, mechanically deformed massive rocks and their schistosity in no way relates to true bedding, as has been already noted (p. 156).

The varietal names given are dependent mainly upon the character of the prevailing ferro-magnesian silicate. We thus

[1] Bull. Geol. Soc. of America, Vol. VII, p. 122.

have *mica schists, chlorite schists, talc schists, hornblende, actinolite, glaucophane schists,* etc. The term *slate* was originally applied to these and other types of rocks of schistose or fissile character. In the arrangement here adopted this term is restricted to the argillaceous fragmental or semi-crystalline and foliated rocks next to be described.

Of the above-mentioned varieties the mica schists are the most common and widely distributed, the mica being in some cases biotite, in others muscovite, or perhaps a mixture of the two. The principal accessories sufficiently developed to be conspicuous are staurolites, chiastolites, garnets, and tourmalines. In the sericite schists the hydrous mica sericite prevails; *paragonite schist* carries the hydrous sodium-mica paragonite; *ottrelite schist* carries the accessory mineral ottrelite.

The name *phyllite* is used by German petrographers to designate a micaceous semi-crystalline rock standing intermediate between the true schists and clay slates. *Quartzite* is a more or less schistose or banded rock consisting essentially of crystalline granules of quartz. Such originate from the induration of siliceous sandstones. This induration is brought about through a deposition of crystalline silica in the form of a binding material or cement around each of the sand particles of which the stone is composed. Each of these granules then forms the nucleus of a more or less perfectly outlined quartz crystal. This structure is shown in Fig. 16, drawn from a thin section of a Potsdam quartzite from St. Lawrence County, New York. The rounded, more or less shaded, portions represent the original grains of quartz sand, and the clear, colorless, interstitial portions the secondary silica.

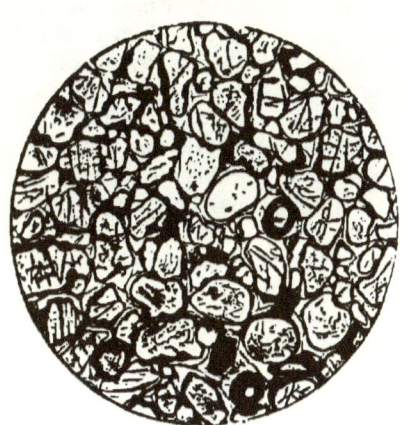

Fig. 16. — Microstructure of quartzite.

The quartzites consist, as a rule, only of silica, or silica colored brown and red by iron oxides. At times a greenish tinge is imparted through the development of chloritic minerals; accessory minerals are not, as a rule, abundant.

Among the hornblende schists there are but few needing especial attention. These are, as a rule, less finely schistose than are the mica-bearing schists, owing to the fact that the mineral hornblende itself has not a platy structure. The *glaucophane schists* are perhaps the least abundant of the hornblendic varieties. Such have been described from the Isle of Syra, in the Mediterranean Sea, Switzerland, Wales, and Italy; a more massive form, probably an altered eruptive, is found near the mouth of Sulphur Creek, Sonoma County, California. *Amphibolite* is the name given to an extremely tough and often massive rock of obscure origin, and consisting essentially of the mineral amphibole or hornblende. In some instances the varieties of amphibole, actinolite, and tremolite take the place of the common hornblende. The tremolite rock may undergo alteration into serpentine under proper conditions. *Eclogite* is a tough, massive, or slightly schistose rock, consisting of the grass-green variety of pyroxene, *omphacite*, and small red garnets, with which are frequently associated bluish kyanite, green hornblende (smaragdite), and white mica. *Garnet rock*, or *garnetite*, is a crystalline granular aggregate of garnets with black mica, hornblende, quartz, and magnetite. *Kinzigkite* is a somewhat similar, though fine-grained and compact, rock consisting of garnets, plagioclase feldspar, and black mica, and which is found in Kinzig and the Odenwald.

Many of the rocks of this group are but products of dynamic or contact metamorphism, as is the case with many of the chiastolite and argillaceous schists or roofing slates. Rocks of the latter group pass by insensible gradations into clastic argillites. They owe their cleavable property to shearing, as already explained. Under the microscope these rocks are found to be quite variable. Hawes describes clay slate from Littleton, New Hampshire, as consisting of a mixture of quartz and feldspar, in particles as fine as dust. They contained also amorphous carbonaceous matter and little needles of a mineral assumed to be mica. A slate from Hanover, in the same state, contained garnets and staurolites. Wichman found slates from Lake Superior to consist of a colorless, isotropic ground-mass carrying quartz and feldspar particles, scales of iron oxide, carbonaceous matter, minute tourmalines, and mica fragments.

The red slates of New York state are composed of an impalpable red, dust-like ground-mass, carrying grains of quartz and feldspar, all arranged with their longer axes parallel to the plane of schistosity. These can scarcely be considered as other than clastic rocks, the dynamic action not having been sufficient to produce crystallization in more than incipient stages. In this case the plane of schistosity is very nearly parallel with that of bedding, but in many cases, as in the roofing slates of Pennsylvania, the schistose structure is developed at a very considerable, though ever-varying, angle with the bedding. In such cases the true bedding plane is often determinable only by the dark bands, or ribbons, by which the split slates are traversed.

Chemical Composition.—As may be readily imagined, the schists vary almost indefinitely in composition, approximating pure quartzite on the one hand and the gneisses on the other. The table given below is intended to show the composition of a few characteristic types only. All gradations, from the most acid of quartzites to the most basic of the amphibolites, may readily be found.

Constituents	I	II	III	IV	V	VI
Silica (SiO_2)	82.38%	40.00%	52.30%	49.18%	50.81%	97.1%
Alumina (Al_2O_3)	11.84	23.05	16.33	15.00	4.53	1.30
Ferric oxide (Fe_2O_3)	....	8.07	1.04	12.90	3.52	1.25
Ferrous oxide (FeO)	2.28	....	1.44	....	4.20	....
Lime (CaO)	....	0.63	8.76	10.59	....	0.18
Magnesia (MgO)	1.00	0.04	4.70	5.22	31.55	0.13
Potash (K_2O)	0.83	0.11	1.42	1.51	....	....
Soda (Na_2O)	0.38	1.75	2.59	3.04	....	....
Ignition	0.77	3.41	0.17	1.87	4.42	....

I. Mica schist: Monte Rosa, Switzerland. II. Sericite schist: Wisconsin. III. Hornblende schist: Grand Rapids, Wisconsin. IV. Chlorite schist: Klippe, Sweden. V. Talc schist: Gastein, Austria. VI. Quartzite: Chickies Station, Pennsylvania. All analyses quoted from J. F. Kemp's Lecture Notes on Rocks.

PART III

THE WEATHERING OF ROCKS

"In the economy of the world, I can find no traces of a beginning, no prospect of an end." — HUTTON.

THE stability of chemical compounds is governed by prevailing conditions. A form of combination stable under conditions existing to-day may, under those of to-morrow, become impossible. As was suggested in the introductory chapter, the conditions under which the more superficial portions of the earth's crust exist are ever changing, and as a result old compounds are broken up and new continually formed. All over the earth rocks laid down as sediments on oceanic floors have been folded, faulted, and pushed out of place until brought under influences as different from those under which they were formed as it is possible to conceive. Molten magmas cooling suddenly on the immediate surface formed compounds in which mere loss of heat was the controlling factor, but which time proves to be unstable. Slow cooling, deep-seated magmas have been, and are being, continually exposed by denudation, and thus brought under new influences and environments. Hence a constant readjustment is everywhere going on, which, as we shall see, is manifold in its physical manifestations. As where an entire building is razed to the ground, and another of quite different architectural features constructed from the old materials; or again, where, without change of general plan, old timbers are here and there replaced by new, so here we have at work a series of processes in part seemingly destructive and in part constructive, but all tending toward one end.

The firm and everlasting hills we must learn to regard as neither firm nor everlasting. Whole mountain chains of the geological yesterday have disappeared from view, and as with

PLATE 14

Weathered granite, District of Columbia.

the ancient cities of the East, we read their histories only in their ruins. Yet, in all this seemingly destructive process of breaking down, decomposition, and erosion, there is traceable the one underlying principle of transformation from the unstable toward that which is to-day more stable. Nothing is lost or wasted: It is a change which began with the beginning of matter; which will end only with the blotting out of matter itself. There are no traces of a beginning, there is no prospect of an end.

I. THE PRINCIPLES INVOLVED IN ROCK-WEATHERING

The processes involved in this readjustment from unstable to stable compounds, as above outlined, and of incidental soil formation, are in part physical and in part chemical in their nature; they operate under ever-varying conditions, and through processes at times simple, or again complex. What these processes are, and how they operate, it must be our purpose to now consider.

It may be said at the outset, that whatever the forces engaged, they are, with a few isolated exceptions, superficial, — they work from without downwards. However much they may have accomplished since the first rock masses appeared above the primeval ocean, in no case can the actual amount of débris *in situ* have formed at one time more than a scarcely appreciable film over the underlying and unchanged material. The decomposing forces early lose their active principles and become quite inert at depths comparatively insignificant. It is only where through erosion the results of the disintegration are gradually removed, that the processes have gone on to such an extent as to perhaps quite obliterate thousands of feet of strata or of massive rock, and furnished the necessary débris for the vast thicknesses of sandstone, slate, and shale which characterize the more modern horizons. In certain isolated cases, it is true, ascending steam and heated waters, arising from depths unknown, have been instrumental in promoting decomposition, as is well illustrated in the areas of decomposed rhyolites in the Yellowstone National Park. Nevertheless, it is to the almost incalculably slow process of superficial weathering that we owe a very large share of the apparent rock decomposition and incidental soil formation.[1]

This transformation, as already noted, takes place through processes that may be simple, or again complex. It is but rarely that one, alone, prevails for any length of time, and as a rule several or many go merrily on together. Were it possible, it might be well to consider briefly each of these in its turn and by itself. From the fact, however, as above stated, that any one, either physical or chemical, rarely goes on alone, it is thought best to treat the subject as below, and describe in more or less detail the action, first, of the atmosphere, second, of water, in both the solid and liquid form, and third, that of plant and animal life, finally considering the combined action of all these forces, as manifested on the various types of rock which go to make up the earth's crust.

So striking a phenomenon as the breaking down, or *degeneration* as we may call it, of a mass of firm rock, naturally did not escape the observation of the earlier workers in this and allied branches of science, and the older literature from the time of Hutton contains numerous references to it, though the full significance of atmospheric agencies in bringing about the results, seems not at first to have been fully realized.

The exciting cause of the degeneration, particularly in warm latitudes, where phenomena of this nature are, as a rule, more apparent, has been a matter of some speculation, and at the outset it may be well to indicate in brief their tendencies.

---

[1] The term *weathering*, as here used, is applied only to those superficial changes in a rock mass brought about through atmospheric agencies, and resulting in a more or less complete destruction of the rock as a geological body, as where granitic rocks are resolved into sand, and kaolinic material, with liberation of carbonates of the alkalies and of lime, and oxides of iron. It does not include those deeper-seated changes — changes taking place below the zone of oxidation and which result mainly in hydration and the production, it may be, of new mineral species, as chlorite, sericite, zeolites, etc., but during which the rock mass as a whole retains its individuality and geological identity. The distinction is not one that has been sharply insisted upon, and indeed geologists and petrologists as a rule have been extremely careless in their use of such terms as *alteration*, *decomposition*, and *weathering*. The distinction drawn here is essentially that made by Roth (Allgemeine u. Chemische Geologie), between *Verwitterung* and *Complicirte Verwitterung*. For reasons above stated and others given on p. 161, it seems best to limit the terms *weathering* and *decomposition* to processes involving the destruction of the rock mass as a geological body, and to designate the purely mineralogical deeper-seated changes as *alteration*, which may or may not be due wholly to hydrometamorphism.

Fournet, as quoted elsewhere, writing as early as 1833, insisted upon the efficacy of water containing carbonic acid in promoting the decomposition of igneous rocks, while Brogniart, writing with particular reference to feldspathic decomposition and the origin of kaolin, laid great stress on the acceleration of the ordinary process of decay through the electric currents resulting from the contact of heterogeneous rock masses. Darwin [1] believed the extensive decomposition observed by him in Brazil, to have taken place under the sea, and before the present valleys were excavated. Hartt [2] gave it as his opinion that the decomposition was due to the action of warm rain water soaking through the rock, and carrying with it carbonic acid derived not only from the air, but from the vegetation decaying in the soil as well, together with organic acids, nitrate of ammonium, etc. Further, that the decomposition had gone on only in regions once covered by forests. Heusser and Claraz [3] suggested that the decomposition was brought about through the influence of nitric acid. They say "it is without doubt determined by the violence and frequency of the tropical rains, and by the dissolving action of water, which increases with the temperature. It is necessary to observe, moreover, that this water contains some nitric acid, on account of the thunder storms which follow each other with great regularity during many months of the year."

Belt,[4] in discussing the extensive decomposition observed by him in Nicaragua, says: "This decomposition of the rocks near the surface prevails in many parts of tropical America, and is principally, if not always, confined to the forest regions. It has been ascribed, and probably with reason, to the percolation through the rocks of rain water charged with a little acid from the decomposing vegetation."

The elder Agassiz laid much stress on the decomposing effects of the hot water from rainfall,[5] while Mills and Branner,[6] in addition, attributed no insignificant amount of the decomposition to the action of decomposing organic matter carried into

---

[1] Geological Observations, p. 417.
[2] Phys. Geog. and Geol. of Brazil.
[3] Ann. des Mines, 5th series, 17, 1860, p. 291.
[4] The Naturalist in Nicaragua, 1874.
[5] Journey in Brazil, p. 89.
[6] Bull. Geol. Soc. of America, Vol. VII, 1896.

the ground by ants, and also to the acid secretions of the ants themselves.

The chemical changes involved in the process of decomposition received attention from several of the earlier workers, among whom the names of Berthier, J. G. Forschammer, Brogniart, Gustav Bischof, and Ebelmen stand out in greater prominence. More recently the name of Sterry Hunt becomes conspicuous, while the purely geological side of the question has been ably set forth in numerous papers by L. Agassiz, R. Pumpelly, N. S. Shaler, O. A. Derby, R. Irving, J. C. Branner, and others, to whom reference is frequently made in these pages.

1. ACTION OF THE ATMOSPHERE

Atmospheric air, as is well known, consists in its normal state of a mechanical admixture of free nitrogen and oxygen in the proportion of four volumes of the former to one of the latter. In addition are small and comparatively insignificant amounts of various combined gases and salts, of which carbonic acid is by far the most abundant, constant, and, from our standpoint, important. Still smaller quantities of ammoniacal vapors exist, and in volcanic regions there have been detected appreciable but variable quantities of sulphuric and hydrochloric and nitric acids as well. With rare exceptions these last exist in combination as sulphates, chlorides, and nitrates and with the exception of the last-named need little consideration.

(1) **Nitrogen, Nitric Acid, and Ammonia.** — Nitrogen, by itself, is believed to be wholly inoperative in promoting rock decomposition. In works on agricultural chemistry, much has, however, been written concerning the presence in the atmosphere of the compounds of nitrogen, nitric acid, and ammonia, and it will be well to devote a little space to a consideration of the facts as known, and their possible application to the subject under consideration.

The well-known experiments of Cloez, Boussingault, De'Luca, Kletzinsky, and Way, as well as the recent ones of G. H. Failyer,[1] prove conclusively the existence of ammonia and nitric acid in the air, from whence it is brought to the surface of the earth in the water of rainfalls.

[1] Ammonia and Nitric Acid in Atmospheric Waters, 2d Ann. Rep. Kansas Experiment Station, 1880.

In nearly every case, however, the percentage of ammonia, as determined, equalled or exceeded the amount necessary to combine with the acid, forming thus the salt *ammonium nitrate*. Failyer's experiments in Kansas, carried on for a period of four years, during which time water was collected from 266 rainfalls, showed in but seven instances nitric acid equalling or exceeding the ammonia. In all other cases the amount is less, with the possible exception of the reported occurrence (at Nismes, in 1845) of a fall of hail sufficiently acid to be sour to the taste. As direct promoters of rock decomposition, neither atmospheric nitrogen nor free nitric acid need, then, serious attention. The following tables are, however, of interest, the first being abridged from Johnson's How Crops Feed, and the second from Professor Failyer's paper above quoted.

AMOUNTS OF RAIN AND OF AMMONIA, NITRIC ACID, AND TOTAL NITROGEN THEREIN, COLLECTED AT ROTHAMSTEDD, ENGLAND, IN THE YEARS 1855-56, CALCULATED PER ACRE, ACCORDING TO MESSRS. LAWES, GILBERT, AND WAY.

	Quantity of rain in Imperial gallons. 1 gal. = 10 ℔. water		Ammonia (in pounds)		Nitric Acid (in pounds)		Total Nitrogen (in pounds)	
	1855	1856	1855	1856	1855	1856	1855	1856
Total . .	663.332	616.051	7.11	9.53	2.08	2.80	6.63	8.31

AMOUNTS OF RAIN AND OF AMMONIA, NITRIC ACID, AND NITROGEN THEREIN, COLLECTED AT MANHATTAN, KANSAS, 1887-90, ACCORDING TO G. H. FAILYER.

	Total Nitrogen. Means for 4 years	Nitrogen in ammonia. Means for 3 years	Nitrogen in nitric acid. Means for 3 years
Parts per million of water	0.522	.388	0.150
Grammes per acre	1563.0	1190.0	480.0
Pounds per acre	3.44	2.63	1.06

It has been demonstrated, however, that nitrogen compounds and nitrogenous matter in the soil may become subject to nitrification through the action of bacteria, whereby ammonia, nitrous or nitric acid, carbon dioxide, and water are formed, though, as Wiley says, "The ammonia and nitrous acid may not appear in the soils, as the nitric organism attacks the latter at once and converts it into nitric acid."[1] (See further under influence of plant and animal life, p. 203.)

In considering the possible efficacy of these compounds, one must not lose sight of the fact that the amount of nitrogen in the soils is as a rule far too small to supply the demands of growing plants, and it is probable that a very large proportion of that which finds its way there is quickly taken up again by these organisms. It is possible that other salts of ammonium than the nitrate may be locally efficacious. Thus M. Beyer, as quoted by Van Den Broeck,[2] has shown that the feldspars decompose very rapidly under the influence of water containing ammonium sulphate or even sodium chloride, either of which substance may be found in vegetable soil. Daubrée, who experimented by means of revolving iron cylinders (see p. 197), found, however, that the presence of sodium chloride retarded decomposition.

(2) **Carbonic Acid.** — The amount of carbonic acid in the air under natural conditions is not a widely variable quantity, excepting near volcanoes and the immediate vicinity of gaseous springs. In the vicinity of large cities and manufactories consuming great quantities of coal, the amount is naturally increased. Although carbonic acid is the most abundant gas given off by decomposing vegetable matter, it has apparently been definitely ascertained that the amount of this gas in regions of abundant vegetation is no greater than elsewhere. This has been accounted for on the assumption that, as fast as liberated, it is taken up by growing organisms or carried by rains into the soil.[3]

[1] Wiley, Principles and Practice of Agricultural Analysis.
[2] Mem. sur les phénomènes d'Alteration des Dépôts Superficial, p. 16.
[3] The researches of Boussingault and Lewey (Mem. de Chemie Agricole, etc.), as quoted by Johnson (How Crops Feed, p. 139), showed the following proportions existing between the CO_2 of the air of the atmosphere and of various soils: —

	CO_2 IN 10,000 PARTS BY WEIGHT
Ordinary atmosphere	6 parts
Air from sandy subsoil of forest	38 parts
Air from loamy subsoil of forest	124 parts
Air from surface soil of forest	130 parts
Air from surface soil of vineyard	146 parts
Air from pasture soil	270 parts
Air from soil rich in humus	543 parts

Twenty-one tests of the air in various parts of Boston, during the spring, 1870, showed the presence of 385 parts of carbonic acid in 1,000,000. Eleven tests of the winter air in Cambridge yielded 337 parts in 1,000,000.[1] Dr. J. H. Kidder found the out-door air of Washington to contain 387 to 448 parts in 1,000,000, while Dr. Angus Smith, after an elaborate series of experiments, reported the atmosphere of Manchester (England) as containing 442 parts in 1,000,000.[2]

These amounts are considerably in excess of those reported by Müntz and Aubin,[3] who give the following figures relative to the proportional amounts in 10,000 by volume, as determined at the various widely separated stations. The amount, it will be perceived, is slightly greater during the night than during the day.

	Day	Night
Haytí	2.704	2.920
Florida	2.897	2.947
Martinique	2.735	2.850
Mexico	2.665	2.800
Santa Cruz, Patagonia	2.664	2.670
Chubut, Patagonia	2.790	3.120
Chili	2.665	2.820

The general mean is then 2.78 parts in 10,000, that for the night alone being 2.82. For the north of France the mean is given as 2.962, for the plain of Vincennes 2.84, and for the summit of the Pic du Midi 2.86.

Fischer, as quoted by Branner,[4] has shown that in rain and snow water the amount of carbonic acid varies between 0.22 % and 0.45 % by volume of water. Assuming that the mean of these figures fairly represents the general average, it is easy, knowing the rainfall of any region, to calculate the amount of the gas thus annually brought to the surface. Professor Branner has thus calculated that from 3.21 to 11.80 millimetres of carbonic acid (CO_2) are annually brought to the surface in certain parts of Brazil. The same method of calculation applied to the various parts of the United States, would give us for the Atlantic coast states 3.75 mm.; for the upper Mississippi valley, 2.50 mm.; for the lower Mississippi valley, 4.50 mm.; and

---

[1] 2d Ann. Rep. Mass. State Board of Health, 1871.
[2] Air and Rain, p. 52.
[3] Comptes Rendus, Vol. XCIII, 1881, p. 797; also XCVI, 1883, pp. 1793-97.
[4] Op. cit.

for the northern Pacific states, 6.25 mm. As it is mainly when this carbonic acid is thus brought to the surface by the rain and snows that its effects become of direct significance in our present work, the matter may be dropped here, to be taken up again when considering the chemical action of water.

(3) **Oxygen.** — Under ordinary conditions oxygen is the most active principle in atmospheric air, and it is to this agent that is due the process of oxidation which almost invariably characterizes the decomposition of silicates and other minerals containing iron in the protoxide state. Such oxidation is, however, almost inactive unless aided by moisture, and a further discussion of the subject may well be deferred, to be taken up again when discussing the action of water.

(4) **Heat and Cold.** — The ordinarily feeble action of the air is greatly augmented through natural temperature variations. That heat expands and cold contracts is a fact too well known to need elaboration. That, however, the constant expansion and contraction due to diurnal temperature variations may be productive of weakness and ultimate disintegration in so inert a body as stone, seems not so generally understood, or is, at least, less well appreciated, and hence a little space is devoted to the subject here. Rocks, it must be remembered, as the writer has noted elsewhere,[1] are complex mineral aggregates of low conducting power, each individual constituent of which possesses its own ratio of expansion, or contraction, as the case may be. In crystalline rocks these various constituents are practically in contact. In clastic rocks they are, on the other hand, frequently separated from one another by the interposition of a thin layer of calcareous, ferruginous, or siliceous matter which serves as a cement. As temperatures rise, each and every constituent expands and crowds with almost resistless force against its neighbor; as temperatures fall, a corresponding contraction takes place. Since in but few regions are surface temperatures constant for any great period of time, it will be readily perceived that almost the world over there must be continuous movement within the superficial portions of the mass of a rock.

The actual amount of expansion and contraction of stone under ordinary temperatures has been a matter of experiment. W. H. Bartlett[2] has shown that the average rate of expansion

---

[1] Stones for Building and Decoration, Wiley & Sons, New York.
[2] Am. Jour. of Science, Vol. XXII, 1832, p. 136.

for granite amounts to .000004825 inch per foot for each degree Fahrenheit; for marble .000005668 inch, and for sandstone .000009532 inch. Adie, in a series of similar experiments, found the rate of expansion for granite to be .00000438 inch, and for white marble .00000613 inch.[1] Slight as these movements may seem, they are sufficient to in time produce a decided weakening and afford a starting-point for other physical and chemical agencies, such as are ever lying in wait for an opportunity to get in their work. The writer well remembers the peculiar impressions produced during one of his earlier trips into the comparatively arid regions of Montana, at finding, at a certain place, the slopes and valley bottoms strewn with small, beautifully fresh, concave and convex chips of a dense, coal-black, andesitic rock that occupied the crest of one of the higher hills. So fresh were the fractures, so free were they from oxidation or other signs of decomposition, it was at first felt that they must be of human origin, that they were chips flaked off by aboriginal workmen in making stone implements, and some time was wasted in seeking for the more complete results of their handiwork. It, however, did not take long to convince him that the flakes were far too abundant and too widely spread to have originated in any such way, while the finding, on the top of the hill, of the coal-black rock, broken into larger columnar blocks, each with its angles rendered more obtuse or even fluted by the springing off of just such flakes, — this, coupled with the knowledge that during the day, exposed under a cloudless sky, the rocks became so highly heated as to be uncomfortable to the touch, whilst at night the temperature sank nearly to the freezing-point, sufficed to teach, as it must have taught the most obtuse, that the ordinary daily temperature variations were amply sufficient to account for the phenomenon.

Shaler states[2] that rock surfaces in the eastern United States may be subjected to temperatures varying from 150° F. at midday in summer to 0° and below in winter. This change of 150° in a sheet of granite 100 feet in diameter would produce a lateral expansion of about one inch of surface. That this expansion must tend to lessen the cohesion and tear the upper from the deeper lying layers, is self-evident. As exemplifying

[1] Trans. Royal Soc. of Edinburgh, Vol. XIII, p. 366.
[2] Proc. Boston Soc. of Nat. History, XII, 1869, p. 292.

this, Professor Shaler states that there are on Cape Ann (Massachusetts) hundreds of acres of bare rock surface completely covered with blocks of stone, which have been separated from the mass beneath by just this process.[1]

The size of such flakes may vary from those of microscopic proportions to masses of several tons' weight. The higher slopes of Lone Mountain, east of the Madison, in Montana, are covered above timber line with thousands upon thousands of these loose flakes of all sizes up to ten or more feet in diameter. Such, here, as in general, are characterized by a roughly lenticular outline in cross-section, possessing a large superficial area in proportion to their thickness, and are further distinguished from boulders of decomposition by the entire freshness of their materials even to the very surface. In close-grained, black andesitic and basaltic rocks the chip or flake not infrequently shows a beautiful concave and convex form and is greatly elongated in proportion to its breadth, resembling the long and slender chips of obsidian or flint found on the sites of aboriginal workshops. The surface left by the springing off of these flakes is of course fluted as though the work were done with a carpenter's gouge.

It is natural that this form of disintegration should be most pronounced in massive, close-grained rocks. In regions of great extremes of daily temperature the rupturing of these masses from the parent ledge is frequently attended by gun-like reports sufficiently loud to be heard at a considerable distance. H. von Streeruwitz states[2] that the rocks of the Trans Pecos (Texas) region undergo a very rapid disintegration from diurnal temperature variations, which here amount to from 60° to 75° F. He says: "I frequently observed in summer, as well as in winter time, on the heights of the Quitman Mountains a peculiar crackling noise and occasionally loud reports, . . . and careful

[1] The rifting action of heat upon granitic masses is said to have been made a matter of quarry utility in India. It is stated (Nature, January 17, 1895) that a wood fire built upon the surface of the granite ledge and pushed slowly forward causes the stone to rift out in sheets six inches or so in thickness, and of almost any desired superficial area. Slabs 60 × 40 feet in area, varying not more than half an inch from a uniform thickness throughout, have been thus obtained. In one instance mentioned, the surface passed over by the line of fire was 460 feet, setting free an area of stone of 740 square feet of an average thickness of five inches. This stone is undoubtedly one of remarkably easy rift, but the case will, nevertheless, serve our present purposes of illustration.

[2] 4th Ann. Rep. Geol. Survey of Texas, 1892, p. 144.

research revealed the fact that the crackling was caused by the gradual disintegration and separation of scales from the surface of the rock, and the loud reports by crackling and splitting of huge boulders." The scales thus split off, he says, vary in thickness from one-half to four inches, and their superficial area from a few square inches to many feet. This form of disintegration is necessarily confined to slopes unprotected by vegetation, and is the more pronounced the greater the diurnal variations.

In Arabia Petrea, according to Marsh,[1] "when a wind powerful enough to scour down below the ordinary surface of the desert and lay bare a fresh bed of stones is followed by a sudden burst of sunshine, the dark agate pebbles are often cracked and broken by the heat." According to Livingstone, the rock temperatures in certain parts of Africa, on the immediate surface, rise during the day as high as 137° F. and at night fall so rapidly as to throw off by their contraction sharp, angular masses in sizes up to 200 pounds' weight. Stanley, in his reports, is inclined to lay considerable stress on the effects of cold rains upon the heated rock surfaces, though it is doubtful if this is as powerful an agent as his descriptions would give us to understand. (See further under action of water.) Throughout the desert regions of lower California, as observed by the writer, the granitic and basic eruptive rocks subject to very little rainfall, and hence almost completely bare of vegetation, under the blistering heat of the desert sun have weathered down into dome-shaped masses, their débris in the form of angular bits of gravel being strewn over the plain. Particles of this gravel, when compared with those which are a product of chemical agencies, are found to differ in that each, however friable, is a complex molecule of quartz, feldspar and mica or other mineral that may have composed the rock from which it was derived. Aside from a whitening of the feldspathic constituent, due to the reflection of the light from its parted cleavage planes, scarcely any change has taken place, and indeed it more resembles the finely comminuted material from a rock-crusher than a product of natural agencies.

Owing, however, to the low conducting power of rocks, disintegration from this cause alone can go on to any extent only at the immediate surface, and on flat and level plains, where

[1] The Earth as modified by Human Action, p. 552.

the débris is allowed to accumulate, must in time completely cease.[1] It is only on hillsides and slopes, or where by the erosive action of running water, or by wind, the débris is removed as fast as formed, that such can have any geological significance, although the rate of such disintegration is sufficiently rapid in exposed places to be of serious consequence in stone used for architectural application. (See further on p. 198, Action of Ice.)

(5) **Wind.** — But it is to the action of the air when in motion — to the wind — that is due a very considerable part of atmospheric work. Particles of sand drifting along before the wind become themselves agents of abrasion, filing away on every hard object with which they come in contact. As a matter of course, this phenomenon is most strikingly active in the arid regions, though the results, when looked for, are by no means wanting in the humid east. It is thought by Professor Egleston that many of the tombstones in the older churchyards of New York City have become illegible by the wearing action of the dust and sand blown against them from the street. There is among

[1] Observations on soil temperatures made at the Orono, Maine, Experimental Station showed that the mean daily range of temperatures from April to October, at a depth below the surface of 1 inch, was 5.62°; at a depth of 3 inches, 5.26°; at 6 inches, 1.9°; and at 9 inches, 1.18°; and at 12 inches very slight. At the depth of 1 inch the temperature was lower than that of the air by 2.4°; at 3 inches by 2.11°; at 6 inches by 3.16°; at 9 inches by 3.94°; at 12 inches by 4.18°; at 24 inches by 5.78°; and at 36 inches by 7.10°.

The following table, compiled by Forbes (Trans. Royal Society of Edinburgh, Vol. XVI, 1849), from observations made near Edinburgh, Scotland, during 1841–42, shows the range of earth temperatures at varying depths in soil, sandstone, and trap rock.

Depth	Trap Rock			Sand of Garden			Chaigleith Sandstone		
	Max.	Min.	Range	Max.	Min.	Range	Max.	Min.	Range
3 feet	52.85°	38.88°	13.97°	54.50°	37.85°	17.65°	53.15°	38.25°	14.90°
6 feet	51.07	40.78	10.29	52.95	39.55	13.40	51.90	38.95	12.95
12 feet	49.00	44.20	4.80	50.40	43.50	6.90	50.30	41.60	8.70
24 feet	47.50	46.12	1.38	48.10	46.10	2.00	48.25	44.35	3.90

It has been shown that the thermal conductivity of rocks varies in direction according to their structure, being greatest in the direction of their schistosity, where such exists. In massive, homogeneous rocks the conductivity is the same in all directions. In finely fissile rocks, on the other hand, it may be four times as great in the direction of their fissility as at right angles thereto.

ACTION OF THE ATMOSPHERE

the heterogeneous collections of the National Museum at Washington a large sheet of plate glass, once a window in a lighthouse on Cape Cod. During a severe storm, of not above forty-eight hours' duration, this became on its exposed surface so ground from the impact of grains of sand blown against it as to be no longer transparent, and to necessitate its removal. Window panes in the dwelling-houses of the vicinity are, it is stated, not infrequently drilled quite through by the same means.

Apply now this agency to a geological field in a dry region. The wind, sweeping across a country bare of verdure and parched by drought, catches up the loose particles of dust and sand and drives them violently into the air in clouds, or sweeps them along more quietly close to the surface, where they are at first scarce noticeable. The impact of a single one of these moving grains on any object with which it may come in contact is far too small to be appreciable; but the impact of millions, acting through days, weeks, and years, produces results not merely noticeable, but strikingly conspicuous. We have here, in fact, a natural sand blast, an illustration on a grand scale of a principle in common use in glass-cutting, and to a small extent in stone-cutting also. Constantly filing away on every object with which they come in contact, the grains go sweeping on, undermining cliffs, scouring down mountain passes, wearing away the loose boulders, and smoothing out all inequalities. Naturally the abrading action on exposed blocks of stone is most rapid near the ground, as here the flying sand grains are thickest. First the sharp angles and corners are worn away, and the masses gradually become pear-shaped, standing on their smaller ends. Finally the base becomes too small for support, the stone topples over, and the process begins anew without a moment's intercession, and continues until the entire mass disappears, — becomes itself converted into loose sand drifted by the wind and an agent for destruction. Professor W. P. Blake was the first, I believe, to call public attention to this phenomenon, having observed it while in the Pass of San Bernardino (California) in 1853. G. K. Gilbert has also published some interesting facts as noted by himself while geologist of the Wheeler Expedition west of the 100th meridian, in 1878.[1] In acting on the hard rocks, the sand cuts

---

[1] It should be noted that the "sand-blast carving" described by Gilbert in this report is not due wholly to the action of wind-blown sand. The rock is fine

so slowly as at times to produce only grooved or fantastically carved surfaces, often with a very high polish. The geologists of the 40th Parallel Survey in 1878 described like interesting phenomena as observed on the western faces of conglomerate boulders exposed to the sand blast of the desert regions of Nevada. The surface of the otherwise light-colored rock was found to have assumed a dark lead-gray hue and a polish equal to that of glass, while the sand had drilled irregular holes and grooves, often three-fourths of an inch deep and not more than an eighth of an inch in diameter, through pebbles and matrix alike. Professor W. M. Davis,[1] G. H. Stone,[2] and J. B. Woodward[3] have described pebbles occurring in the glacial deposits of Cape Cod and of Maine, carved and facetted by the same agencies.

2. CHEMICAL ACTION OF WATER

Pure water, although an almost universal solvent, nevertheless acts with such slowness upon the ordinary materials of the earth's crust, that its results are scarcely appreciable to the ordinary observer. But it by no means follows that its effects are not worthy of our consideration here. This is particularly true when we reflect that the results being discussed are not merely those of days and weeks, but of years even when counted by the tens of thousands and millions. Moreover, absolutely pure water, as a constituent of our sphere, presumably does not exist. We have to consider its action as well when contaminated with sundry salts and acids which it almost universally holds, having taken them up in passing through the atmosphere, and in filtering through the overlying layer of organic matter and decomposition products which cover so large a portion of the surface of the land. It is when thus contaminated that are manifested the wonderful solvent and other chemical reactions which have been instrumental in promoting rock destruction, and it is here, then, that will be considered the complex chemical

calcareous shale. Through the solvent action of meteoric water the calcareous cement is removed, the fine, argillaceous interstitial material mechanically eroded, while the more resisting granules of quartz sand stand in relief, giving rise to elevated points and ridges.

[1] Proc. Boston Soc. of Natural History, Vol. XXVI, 1893, p. 166.
[2] Am. Jour. Science, Vol. XXXI, 1886, p. 133.
[3] Ibid., Jan., 1894, p. 63.

processes commonly grouped under the head of oxidation, deoxidation, hydration, and solution.

 (1) **Oxidation.** — Oxidation is perceptibly manifested only in rocks carrying iron either as sulphide, protoxide carbonate, or silicate. The sulphides, in presence of water and when not fully protected from atmospheric influences, readily succumb, producing sulphates which, being soluble, are removed in solution, or hydrated oxides, sulphuretted hydrogen, and perhaps free sulphur, as already noted (p. 29). Such an oxidation is attended by an increase in bulk, so that if nothing escapes by solution, there may be brought to bear a physical agency to aid in disintegration. Weathered rocks, containing iron sulphides, may not infrequently be found with cubical cavities quite empty or partially filled with the brownish, yellow, or red product of its oxidation in a more or less powdery condition. Pyrites, though a wide-spread constituent, is, nevertheless, a less conspicuous agent in promoting rock decomposition than the protoxide carbonates and silicates. In these the iron passes also over to the hydrated sesquioxide state, as is indicated by the general discoloration, whereby the rock becomes first streaked and stained, and finally uniformly ochreous. The more common minerals thus attacked are the ferruginous carbonates of lime and magnesia, and silicates of the mica, amphibole, and pyroxene groups. As the oxidation progresses, the minerals become gradually decomposed and fall away into unrecognizable forms. The red and yellow colors of soils are due invariably to the iron oxides contained by them. In many cases, the mineral magnetite, a mixture of proto- and sesqui-oxides, undergoes further oxidation and also loses its individuality.

(2) **Deoxidation** is a less common feature than oxidation. Water, carrying small quantities of organic acids, may take away a portion of the combined oxygen of a sesquioxide, converting it once more into the protoxide state. The local bleaching of certain ferruginous sands and sandstones is due to this action and a partial removal of the ferriferous salt in solution. Through a similar process of deoxidation, ferrous sulphates may be converted into sulphides, a process which undoubtedly takes place in marine muds protected from atmospheric action.

(3) **Hydration** — the assumption of water — more commonly accompanies oxidation, and, indeed, is an almost constant accom-

paniment of rock decomposition, as may be observed in comparing the total percentages of water in fresh and decomposed minerals and rocks, as given in the analyses.

This assumption, provided it be not accompanied by a loss of constituents, either by solution or erosion, must be attended by an increase in bulk, such as may be quite appreciable. The Comte de la Hure, as quoted by Branner,[1] has expressed the opinion that some of the hills of Brazil have actually increased in height through this means. The present writer has calculated that the transition of a granitic rock into arable soil, provided the same took place without loss of material, must be attended by an increase in bulk amounting to 88%.

Hydration as a factor in rock disintegration is, in the writer's opinion, of more importance than is ordinarily supposed. Granitic rocks in the District of Columbia have been shown[2] to have become disintegrated for a depth of many feet with loss of but comparatively small quantities of their chemical constituents and with apparently but little change in their form of combination. Aside from its state of disintegration, the newly formed soil differs from the massive rock, mainly in that a part of its feldspathic and other silicate constituents have undergone a certain amount of hydration. Natural joint blocks of the rock brought up from shafts were, on casual inspection, sound and fresh. It was noted, however, that on exposure to the atmosphere such shortly fell away to the condition of sand. Closer inspection revealed the fact that the blocks when brought to the surface were in a hydrated condition, giving forth only a dull, instead of clear, ringing sound, when struck with a hammer, and showing a lustreless fracture, though otherwise unchanged. That such had not previously fallen away to the condition of sand was evidently due to the vice-like grasp of the surrounding rock masses. These observations seem to have since received confirmation from Professor Derby,[3] who states that the sedimentary rocks of Sao Paulo, Brazil, as seen in the deep railway cuttings, "are almost invariably soft even when they show no signs of decay, and go to pieces by a kind of slaking process when broken up and exposed to the air, though they may have required blasting in the original opening of the cuttings."

[1] Op. cit., p. 284.
[2] Bull. Geol. Soc. of America, Vol. VI, p. 321.
[3] Decomposition of Rocks in Brazil, Jour. of Geol., Vol. IV, 1896, p. 205.

Professor W. O. Crosby[1] gives it as his opinion that the disintegration of the Pike's Peak (Colorado) granite is due mainly to hydration, the mica particularly being affected.

Professor Alexander Johnstone showed[2] by experimentation that normal muscovites, when submitted to the action of pure and carbonated waters for the space of a year, underwent very little change other than hydration, and a diminution in lustre, hardness, and elasticity. They appeared, in fact, to be converted merely into hydromuscovites, the hydration in pure water having gone on nearly as rapidly as in that which was carbonated. Biotite, when similarly treated, showed a slight discoloration or bleaching on the edges, accompanied also by hydration, and, in the case of that in carbonated water, a distinct loss of iron and magnesia through solution. Lepidolite, voigtite, vermiculite, and pyrosclerite were similarly acted upon, the iron and magnesia being removed in the form of carbonates. The fact was noted "that whenever anhydrous micas, or lower hydrated micas, become hydrated, they always at the same time increase in bulk." This fact he regarded as accounting for the rapid weathering of micaceous sandstones.

(4) **Solution.** — The solvent action of water is perhaps the most important of its immediate effects, though there are many incidental chemical changes set in operation which, in the end, are of equal or even greater significance. It is the solvent action only that concerns us here.

Rain and nearly all superficial waters contain small quantities of carbonic, humic, ulmic, crenic, and apocrenic acids, which greatly increase their solvent capacities. The last-named forms are complex, unstable, and little understood products of plant decomposition,[3] and might logically be considered under effects

[1] Personal Memoranda to the Writer.
[2] Quar. Jour. Geol. Soc. of London, Vol. XLV, 1889.
[3] The following are the chemical formulas of these acids, as commonly given: —

ULMIN AND ULMIC ACID

Carbon 67.1%
Hydrogen . . . . 4.2   } Corresponding to $C_{40}H_{28}O_{12} + H_2O$
Oxygen 8.7

HUMIN AND HUMIC ACID

Carbon 64.4%
Hydrogen . . . . 4.3   } Corresponding to $C_{21}H_{24}O_{12} + 3H_2O$
Oxygen 31.3

of plant and animal life, but that they act only in presence of moisture.

"There is reason to believe that in the decomposition effected by meteoric waters and usually attributed mainly to carbonic acid, the initial stages of attack are due to the powerful solvent capacities of the humus acids. Owing, however, to the facility with which these acids pass into higher stages of oxidation, it is chiefly as carbonates that the results of their action are carried down into deeper parts of the crust or brought up to the surface. Although CO_2 is no doubt the final condition into which these unstable organic acids pass, yet during their existence they attack not merely alkalies and alkaline earth, but even dissolve silica."[1] P. Thernard found that the solvent power of these acids was largely controlled by the amount of nitrogen they contained.[2]

CRENIC ACID

Carbon 44.0%
Hydrogen 5.5
Nitrogen 3.0
Oxygen 46.6

Corresponding to $C_{12}H_{12}O_8$?

APOCRENIC ACID

Carbon 34.4%
Hydrogen 3.5
Nitrogen 3.0
Oxygen 39.1

Corresponding to $C_{24}H_{24}O_{12}$?

Berthelot and Andre (Comptes Rendus Academie de Paris, 114, 1892, pp. 41-43) have shown that the brown substance of humus and analogous compounds undergo direct oxidation under the influence of the air and sunlight, forming carbonic acid. These reactions are purely chémical, taking place without the intervention of microbes, and are accompanied by a change in color of the original humus. The oxidation is rendered more active through the division and mellowing of the humus by cultivation. Through chemical union of the carbonic acid with certain bases, as lime soda and potash, there are found soluble carbonates which may be leached out by meteoric waters.

[1] Geikie, Text-book of Geology, 3d ed., p. 472.
The writer was shown not long since, by Professor Charles E. Munroe, a very practical illustration of the remarkable corrosive power of organic acids. A highly ornate French clock, with case of black marble, was packed for storage in excelsior which was a trifle damp. The clock remained in storage from the last of May until about the first of October of the same year. When the packing material was removed, the marble was found to be so corroded as to need rehoning and polishing. The roughness could be easily felt by passing the finger over the surface, and long lustreless lines indicating the contact of excelsior fibres traversed the surface in every direction.

[2] Julien, The Geological Action of Humus Acids, Proc. Am. Assoc. Adv. of Science, 1879, p. 324.

It is stated by Storer[1] that "on the tops of the higher hills of New Hampshire, and on the coast of Maine also, a cold, sour black earth will often be noticed at the surface of the ground, immediately beneath which is sometimes a layer of remarkably white earth. The whiteness is due to the solvent action of acids that soak out from the black humus, and which leach out from the underlying clay and sand the oxides of iron that formerly colored them."

As long ago as 1848 the Rogers brothers showed[2] that pure water partially decomposed nearly all the ordinary silicate minerals which form any appreciable part of our rocks. The action of carbonated water was recognizable in less than ten minutes, but pure water required a much longer time before its effect was sufficient for a qualitative determination. So pronounced was the action of carbonated water that the presence of the alkalies of lime and magnesia could be recognized in a single drop of the filtrate from the liquid in which the powdered minerals were digested. By digestion for forty-eight hours they obtained from hornblende, actinolite, epidote, chlorite, serpentine, feldspar, etc., a quantity of lime, magnesia, oxide of iron, alumina, silica, and alkalies amounting to from 0.4 % to 1 % of the whole mass. The lime, magnesia, and alkalies were obtained in the form of carbonates; the iron, in the case of hornblende, epidote, etc., passing from the state of carbonate to that of peroxide during the evaporation of the solutions. Forty grains of finely pulverized hornblende, digested for forty-eight hours in carbonated water at a temperature of 60°, with repeated agitation, yielded — silica, 0.08 %; oxide of iron, 0.095 %; lime, 0.13 %, and magnesia, 0.095 %, with traces of manganese. Commenting on these results, Bischof remarks[3] that "by repeating this treatment 112 times with fresh carbonated water, a perfect solution might be effected in 224 days. If now," he says, "40 grains of hornblende, unpowdered, in which, according to the above assumption, the surface is only one millionth of the powdered, were treated in the same way, and the water renewed every two days, the time required for perfect solution would be somewhat more than six million years." In considering these figures and their practical bear-

[1] Chemistry as applied to Agriculture.
[2] Am. Jour. of Science, Vol. V, 1848.
[3] Chemical and Physical Geology, Vol. I, p. 61.

ing, it must be remembered that while in nature the quantity of water coming in contact with a crystal embedded in a rock during a given time is much less than that assumed above, the mineral is undergoing a gradual splitting up, becoming more and more porous, so that the process is gradually accelerated.

To quote Bischof again, it is probably admissible to assume that the time in which water produces similar effects of decomposition or solution on minerals, is inversely as the magnitude of the surface of contact. If, therefore, a mineral were so far subdivided that the surface was increased ten million-fold, the quantity then dissolved during a certain time would be the same as that dissolved during a period ten million times as long.

Richard Müller has also shown[1] that carbonic acid waters will act even during so brief a period as seven weeks upon the silicate mineral with such energy as to permit a quantitative determination of the dissolved materials. The accompanying table shows (1st) the percentages of the various constituents thus taken out by the carbonated water, and (2d) the total percentages of the materials dissolved. That is to say, the figures 0.1552 given for adular under SiO_2, indicate that 0.1552% of the total 65.24% of the silica contained by the mineral have been removed, and so on. The last column gives the total per cent of all the constituents extracted.

Mineral	SiO_2	Al_2O_3	K_2O	Na_2O	MgO	CaO	P_2O_5	FeO	Total
	%	%	%	%	%	%	%	%	%
Adular . . .	0.1552	0.1308	....	....	....	....	....	trace	0.328
Oligoclase . .	0.237	9.1713	....	2.367	....	3.213	....	trace	0.533
Hornblende .	0.419	trace	....	....	....	8.528	....	4.829	1.536
Magnetite . .	trace	....	....	....	....	....	....	0.042	0.307
Apatite . . .	....	....	....	....	....	2.108	1.822	....	2.018
Olivine . . .	0.873	trace	....	....	1.291	trace	....	8.733	2.111
Serpentine .	0.354	....	....	....	2.649	....	....	1.527	1.211

● The summary of his investigations he gives as below: —

(1) All the minerals tested were acted upon by the carbonated water.

(2) In this process there were formed carbonates of lime, iron, manganese, cobalt, nickel, potash, and soda.

[1] Untersuchen über die Einwirkung des kohlensäurehaltigen Wassers auf einige Mineralien und Gesteine, Tschermaks Min. Mittheilungen, 1877, p. 25.

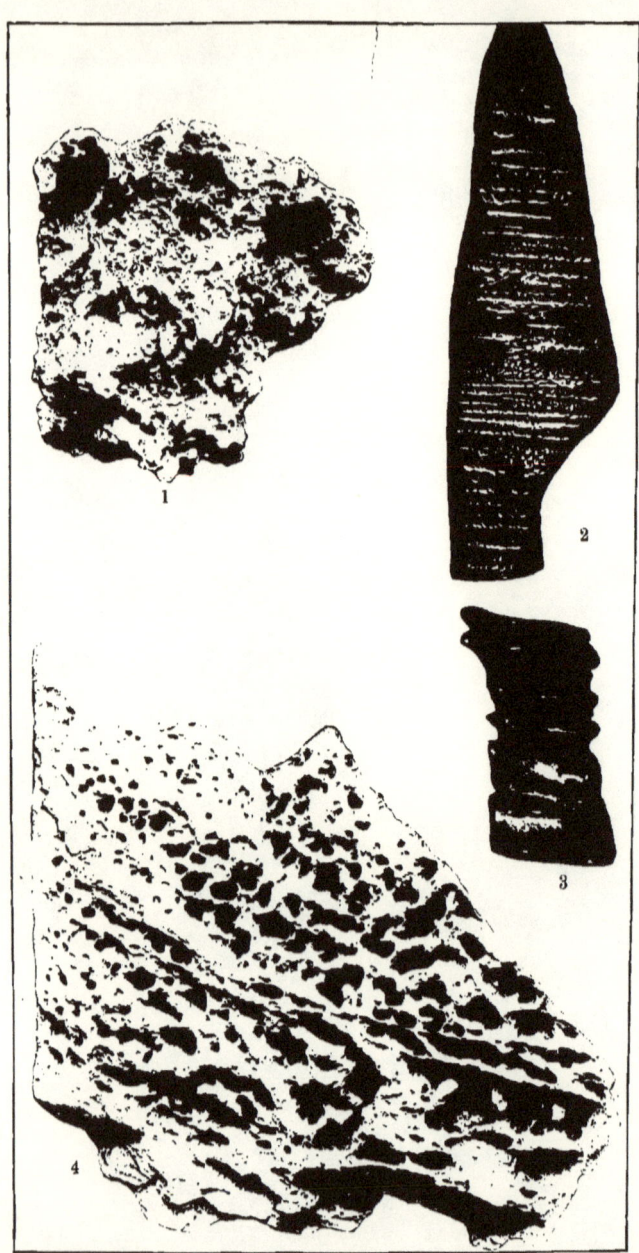

Corroded limestones.

(3) In the action of the carbonated waters upon the alkaline silicates, like the feldspars, a small amount of silica went always into solution, presumably in the form of hydrate.
(4) Even alumina was dissolved in appreciable quantities.
(5) Adular proved more resisting to the action of the acid than did the oligoclase.
(6) The first stage of decomposition in the feldspars is a reddening process; the second, kaolinization.
(7) Hornblende was more easily decomposed than feldspar.
(8) Increase of pressure on the solution was productive of more energetic action than prolonging the time.
(9) Of all the minerals tested, the magnetic iron was least affected.
(10) Apatite was readily acted upon, as could be detected by its appearance under the microscope.
(11) Olivine was the most readily attacked of all the silicates tested, probably twice as easily decomposed as the serpentine.
(12) Magnesian silicates were attacked by the carbonated waters. Hence serpentine cannot be considered a final product of decomposition.[1]

Of all the materials forming any essential part of the earth's crust the limestones are most affected by the solvent power of water. It is stated that pure water will dissolve lime carbonate in the proportions of one part in 10800 when cold and one part in 8875 when boiling.

Since rock-weathering is, as already stated, a superficial phenomenon, we have to do only with waters of ordinary temperatures and under ordinary conditions of pressure, though this expression must not be taken as necessarily meaning *cold* waters, since, if we accept the statements of Caldcleugh,[2] rain waters falling upon the heated rocks may have their temperatures raised as high as 140° F. The enormously destructive effect of carbonated waters on limestone is scarcely apparent on casual inspection, owing to the fact that the material is carried away in solution, leaving only the insoluble impurities behind. In such cases it is possible to estimate the amount of corrosion through a comparison of the proportional amounts of various constituents in this residue with those in the fresh rock

[1] Serpentine, however, cannot be properly considered a decomposition product. It is rather a product of *alteration*.
[2] Trans. Geol. Soc. of London, 1820.

(see p. 209 *et seq.*), and the time limit of corrosion through determining the percentage amounts of the constituents in the water which annually drains from any given area. By such methods it has been estimated[1] that some 275 tons of calcium carbonate are annually removed from each square mile of Calciferous limestone exposed in the Appalachian region alone ; while a well-known English authority[2] has calculated that with an annual rainfall of 32 inches, percolating only to a depth of 18.3 inches, there are annually removed by solution from the superficial portions of England and Wales an average of all constituents amounting to 143.5 tons per square mile of area. He further calculates that the average amount of carbonate of lime annually removed from each square mile of the entire globe amounts to 50 tons.[3] It is to this corrosive action of meteoric waters that still another authority[4] would attribute the slight thickness and nodular condition of many beds of Palæozoic limestone. He argues that originally thick-bedded limestones have, during the ages subsequent to their formation and uplifting, become so impoverished through the dissolving out and carrying away in solution of the lime carbonate, as to have been quite obliterated, or reduced to mere nodular bands, and given rise to important palæontological breaks in the geological record. Other than organic acids may locally exert a potent influence. Thus Robert Bell has described the dolomitic limestones underlying the waters along Grand Manitou Island, the Indian peninsula, and adjacent portions of Lake Huron and the Georgian Bay, as pitted and honeycombed in a very peculiar and striking manner. This corrosion, it is believed, is produced through the solvent action of sulphuric acid in the water, the acid itself arising from the decomposition of the sulphides of iron, pyrites and pyrrhotite, which exist in great quantities in the Huronian rocks to the northward.[5]

---

[1] A. L. Ewing, Am. Jour. of Science, 1885, p. 29.

[2] T. Mellard Reade, Chemical Denudation in Relation to Geological Time.

[3] The total dissolved constituents thus removed are divided up as follows: Carbonate of lime, 50 tons ; sulphate of lime, 20 tons ; silica, 7 tons ; carbonate of magnesia, 4 tons ; peroxide of iron, 1 ton ; chloride of sodium, 8 tons ; alkaline carbonates and sulphates, 6 tons.

[4] F. Rutley, The Dwindling and Disappearance of Limestones, Quar. Jour. Geol. Soc. of London, August, 1893.

[5] Bull. Geol. Soc. of America, Vol. VI, pp. 47–304. Messrs. C. W. Hayes and M. R. Campbell, of the United States Geological

3. MECHANICAL ACTION OF WATER AND OF ICE

Aside from its solvent capacity, water acts as a powerful erosive agent, as well as an agent for the transportation of the eroded materials. It is only its erosive power that need concern us here, though, as will be seen, this is to a considerable extent dependent upon its power of transportation. Every raindrop beating down upon a surface already sorely tried by heat and frost serves to detach the partially loosened granules, and, catching them up in the temporary rivulets, carries them to the more permanent rills, to be spread out over the valley bottoms, or perhaps if the slopes be steep and the current ac-

Survey, have recently reported some remarkable examples of corroded quartz pebbles which should be mentioned here, although a satisfactory explanation for the phenomenon has not yet been given.

Dr. Hayes, in a personal memorandum to the writer, describes the occurrence as follows:—

"At three rather widely separated points in the South, conglomerates have been observed in which the projecting portions of the pebbles have been etched or partly dissolved.

"The first, observed by Mr. Campbell, is at Nuttall, West Virginia. The conglomerate in question, which belongs to the coal measures, is composed of rather coarse quartz sand with slightly yellowish cement, in which are embedded well-worn pebbles of white vein quartz. The latter vary in size up to three-quarters of an inch in diameter, and are somewhat irregularly distributed. Ordinarily the pebbles, wholly unaltered, weather out by the chemical or mechanical disintegration of the sandy matrix. In the case observed, however, where the conglomerate received the drip from an overhanging cliff, the projecting portions of the pebbles are deeply pitted, evidently by solution. Mechanical wear is precluded by the form of the resulting surface, which is not smooth like the portions of the pebble still protected by the matrix, but is rough and irregular. The outer portion of the pebbles is evidently less easily affected by the solvent than the interior, and forms a sharp rim about the irregular cavities hollowed out within. In some cases a third of the pebble has thus been removed. The surface of the sandstone matrix in which the pebbles are embedded is also pitted, possibly by the same process of solution as that which has affected the pebbles, but such a surface might also be produced by mechanical means in case the cement were less indurated in some places than in others.

"The second case is on Clifty Creek, White County, Tennessee. The conglomerate, also a member of the coal measures, forms the bottom of a small cañon, and is covered by the creek at high water, but uncovered throughout the greater part of the year. The matrix is a coarse white sandstone which weathers yellow by the oxidation of the slightly ferruginous cement. Embedded in this are rather abundant pebbles, varying in size up to two inches in diameter, and composed chiefly of quartz, with a few of chert and possibly of quartzite. The projecting portions of these pebbles have been in part removed, though they still project somewhat above the enclosing matrix. As in case of the Nuttall conglomerate, the exterior portions of the pebbles are less easily affected than

cordingly strong, to the rivers and thence to the sea. The amount of detrital matter thus mechanically removed from the hills and spread out over valley and sea-bottoms quite exceeds our comprehension, but it is estimated that at the rate the Mississippi River is now doing its work, the entire American continent might be reduced to sea-level within a period of four and one-half million years. The Appalachian Mountain system, whose uplifting began in early Cambrian times and terminated at the close of the Carboniferous, has already through this cause lost more material than the entire mass of that which now remains. But the rivers, like the winds and glaciers, in virtue of this load they bear, become themselves converted into agents of erosion, filing away upon their rocky beds, undermining their banks, and continually wearing away the land by their ceaseless activity. The pot-holes in the bed of a stream, formed by the constant swirl of sand and gravel in an eddy, furnish on a small scale striking illustrations of this cutting power, while the rocky cañons of the Colorado of the West, where thousands of feet of horizontal strata have been cut through as with a file, show the same thing on a scale so gigantic as to be at first scarce comprehensible.[1] An item of no insignificant importance to be considered here is the possibility,

the interiors, and when the pebble has been a third or half removed the outer shell forms a rim within which is a depression with a slight elevation in the centre. The chert pebbles show less evidence of corrosion by a solvent than those composed of quartz. Their upper surfaces are somewhat worn down and even slightly hollowed, but this might easily have been produced by mechanical means, which is not the case with the quartz.

"The third case is a block of conglomerate from Starrs Mountain, Tennessee, collected by Mr. Bailey Willis. This is of Lower Cambrian age. The matrix is a coarse feldspathic sandstone containing layers of well-rounded pebbles, mostly quartz, with a few probably of some feldspar. The former are between one-half and one inch in diameter and the latter somewhat larger. The projecting portions of the quartz pebbles on one side of the block are almost entirely removed, and as in the other cases evidently by solution. A slight rim projects above the matrix in which the pebbles are embedded; within this is a depression, while a slight elevation occupies the centre.

"'The projecting portions of the feldspathic pebbles also are partly removed, but this may be due to corrasion instead of corrosion, that is, to the action of mechanical rather than chemical agents. The pebbles on the lower side of the block have their original water-worn surfaces without any trace of etching."

[1] Captain C. E. Dutton has estimated (Tertiary History of the Grand Cañon of the Colorado) that from over an area of 13,000 to 15,000 square miles drained by the Colorado River, an average thickness of 10,000 feet of strata have been removed.

indeed probability, of an incidental chemical decomposition taking place during this abrasive action. Daubree showed [1] that when feldspathic fragments were submitted to artificial trituration in a revolving cylinder containing water, a decomposition was effected whereby the alkalies were liberated in very appreciable amounts. He found further that the principal product of mutual attrition of feldspar fragments in water was not sand, but an impalpable mud (*limon*). This mud was of such tenuity as to remain for many days in suspension, and on desiccation became so hard as to be broken only with the aid of a hammer, resembling in many respects the argillites of the coal measures, but differing in that it carried a high percentage of alkalies. Granitic rocks thus treated yielded angular fragments of quartz and very minute shreds of mica, while the feldspars ultimately quite disappeared in the form of the impalpable mud above mentioned. It was noted that after the quartzose particles had reached a certain degree of fineness further diminution in the size ceased, owing to the buoyant action of the water, which in the form of a thin film between adjacent particles acted as a cushion and prevented actual contact to the extent necessary for mutual abrasion. It is to a similar action on the part of sea-water that Shaler [2] would attribute the lasting qualities of the sand grains upon our sea beaches. Indeed the conditions of Daubree's experiments as a whole were not so different from those existing in nature that we need hesitate, as it seems to the writer, to conclude similar action, both chemical and physical, may be going on wherever abrasion takes place in the presence of continual moisture, as in the bed of a river or glacier.

[1] It will be remembered that this authority placed rock fragments in stone and iron cylinders containing water and made to revolve horizontally at a measured rate of speed, so that the actual distance travelled by any of the particles during a given time could be readily calculated. The product of this disintegration, even when carried to the condition of fine silt, was always sharply angular. His experiments further showed that when feldspathic fragments were thus treated, there was always a certain amount of decomposition, whereby salts of potash were liberated; in one instance, when 3 kilogrammes of feldspar were revolved for 192 hours in iron cylinders containing 5 litres of water, 2.72 kilogrammes of finely comminuted mud were obtained, and in solution in the water, 12.6 grammes of potash, or 2.52 grammes per litre. The presence of carbonic acid in the water increased the amount of potash. When the feldspar was triturated dry and then treated with water, no such solvent action could be detected. — Geologie Experimentale, p. 268.

[2] Bull. Geol. Soc. of America, Vol. V, p. 208.

The hammering action of waves upon the sea-coast exerts a powerful erosive action, particularly upon particles of rock of such size as to be lifted or moved by wave action, but too heavy to be protected from attrition by the thin film of water above alluded to. Shaler's observations[1] at Cape Ann were to the effect that ordinary granitic paving blocks (weighing perhaps twenty pounds) were, when exposed to surf action, worn in the course of a year into spheroidal forms such as to indicate an average loss of more than an inch from their peripheries. Experiments made with fragments of hard burned brick showed that in the course of a year they would be reduced fully one-half their bulk. Even the crystallization of the salt thrown up by wave action and absorbed into the pores of rocks[2] serves in its way the purposes of disintegration.

The Action of Freezing Water and of Ice. — The action of dry heat and cold in disintegrating rocks has already been described. The effects of such temperature changes upon stone of ordinary dryness are, however, slight in comparison with the destructive agencies of freezing temperatures upon stones saturated with moisture. The expansive force of water passing from the liquid to the solid state has been graphically described as equal to the weight of a column of ice a mile high (about 150 tons to the square foot). Otherwise expressed, 100 volumes of water expand, on freezing, to form 109 volumes of ice. Provided, then, sufficient water be contained within the pores of a stone, it is easy to understand that the results of freezing must be disastrous. That stones as they lie in the ground do contain moisture, often in no inconsiderable amounts, is a well-known and well-recognized fact by all those engaged in quarrying operations, and indeed no mineral substance is absolutely impervious to it. The amount contained, naturally varies with the nature of the mineral constituents and their state of aggregation. According to various authorities, granite may contain some 0.37 % by weight; chalk, 20 %; ordinary compact limestone, 0.5 % to 5 %; marble, about 0.30 %; and sandstones, amounts varying up to 10 % or 12 %, while clay

---

[1] Bull. Geol. Soc. of America, Vol. V, p. 208.
[2] According to Dana (Wilkes' Exploring Expedition, Geology, p. 529), the sandstones along the coast of Sydney, Australia, are subjected to a mechanical disintegration through the crystallization of salt which is absorbed from the saline spray of the ocean waves.

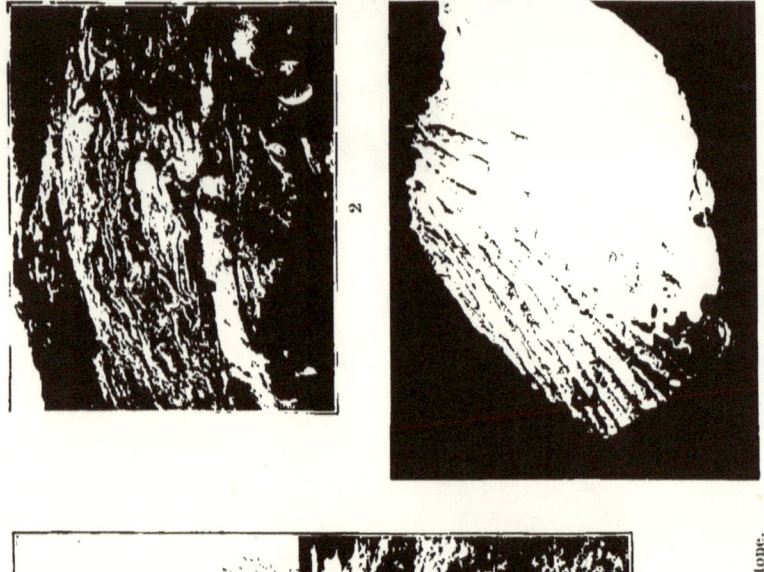
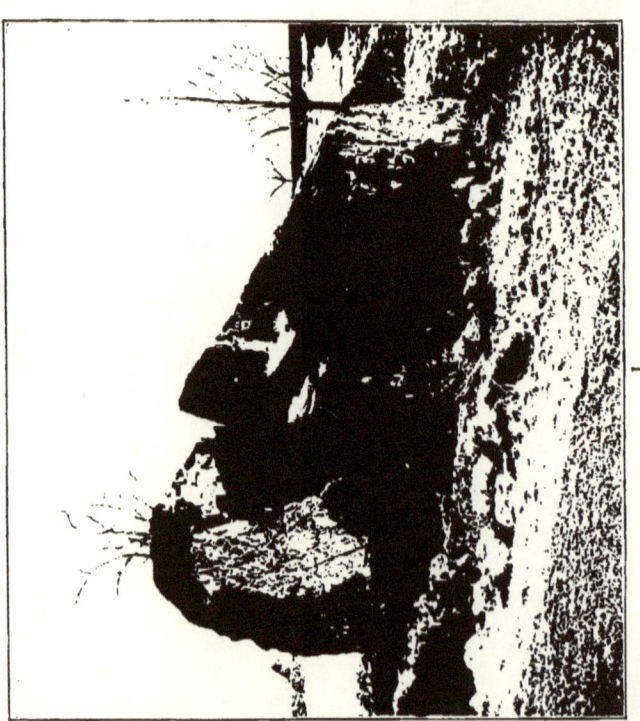

PLATE 16

Fig. 1. Diorite boulder split along joint planes by frost. Fig. 2. Corroded surface of pyroxenic limestone. Fig. 3. Corroded limestone.

may contain nearly one-fourth its weight. This water is largely interstitial — the *quarry water*, as it is sometimes called. In addition to this, the quartz, particularly of granitic rocks, almost universally contains innumerable minute cavities partially filled with water, and which are, in extreme cases, so abundant as to make up, according to Sorby, at least 5 % of the whole volume of the mineral.

That the passage of this included moisture from the liquid to the solid state, must be attended with results disastrous to the stone is self-evident, though the rate of disintegration may be so slow under favorable circumstances as to be scarce noticeable. Freezing of the absorbed water is one of the most fruitful sources of disintegration in stones confined in the walls of a building, and even in the quarry bed it is by no means uncommon to have stone so injured as to render it worthless. However slight may be the effects of a single freezing, constant repetition of the process cannot fail to open up new rifts, and still further widen those already in existence, allowing further penetration of water to freeze in its turn and to exert a chemical action as well. So year in and year out, through winter's cold and summer's heat, the work goes on until the massive rock becomes loose sand to be caught up by winds or temporary rivulets and spread broadcast over the land. In some instances, it may be, the rock is of sufficiently uniform texture to be affected in all its mass alike. More commonly, however, it is traversed by veins, joints, or other lines of weakness along which the rifting power is first made manifest, as in our illustration. Naturally disintegration of this kind is confined to frigid and temperate latitudes. As bearing upon the extreme rapidity with which such disintegration may take place, the following is quoted from a letter of Dr. L. Stejneger, of the United States National Museum, who passed several months among the islands of Bering Sea.

"In September, 1882, I visited Tolstoi Mys, a precipitous cliff near the southeastern extremity of Bering Island. At the foot of it I found large masses of rock and stone which had evidently fallen down during the year. Most of them were considerably more than six feet in diameter, and showed no trace of disintegration. The following spring, April, 1883, when I revisited the place, I found that the rocks had split up into innumerable fragments, cube-shaped, sharp-edged, and of a very uniform size, — about two inches. They had not yet fallen to pieces, the rocks still retaining their original shape. I may remark, however, that the weather was still freezing when I was there. The winter was not one of great severity, and several thawing spells broke its continuity. These cubic fragments did not seem to split up any further, for everywhere on the islands where the rock consisted of the coarse sandstone, as in this place, the talus consisted of these sharp-edged stones."

Ice acts as a disintegrating agent in still other ways than that mentioned. The phenomenon of the glacier is now so well known that we need dwell upon it but briefly here. Long-continued precipitation of snow upon regions of such elevation, or in such latitudes as to preclude anything like an equally rapid melting, gives rise to deep fields of snow, compacted in the lower portions into the condition of ice. These, in virtue of the weight of the overlying mass, and perhaps the steepness of the slopes, aided by a certain amount of plasticity possessed in some degree by even the most rigid of so-called solids, creep slowly down the slopes in the form of glaciers or rivers of ice. Advancing, it may be, but an inch or several feet a day, now scarce moving at all, or even retreating temporarily through a diminution in the amount of their supplies, or an increase in the sun's heat, these bring, either upon their surfaces as moraines, or frozen into their mass, large quantities of fragmental rock material fallen upon them from above, or picked up from the surfaces over which they flow. Those fragments which remain upon the upper surface, or frozen into the upper portions, are but transported to the lower levels where, the temperature being sufficient, the ice is melted and the load deposited in the form of a moraine.

Beneath, and frozen into the lower portion of the ice sheet, there is, however, a variable amount of rock material, which, as the glacier moves along, is crowded with all the weight of the overlying mass, and all the resistless energy of the ice behind, over the surface of the underlying rock. In virtue of this material, this sand, gravel, and boulder aggregate, the glaciers become converted into what we may compare to extremely coarse files, to tear away the rocks over which they pass, and grind and crush them into detritus of varying degrees of fineness. The small streams which originate from the melting of these glaciers become, hence, not infrequently charged to the point of turbidity with the fine silt-like detritus ground from the ledges and in part from the boulders themselves. Figure 3 of plate 19 shows a slab of limestone still bearing upon its surface the evidences of the severity of the onslaught. A consideration of the amount of detritus thus brought down either merely as transported or as abraded material belongs properly to the chapter on transportation, but a few illustrations are not without interest here. The Aar in Switzerland is stated by Geikie to discharge every day in August some 440,000,000 gallons of water, carrying some 280 tons of sand. A portion of this is in a state of such minute subdivision as to remain a long time in suspension, and give the water a milky appearance for several miles. I. C. Russell has described[1] the Tuolumne River, issuing from the foot of the Lyell Glacier in the Sierras of California, as turbid with silt which has been ground by the moving ice.

At the foot of the Dana Glacier there is a small lakelet whose waters are of a peculiar greenish yellow color from the silt held in suspension, and which, when submitted to microscopic examination, is found to be made up of fresh angular fragments of various silicate minerals of all sizes from 0.35 mm. in diameter down to impalpable silt.

4. ACTION OF PLANTS AND ANIMALS

Both plants and animals aid to some extent in the work of rock disintegration. Plants are also not infrequently an important factor in promoting sedimentation, while burrowing insects and animals may exert an important influence upon the texture of soils and in bringing about a more general admixture by transferring to the surface that which is below.

The lowest forms of plant life,— the lichens and mosses,— growing upon the hard, bare face of rocky ledges send their minute rootlets into every crack and crevice, seeking not merely foot-hold, but food as well.

Slight as is the action, it aids in disintegration. The plants die, and others grow upon their ruins. There accumulates thus, it may be with extreme slowness, a thin film of humus, which serves not merely to retain the moisture of rains and thus bring the rock under the influence of chemical action,

[1] 5th Ann. Rep. U. S. Geol. Survey, 1883–84.

but supplies at the same time small quantities of the humic and other organic acids to which reference has already been made.[1] These act both as solvents and deoxidizing agents. As time goes on, sufficient soil gathers for other, larger and higher types of life, which exert still more potent influences. It may be the rock is in a jointed condition. Into these joints each herb, shrub, or sapling pushes down its roots, which, in simple virtue of their gain in bulk, day by day, serve to enlarge the rifts and furnish thereby more ready access for water, and the wash of rains, to still further augment disintegration.

This phase of root action is often well shown in walls of ancient masonry, either of brick or stone, whereby the usual rate of destruction is greatly accelerated. The depth to which such roots may penetrate has often been noted, varying, as is to be expected, with the nature of the soil.[2] In the limestone caverns of the Southern states, the writer has often been impressed by the number of long thread-like rootlets, so fine as to be almost imperceptible, which have found their way through rifts in the rocky roof.

H. Carrington Bolton has shown that very many minerals are decomposed by the action of cold citric acid for a more or less prolonged period, the zeolites and other hydrous silicates being especially susceptible. Such tests have a peculiar significance when we consider that the roots of growing plants secrete an acid sap, which, by actual experiment, has been found capable of etching marble. The exact nature of this acid is not accurately known, but it is considered probable that in the rootlets of each species of plant there exists a considerable variety of organic acids.[3]

But the effects of plant growth are not necessarily always destructive; such may be conservative or even protective. In glaciated regions, it is often the case that the striated and polished surfaces of the rocks have been preserved only where protected from the disintegrating action of the sun and atmosphere

---

[1] It is stated by Storer (Chemistry as applied to Agriculture) that some lichens have been found to contain half their weight of oxalate of lime.

[2] Aughey has found roots of the buffalo berry (*Sherperdia argophylla*) penetrating the loess soils of Nebraska to the depth of 50 feet.

[3] See Application of Organic Acids to the Examination of Minerals, H. Carrington Bolton, Proc. Am. Assoc. for the Advancement of Science, XXXI, 1883, and Available Mineral Plant Food in Soils, B. Dyer, Jour. Chem. Society, March, 1894.

by a thin layer of turf or moss. As a general rule, however, the manifest action of plant growth is to accelerate chemical decomposition, through keeping the surfaces continually moist, and to retard erosion. (See further on p. 280.)

Action of Bacteria. — The researches of A. Müntz,[1] Widogradsky, Schlösing, and others tend to show that bacteria may exercise a very important influence in promoting rock disintegration and decomposition. Their influence in promoting nitrification has been already alluded to. It would appear that while these organisms secrete and utilize for their sustenance the carbon from the carbonic acid of the atmosphere, as do plants of a higher order, they may also assimilate carbonate of ammonium, forming from it organic matter and setting free nitric acid. Being of microscopic proportions, the organisms penetrate into every little cleft or crevice produced by atmospheric agencies, and throughout long periods of time produce results of no inconsiderable geological significance. The depth below the surface at which such may thrive is presumably but slight, and their period of activity limited to the summer months. They have been found on rocks of widely different character — granites, gneisses, schists, limestones, sandstones, and volcanic rocks — and on high mountain peaks as well as on lower levels. The Pic Pourri, or Rotten Peak, in the Lower Pyrenees of southwestern France, is composed of friable and superficially decomposed calcareous schists, throughout the whole mass of which are found the nitrifying bacteria, which are believed to have been instrumental in promoting its characteristic decomposition. The organism acts even upon the most minute fragments, reducing them continually to smaller and smaller sizes. Each fragment loosened from the parent mass is found coated with a film of organic matter thus produced, and the accumulation begun by these apparently insignificant forces is added to by residues of plants of a higher order, which come in as soon as food and foothold are provided.[2]

Mr. J. E. Mills,[3] and after him J. C. Branner,[4] lay considerable stress on the decomposing effect of vegetable matter

[1] Comptes Rendus de l'Académie des Sciences, CX, 1890, p. 1370.
[2] It is, perhaps, as yet, too early to say to what extent the presence of bacteria may be incidental to decomposition, rather than causative.
[3] American Geologist, June, 1889, p. 357.
[4] Bull. Am. Geol. Soc. of America, Vol. VII.

carried into the ground by ants in certain parts of Brazil, Mills going so far as to describe the ants as continually pouring carbonic acid into the ground. Be this as it may, the evacuations of the ants themselves are undoubtedly of such a nature as to further the processes of decomposition. Certain species of ants, locally known as saubas, or sauvas, live, according to Branner, in enormous colonies, burrowing in the earth, where they excavate chambers with galleries that radiate and anastomose in every direction, and into which they carry great quantities of leaves. Certain species of termites, the *white ants* of Brazil, are also active promoters in bringing about changes in the structure of the soil, and incidentally accelerating decomposition. The organic matter carried by these creatures into the ground, there to decompose, furnishes organic acids to promote further decay in the material close at hand, and by its downward percolation to attack the still firm rocks at greater depths. Indeed, these numerous channels, through affording easy access of air and surface waters with all their absorbed gases or alkaline salts, may serve indirectly a geological purpose scarcely inferior to that of the joints in massive rocks. (See further under soil modified by plant and animal life.)

The mechanical agency which has already been referred to as instrumental in bringing about a certain amount of decomposition in silicate minerals, is greatly augmented when such trituration takes place in connection with organic matter. J. Y. Buchanan has shown,[1] that the mud of sea-bottoms is being continually passed and repassed through the alimentary canals of marine animals, and that in so doing the mineral matter not merely undergoes a slight amount of comminution and consequent decomposition, but a chemical reduction takes place whereby existing sulphates are converted into sulphides. Such sulphides and the metallic constituents of the silicates and other compounds, particularly those of iron and manganese, would on exposure to sea-water become converted into oxides. It is through such agencies that he would account for the presence of sulphur in marine muds, and the variations in color, from shades of red or brown to blue and gray, in the former the iron occurring as oxides, while in the latter it exists as a sulphide. Of course either form may be more or less permanent according

[1] On the Occurrence of Sulphur in Marine Muds, Proc. Royal Soc. of Edinburgh, 1890-91.

as the mud may be devoid of animal life, or protected from oxidizing influences. These reactions, being subaqueous, are somewhat beyond the scope of the present work, but are nevertheless not without interest in this connection.

One' of the most conspicuous results of rock-weathering through the agencies of water and organic acids, as above enumerated, is manifested in the production of carbonates of lime and more rarely of magnesia, iron, and the alkalies. Thus in the decomposition of lime-bearing silicates, as the feldspars, pyroxenes, and amphiboles, the lime almost invariably separates out as calcite or aragonite, and often may be found filling cracks and crevices, as veins of "spar" in the very rock masses from which it was derived. The celebrated verde di Genova and verde di Prato marbles are but secondary rocks derived by hydration from pre-existing pyroxenic masses and in which the lime and magnesia have separated out as carbonates forming the white veins by which the stone is traversed. The almost universality of carbonate formation incident to rock-weathering manifests itself in the ready effervescence of freshly decomposed material when treated with an acid. It is indeed difficult to find weathered rocks of any kind that will not show at least traces of secondary carbonates, of which those of calcium are by far the more abundant.

It is further to be noted that the solvent and general chemical activity of water is often greatly augmented by the salts and acids it acquires through the decomposition of various minerals with which it comes in contact. Thus through the decomposition of iron pyrites there may be formed free sulphuric acid, or through the decomposition of a feldspar, carbonates of the alkalies, any of which, when in solution, are more energetic factors in promoting decomposition than water alone. Hence under certain conditions the process of decomposition once set in operation augments itself, and goes on with increasing vigor until such a depth is reached that the percolating solutions become neutralized and further action, aside from hydration, practically ceases.

THE WEATHERING OF ROCKS (Continued)

II. CONSIDERATION OF SPECIAL CASES

Let us now enter into a consideration of the composition of a few prominent rock types, and note the changes they have undergone in this process of weathering, assuming, as we must for the time being, that they have been all subjected to essentially the same conditions. Inasmuch, as has been noted already, there are divers types of rocks, differing not merely in chemical and mineral composition, but in structure as well, it is an easy assumption that the results of prolonged weathering may be widely divergent. Yet, as will become apparent, the ultimate products from all but the purely quartzose rocks, present striking similarities.

In the tables following are given the results of chemical and mechanical analyses of rocks of various kinds and in varying stages of degeneration. We will begin with a consideration of the granitic rocks of the District of Columbia.[1]

The rock (see Pl. 14) in its fresh condition is a strongly foliated gray micaceous granite showing to the unaided eye a finely granular aggregate of quartz and feldspars arranged in imperfect lenticular masses from 2 to 5 mm. in diameter, about and through which are distributed abundant folia of black mica. In the thin section the structure is seen to be cataclastic. Quartz and black mica are the most prominent constituents, though there are abundant feldspars of both potash and soda-lime varieties, which, owing to their limpidity, can by the unaided eye scarcely be distinguished from the quartz. The potash feldspar has in part a microcline structure. Aside from these minerals, a primary epidote, in small granules and at times quite perfectly outlined crystals, is a strikingly abundant constituent. Small apatites, a few flakes of white mica (sericite),

[1] Disintegration of the Granitic Rocks of the District of Columbia, Bull. Geol. Soc. of America, Vol. VI, 1895, pp. 321, 332.

and widely scattering black tourmalines and iron ores complete the list of recognizable minerals.

The outcrops from which the samples for the analyses to which attention is first called were selected are shown in the plate. At the very bottom, the rock is hard, fresh, and compact, without trace of the decomposition products other than as indicated by minute infiltrations of calcite from above. Just above the level of the small creek which flows at the foot of the bluff, at the point indicated by the first series of right-and-left joints near the centre of the view, the character of the rock changes quite suddenly, becoming brown and friable, though still retaining its form and easily recognizable granitic appearance. A few feet above a third zone begins, in which the rock is converted into sand and gravel and which becomes more and more soil-like to the top of the bank, where it becomes admixed with organic matter from the growing plants. The amount of organic matter is quite small, however, and in making the analyses care was taken to remove such as was recognizable in the form of rootlets, leaves, and twigs.

Bulk analyses of these three types, (I) fresh gray granite, (II) brown but still moderately firm and intact rock, and (III) the residual sand, yielded the results given in the columns correspondingly numbered below: —

Constituents	I	II	III
Ignition	1.22 %	3.27 %	4.70 %
Silica (SiO_2)	69.33	66.82	65.69
Titanium (TiO_2)	not det.	not det.	0.31
Alumina (Al_2O_3)	14.33	15.62	15.23
Iron protoxide (FeO)	3.60[1]	1.69	....
Iron sesquioxide (Fe_2O_3)	....	1.88	4.30
Lime (CaO)	3.21	3.13	2.63
Magnesia (MgO)	2.44	2.76	2.64
Soda (Na_2O)	2.70	2.58	2.12
Potash (K_2O)	2.67	2.04	2.00
Phosphoric acid (P_2O_5)	0.10	not det.	0.06
	99.60 %	99.79 %	99.77 %

In glancing over these figures it is at once apparent that there is a surprisingly small difference in ultimate composition

[1] 4.00 % when calculated as Fe_2O_3.

between the sound rock and the residual sand, the more marked differences being a slightly smaller amount of silica, more alumina, and slightly diminished amounts of lime, magnesia, potash, and soda, with a considerable increase in the amount of water. The ferrous salts have moreover been converted into ferric forms. It does not necessarily follow, however, that no more actual gain or loss of material or change in manner of combination than is here indicated may not have taken place, and at the very outset it may be well to enter into a discussion of the manner in which the results of such analyses are to be considered.

We must first of all remember that any indicated loss or gain of a constituent may be only apparent, and that the true relative proportions can be learned only by calculating results of analyses of both fresh and decomposed materials on a common basis. Thus the first glance at analysis III, as given, might lead one to surmise that the decomposed rock had actually lost only some 3.3 % of silica. This, however, is not strictly the case, since this analysis shows 4.7 % volatile constituents against 1.22 % in analysis I of the fresh material. Could we assume that this difference of 3.48 % was due wholly to a uniform absorption of moisture, as by a clay, the problem would resolve itself into simply recalculating all analyses upon a water-free basis.

The results obtained thus are not quite satisfactory, however, and it is thought a more correct view of the changes taking place may be obtained by assuming for one of the constituents a fairly constant value and using this as a basis for comparison.

Of all the essential constituents occurring in appreciable quantities in siliceous crystalline rocks the alumina and the iron oxides are the most refractory and the least liable to be removed by a leaching process, although they may undergo manifold changes in mode of combination. Although not absolutely correct, therefore, we will for our present purposes assume the one or the other of these (in this case the iron as Fe_2O_3) as a constant factor, and in order to show the proportional or actual amount of loss of any constituent will recalculate the analyses upon this basis, a proceeding for which, so far as alumina is concerned, we have already good authority.[1] This method will be adopted, however, only with the siliceous crystalline rocks,

[1] G. Roth, Allegemeine u. Chemische Geologie, 3d ed.

in which, for reasons noted later, the process of decomposition, we have reason to suppose, is more complex than in calcareous and magnesian rocks poor or lacking in the alkalies. The entire discussion is one beset with great difficulties, since we lack definite knowledge as to the exact processes which have been going on and need constantly to guard against assumptions too hastily drawn or based upon insufficient data. Indeed, any assumption based upon the results of chemical analyses alone is likely to lead to grave error.

If, then, we consider the iron in the form of Fe_2O_3 as a constant factor, we may, by proper calculation, obtain the results given in column (IV) below, which represent the proportional gain and loss of the various constituents of the rock in passing from the condition indicated in column (I) above, to that indicated in column (III). Such a comparison is instructive as showing not merely the relative loss and gain, but also the total loss of material, in this case 13.47 %, accompanied by a gain of 2.16%, in volatile matter.

DISINTEGRATED AND DECOMPOSED GRANITE, DISTRICT OF COLUMBIA, SHOWING PROPORTIONAL LOSS OF CONSTITUENTS

CONSTITUENTS	IV	V	VI
	PERCENTAGE LOSS FOR ENTIRE ROCK	PERCENTAGE OF EACH CONSTITUENT SAVED	PERCENTAGE OF EACH CONSTITUENT LOST
Silica (SiO_2)	10.60 %	85.11 %	14.89 %
Alumina (Al_2O_3)	0.46	96.77	3.23
Iron sesquioxide (Fe_2O_3)	} 0.00	100.00	0.00
Iron protoxide (FeO)			
Lime (CaO)	0.81	74.79	25.21
Magnesia (MgO)	0.36	98.51	1.49
Soda (Na_2O)	0.77	71.38	28.62
Potash (K_2O)	0.85	68.02	31.98
Phosphoric anhydride (P_2O_5)	0.04	60.00	40.00
Ignition	2.16 [1]	100.00	0.00
Total loss	13.47 %	....	....

Such results are still far from satisfactory, and it is believed the tables will be more useful and instructive can we show the

[1] Gain.

percentage loss and gain of each constituent as compared with the same constituent in the original rock. This can also readily be accomplished by a process the formula for which is given below,[1] and by which are obtained the results given in columns V and VI.

From a perusal of these figures, it appears that the residual sand retains 85.11% of the original silica; 96.77% of the alumina; all the ferric oxide; 74.79% of its lime; 98.51% of its magnesia, together with 71.38% of its soda and 68.02 of the potash, while there has been an actual gain, as was to be expected, in volatile matter.

Let us not, however, too hastily assume that we have exhausted the subject.

We must remember, further, that while an analysis shows the actual composition of a rock so far as the various elements are concerned, it quite fails to show the manner in which those elements are combined. While the ultimate composition of the fresh and decomposed samples may be closely similar, it is possible, indeed probable, that in some cases at least the manner of combination of these elements is quite different. This is well illustrated in the case of the figures showing the percentages of alumina in analyses I and III and which differ only nine-tenths of one per cent in total amount; yet in the first the alumina exists mainly in the form of anhydrous silicates of alumina, potash, iron, and magnesia (as in the feldspars and mica), while in the last a very considerable proportion, or indeed all in extreme cases of weathering, may exist as a hydrous silicate of alumina only (kaolin). It is in instances of this kind that the microscope may render efficient service, and much may be learned by means of such mechanical analyses as can be made by sifting and washing. Such separations made on this disintegrated rock showed it to consist of particles as given in the following table, the 4.25% silt being obtained by washing the

---

[1] The formula employed in these calculations is as follows: $\dfrac{A}{B \times C} = x$: and $100 - x = y$, in which $A =$ the percentage of any constituent in the residual material; $B =$ the percentage of the same constituent in the fresh rock, and $C =$ the quotient obtained by dividing the percentage amount of alumina (or iron sesquioxide, whichever is taken as a constant factor) of the residual material by that in the fresh rock, the final quotient being multiplied by 100. x then equals the percentage of the original constituent saved, in the residue, and y the percentage of the same constituent lost.

10.75 % of material which passed through fine bolting-cloth of 120 meshes to the lineal inch, and which represents the impalpable mud remaining in suspension while the 6.5 % of fine sand sank quickly to the bottom of the beaker in which the washing was made. The residual sand yielded then: —

Silt	4.25 %	Largest grains	0.1 mm. in diameter
Very fine sand	6.50	"	" 0.18	" "
Fine sand	11.25	"	" 0.25	" "
Medium sand	3.80	"	" 0.65	" "
Sand ⎫	11.00	"	" 1.00	" "
Sand ⎭	23.50	"	" 1.50	" "
Coarse sand	20.50	"	" 2.00	" "
Gravel	10.20	"	" 8.00	" "
Total	100.00 %			

The coarser of these particles, like the gravel and coarse sand, are of a compound nature, being aggregates of quartz and feldspar, with small amounts of mica and other minerals. In the finer material, on the other hand, each particle represents but a single mineral, the process of disaggregation having quite freed it from its associates, excepting, of course, the microscopic inclusions which could be liberated only by a complete disintegration of the host itself. These particles, as seen under the microscope, are all sharply angular, and in many cases surprisingly fresh, though the analyses, as given above, had suggested only a slight change in chemical composition. The mica shows the greatest amount of alteration, the change consisting mainly in an oxidation of its ferruginous constituent, whereby the folia becomes stained and reduced to yellowish brown shreds. The feldspars are, in some cases, opaque through kaolinization, but in others are still fresh and unchanged even in the smallest particles. The finest silt, when treated with a diluted acid to remove the iron stains, shows the remaining granules of quartz, feldspar, and epidote beautifully fresh, and with sharp, angular borders, the mica being, however, almost completely decolorized.

An analysis of the silt, which was found to constitute 4.25% of the entire mass of disintegrated material, as noted above, is given below, and also a partial separation and analysis of the 39.7 % soluble, and 60.3 % insoluble portions.[1]

[1] In all analyses made by or under the direction of the author, the matter tabulated as soluble is that extracted by boiling for three hours in hydrochloric

ANALYSES OF SILT FROM DISINTEGRATED GRANITE

CONSTITUENTS	I BULK ANALYSIS OF SILT	II ANALYSIS OF SOLUBLE PORTION (39.7%) SILT	III ANALYSIS OF INSOLUBLE PORTION (60.3%) SILT
Ignition	8.12%	8.12%	0.97%
Silica (SiO_2)	49.39	In HCl 1.123 / In Na_2CO_3 11.147	37.30
Alumina (Al_2O_3)	23.84	9.21	13.40
Iron sesquioxide (Fe_2O_3)	3.69	4.47	0.82
Lime (CaO)	4.41		2.90
Magnesia (MgO)	4.00	Not det.	Trace
Soda (Na_2O)	3.36		2.75
Potash (K_2O)	2.49		1.07
	99.90%	34.07	59.21
		93.28%	

From these analyses it would appear that of the 17 grammes of silt, representing 4% of the total disintegrated material, only 39.7% is soluble; and, further, that a very considerable proportion of the insoluble residue, as indicated by the high percentages of alkalies and lime, still consist of unaltered soda-lime and potash feldspars, the iron and magnesia alone having been largely removed.

These results are not quite what one would be led to expect from a perusal of the literature bearing upon the subject of rock decomposition. As long since noted by J. G. Forchhammer, G. Bischof, T. Sterry Hunt, and others, the ordinary processes of decay in siliceous rocks containing ferruginous protoxides and alkalies consists in the higher oxidation and separation of the protoxides in the form of hydrous sesquioxides and a general hydration of the alkaline silicates, accompanied by the formation of alkaline carbonates, which, being readily soluble, are taken away nearly as fast as formed. More or less silica is also removed, according to the amount of carbonic acid present, a portion of the alkalies forming soluble

---

acid of one-half normal strength, to which is added the silica set free in a gelatinous form by the acid and subsequently extracted by sodium carbonate solution. All analyses made on material first dried at 100° C.

alkaline silicates when the supply of the acid is insufficient to take them all up in the form of carbonates. The apparent anomaly here shown is partially explained by examination of the various separations with the microscope. Thus the low percentage of silica is found to be in large part due to the fact that the residual quartz granules are, in many cases, too large to pass the 120-mesh sieve, or, if passing, have been largely separated in the process of washing. Further, it is found that the sifting has served to concentrate the small epidotes in the fine sand, and a portion of them have even come over with the silt. The presence of this epidote also explains in part the high percentage of lime shown, since the mineral itself carries some 20 to 24 % of this material. The large percentages of magnesia, soda, and potash cannot, however, be thus accounted for, and we are led to infer that either these elements are there combined in minute amorphous zeolitic compounds, unrecognizable as such under the microscope, or, as seems more probable, the feldspathic constituents, to which the alkalies are to be originally referred, have undergone a mechanical splitting up rather than a chemical decomposition. This view is, to a certain extent, borne out by microscopic studies, but it is difficult to measure by the eye the relative abundance of these constituents with sufficient accuracy to enable one to form any satisfactory conclusion. The magnesia must come from the shreds of mica, many of which, from their small size and almost flocculent nature when decomposed, would naturally be found in the silt obtained as stated.

It is to be noted that the magnesia, together with the iron, exists almost wholly in a soluble form.

It is evident at once that we have had to do here with but the preliminary stages of granitic weathering, that the process is more one of disintegration than decomposition, and it will be well to consider now a case in which the decomposition has gone on to the condition of a residual clay, as found in many of the Southern states. For this purpose a biotite gneiss or gneissoid granite found near North Garden, in Albemarle County, Virginia, is selected. The rock is a coarse gray feldspar-rich variety with abundant folia of black mica. Under the microscope it shows the presence of both potash and soda-lime feldspars, a sprinkling of apatite and iron ores, sporadic occurrences of an undetermined zeolite, and an extraordinary number of minute zircons which are mostly enclosed in the feldspars. There are also present occasional small garnets and aggregates of decomposition products the exact nature of which was not made out. The residual soil resulting from the decomposition of this rock is highly plastic, of a deep red-brown color, and has a distinct gritty feeling in the hand, owing to the presence of quartz and undecomposed silicate minerals. In columns I and III below are given the results of analyses of fresh rock and residual soil, and in II, IV, and V the analyses of the soluble and insoluble portions. In columns VI, VII, and VIII are given the calculated percentage amounts of the various constituents saved and lost, as before.

The particular features to which attention need here be called, are (1) that 30.47 % of the fresh rock and 69.18 % of the decomposed are soluble in hydrochloric acid and sodium carbonate solutions, and that more than half the potash and nearly the same proportion of the soda in the fresh rock is found in the acid extract. (2) That the insoluble portion of the residuary material is mainly in the form of free quartz. (3) That 44.67 % of the original matter has been leached away, and that (4) of the original silica 52.45 % is lost, while 85.61 % of the iron and all the alumina remain. All the lime has disappeared, 83.52 % of the potash, 95.03 % of the soda, and 74.70 % of the magnesia. The total amount of water, as indicated by the ignition, has increased very greatly, as was to be expected. The small original amount of phosphoric acid prohibits our placing too much reliance upon the indicated gain in this constituent, since it may be due to errors in manipulation.

Passing from the acid group of granular crystalline rocks, we will consider next a closely allied form differing mainly in the absence of quartz as an essential constituent, and in the presence of elæolite, the elæolite syenites of the Fourche Mountain region of Arkansas. These are somewhat coarsely crystalline granitic-appearing rocks, in which an orthoclase feldspar in broadly tabular forms is the prevailing constituent, though always accompanied by nepheline, biotite, pyroxene, titanite, and apatite, while fluorite, analcite, and thomsonite, together with calcite, occur as secondary products. The rock weathers away to a coarse gray gravel which ultimately becomes a clay, from which, by washing, may be obtained kaolin in a fair degree of purity.

ANALYSES OF FRESH AND DECOMPOSED GNEISS, ALBEMARLE COUNTY, VIRGINIA

CONSTITUENTS	FRESH GNEISS		DECOMPOSED GNEISS			CALCULATED AMOUNTS SAVED AND LOST		
	I Bulk Analysis	II Analysis of Portion Soluble in HCl and Na$_2$CO$_3$	III Bulk Analysis	IV Analysis of Portion Soluble in HCl and Na$_2$CO$_3$	V Analysis of Insoluble Portion	VI Loss	VII Percentage of Each Constituent saved	VIII Percentage of Each Constituent lost
Silica (SiO$_2$) { in HCl / in Na$_2$CO$_3$ }	60.69 %	{ 1.43 / 8.66 }	46.31 %	{ 0.55 / 17.14 }	28.90 %	31.90 %	47.55 %	52.45 %
Alumina (Al$_2$O$_3$)	16.89	13.54	26.55	24.80	1.55	0.00	100.00	0.00
Iron sesquioxide (Fe$_2$O$_3$)	0.06		12.18	11.80	0.22	1.30	85.66	14.35
Lime (CaO)	4.44	1.64	Trace	0.06	0.07	4.44	0.00	100.00
Magnesia (MgO)	1.06	0.89	0.40	0.37	0.04	0.80	25.30	74.70
Potash (K$_2$O)	4.25	2.40	1.10	0.75	0.31	3.55	16.48	83.52
Soda (Na$_2$O)	2.82	1.10	0.22	0.25	Trace	2.08	4.97	95.03
Phosphoric acid (P$_2$O$_5$)	0.25	0.23	0.47	....	Trace	0.00[1]	100.00	0.00[1]
Ignition	0.02	0.02	13.75	13.40	0.35	0.00[1]	100.00	0.00[1]
	100.08 %	30.47 %	99.98 %	69.18 / 100.62 %	31.44	44.07 %		

[1] Gain.

The following analyses from the work of Dr. J. F. Williams[1] will serve to show the changes which have here taken place in the transformation from (I) fresh syenite through (II and III) intermediate stages of decomposition to (IV) a kaolin-like residue.

ANALYSES OF FRESH AND DECOMPOSED SYENITE, ARKANSAS

Constituents	I	II	III	IV
Silica (SiO_2)	59.70 %	58.50 %	50.05 %	40.27 %
Alumina (Al_2O_3)	18.85	25.71	20.71	38.57
Ferric oxide (Fe_2O_3)	4.85	3.74	4.87	1.36
Lime (CaO)	1.34	0.44	0.02	0.34
Magnesia (MgO)	0.08	Trace	0.21	0.25
Potash (K_2O)	5.97	1.96	1.91	0.23
Soda (Na_2O)	6.29	1.37	0.02	0.37
Ignition (H_2O)	1.88	5.85	8.68	13.01
	99.56 %	97.57 %	94.27 %	101.00 %

Recalculating the numbers given in columns I and IV upon the basis of 100, we may obtain by further calculation, as already described, the figures given in columns V and VI and VII below, which represent the proportional loss of each constituent, as before.

CALCULATED LOSS OF MATERIAL

Constituents	V Percentage Loss for Entire Rock	VI Percentage of Each Constituent saved	VII Percentage of Each Constituent lost
Silica (SiO_2)	37.28 % loss	37.82 %	62.18 %
Alumina (Al_2O_3)	0.00 "	100.00	0.00
Ferric oxide (Fe_2O_3)	4.19 "	13.83	86.17
Lime (CaO)	1.19 "	12.10	87.90
Magnesia (MgO)	0.57 "	17.90	82.10
Potash (K_2O)	5.90 "	18.15	81.85
Soda (Na_2O)	6.15 "	2.89	97.11
Water (H_2O)	0.00 "	100.00	0.00

Total loss of original material, 56.28 %.

[1] Ann. Rep., Vol. II, 1890, Arkansas Geol. Survey.

WEATHERING OF SYENITE AND PHONOLITE 217

Here, as with the granitic rocks, it will be noted we have a gradual increase in the percentage of water as the decomposition advances, and a decrease in the amount of silica even more pronounced. This last, as may be readily imagined, is due to the absence of free quartz in the Fourche Mountain rocks.

The phonolites of Marienfels, near Assig, in Bohemia, have been described by Lemberg [1] as weathering into a bright-colored, porous, friable mass, the composition of which, as compared with the fresh rock, is shown below. Each column, it should be stated, represents an average of three analyses, I being the fresh and II the weathered material, while in III, IV, and V are given the percentage calculations of gain and loss, as before.

ANALYSES OF FRESH AND DECOMPOSED PHONOLITE, BOHEMIA

Constituents	I Fresh Phonolite	II Decomposed Phonolite	III Loss of Constituents	IV Percentage of Each Constituent Saved	V Percentage of Each Constituent Lost
Silica (SiO_2)	55.07 %	55.72 %	4.83 %	91.46 %	8.54 %
Alumina (Al_2O_3)	20.64	22.10	0.37	98.40	1.60
Ferric oxide (Fe_2O_3)	3.14	3.44	0.00	0.00	100.00
Lime (CaO)	1.40	1.28	0.25	83.66	16.34
Magnesia (MgO)	0.42	0.44	0.02	95.65	4.35
Potash (K_2O)	5.56	6.26	0.00 [2]	100.00	0.00
Soda (Na_2O)	7.12	2.65	4.79	34.01	65.99
Ignition	4.33	7.79	0.00 [2]	100.00	0.00
	98.28 %	99.77 %	10.26 %	....	....

This phonolite, it should be remarked, consisted essentially of sanidin feldspars and a soda zeolite, together with accessory augite, black mica, magnetic and titanic iron, and possibly hauyne. The zeolite is assumed to have originated from the alteration of the nepheline. The process of decomposition would seem to consist, then, in the breaking down of this zeolite, and the conversion of the rock into an earthy mass, with little other

[1] Zeit. der Deutschen Geol. Gesellschaft, Vol. 35, 1883, p. 559.
[2] Gain. The calculations for potash in column IV gives: 107.79 % and for ignition 164.77 %.

change, so far as ultimate composition is concerned, than a loss of a considerable proportion of its soda, and an assumption of nearly 3.5 % of water. The decomposed rock yielded 55.44 % of material insoluble in hydrochloric acid, and with essentially the composition of sanidin, showing that this mineral underwent only a physical disintegration, the decomposition proper being limited to the other constituents.[1]

Turning to still more basic rocks, we will next consider a disintegrated diabase occurring in the form of a large dike extending from Granite Street in Somerville, Massachusetts, to Spot Pond in Stoneham, and beyond.[2] The rock at the point selected for study (Medford) is a coarsely granular admixture of lath-shaped feldspar, black mica, augite, and brown basaltic hornblende, with the usual sprinkling of apatite, magnetite, and ilmenite. Secondary uralite, chlorite, biotite, leucoxene, kaolin, calcite, pyrite, and quartz are common.[3]

The rock has undergone extensive disintegration, giving rise to loose sand and gravel of a deep brown color, in which lie rounded boulders of all sizes of the still undecomposed material. These boulders, as is usually the case, show a more or less concentric structure, from without inward, until a solid core of unaltered diabase is met with. (See Pl. 17, and Fig. 2, Pl. 20.)

A mechanical separation of the disintegrated material yielded results as below : —

1.	Coarse gravel	above	2	mm. in diameter	42.300 %
2.	Fine gravel	"	2-1	mm. in diameter	20.355
3.	Coarse sand	"	1-5	mm. in diameter	12.723
4.	Medium sand	"	.5-.25	mm. in diameter	9.507
5.	Fine sand	"	.25-.1	mm. in diameter	4.907
6.	Very fine sand	"	.1-.05	mm. in diameter	4.181
7.	Silt	"	.05-.01	mm. in diameter	1.128
8.	Fine silt	"	.01-.005	mm. in diameter	0.370
9.	Clay	"	.005-.0001	mm. in diameter	1.070
10.	Loss at 110° C.				0.660
11.	Loss on ignition				1.730
					99.691 %

[1] In calculating these analyses, it was found that the loss of alumina had exceeded that of iron oxide, necessitating the assumption of the last-named as a constant for comparison. The apparent gain in potash is presumably due to errors in analysis, since, as will be noted, the analysis of the fresh material, given in column I, foots up only 98.28 %.

[2] See Disintegration and Decomposition of Diabase at Medford, Massachusetts, by G. P. Merrill, Bull. Geol. Soc. of America, Vol. VII, 1896, pp. 349-362.

[3] On the Petrographic Characters of a Dike of Diabase in the Boston Basin, by W. H. Hobbs, Bull. Mus. Comp. Zoölogy, Vol. XVI, No. 1, 1888.

PLATE 17. — Weathered diabase dike in Medford, Massachusetts.

Of the above, the first three sizes could be easily recognized by the unaided eyes, as composed of particles of a compound nature. In number 4 the separation had gone a trifle farther, though even here inspection with a pocket lens revealed the compound nature of many of the granules, somewhat obscured by the prevailing discoloration from the oxides of iron. It forms a gray-brown sand composed of feldspathic particles, dirty brown augites, and lustrous scales of brown mica. Numbers 5 and 6 seemed composed almost wholly of beautifully lustrous, dark mahogany-brown mica scales, while 7 would pass for a finely micaceous umber. Numbers 8 and 9 were uniformly ochreous, the last being several shades lighter than number 8, and without appreciable grit.

The chemical nature of the fresh and decomposed rock is shown in the accompanying table, the results being in nearly every case averages obtained from two or more analyses. The "fresh" material, obtained from the interior of one of the boulders, is firm in texture, has a bright clean fracture, and shows to the unaided eye no signs of decomposition. When pulverized and treated with acid, however, it effervesces distinctly, indicating the presence of free carbonates, which are also observable as secondary calcite when thin sections are examined under the microscope. Some of this calcite is evidently a deposit from infiltrated waters, being derived from the surrounding decomposed material, while a portion results from the decomposition of the silicate minerals in place. Aside from a slight kaolinization of the feldspars and development of chlorite from the ferruginous silicates, there are no other observable signs of decomposition, though the presence of a soda-bearing zeolite is indicated by cubes of chloride of sodium, which separate out when an uncovered slide is treated with a drop of hydrochloric acid.

A glance at this table is sufficient to show that the disintegration is accompanied by decomposition and a leaching action which has resulted in the removal of a portion of the more soluble constituents. The fact that the fresh rock yields the larger percentages of its constituents to the solvent action of acid and alkaline solutions is readily explained on this ground, though it may be doubted if the full significance of the fact, so far as it relates to siliceous crystallines, is as yet appreciated. It will be observed that 36.23 % of the fresh rock and 82.28 % of the decomposed is thus extracted.

ANALYSES OF FRESH AND DISINTEGRATED DIABASE FROM MEDFORD

CONSTITUENTS	Fresh Diabase		Disintegrated Diabase		Silt from Disintegrated Diabase, Nos. VII, VIII, and IX of Table, on P. 218		
	I	II	III	IV	V	VI	VII
	Bulk Analysis	Analysis of Portion Soluble in HCl and Na₂CO₃	Bulk Analysis	Analysis of Portion Soluble in HCl and Na₂CO₃	Analysis of 77.87 per cent Soluble in HCl and Na₂CO₃	Analysis of 22.13 per cent Insoluble in HCl and Na₂CO₃	Total
	%	%	%	%	%	%	%
Silica (SiO₂) in HCl / in Na₂CO₃	47.28	{ 1.19 / 0.66 }	}44.44	{ 0.85 / 8.65 }	0.47 / 22.03	}13.51	36.61
Alumina (Al₂O₃)	20.22	4.74	23.10	4.86	21.98		
Ferric oxide (Fe₂O₃)	3.66	}10.91	12.70	10.00	12.83	5.88	40.68
Ferrous oxide (FeO)	8.80						
Lime (CaO)	7.09	3.09	6.03	1.50	3.32	0.12	3.44
Magnesia (MgO)	3.17	2.20	2.82	1.84	3.23	0.79	4.02
Manganese oxide MnO	0.77	Not det.	0.52	Not det.	Not det.	Not det.	Not det.
Potash (K₂O)	2.16	1.21	1.75	0.68	1.30	0.52	1.82
Soda (Na₂O)	3.04	0.50	3.93	0.17	0.90	1.24	2.14
Phosphoric acid (P₂O₅)	0.68	Not det.	0.70	Not det.	Not det.	Not det.	....
Ignition	2.73	2.73	3.73	3.73	10.86	0.11	10.97
	100.59	36.23	99.81	32.28	77.52	22.17	99.68

Of the material classed as silt in columns V, VI, and VII, or as silt and clay, on p. 218, and which constitutes only some 3.17 % of the entire residual débris, 77.87 % is soluble in dilute hydrochloric and sodium carbonate solutions. The insoluble portion, constituting 22.13 % of the silt, consists of unaltered feldspar and iron, lime and magnesian silicates, which are easily recognized under the microscope, in the form of minute, sharply angular particles. Recalculating, as before, the matter in columns I and II on the basis of 100 and considering the alumina as a constant factor, we get the results given in columns VIII to XII inclusive, representing, so far as it can be obtained by this

WEATHERING OF DIABASE 221

method, the actual percentage loss of materials attending the
breaking down.

CALCULATED LOSS OF MATERIAL

CONSTITUENTS	VIII	IX	X	XI	XII
	RECALCULATED ON BASIS OF 100		Percentage Loss for Entire Rock	Percentage of Each Constituent saved	Percentage of Each Constituent lost
	Fresh Diabase	Decomposed Diabase			
Silica (SiO$_2$)	47.01%	44.51%	8.48	81.97%	18.03%
Alumina (Al$_2$O$_3$) . . .	20.11	23.24	0.00	100.00	0.00
Ferric oxide (Fe$_2$O$_3$) . .	3.63	} 12.71	2.42	81.90	18.10
Ferrous oxide (FeO) . .	8.83				
Lime (CaO)	7.06	6.04	1.83	74.11	25.89
Magnesia (MgO) . . .	3.15	2.85	0.68	78.30	21.70
Manganese (MnO) . . .	0.77	0.52	0.32	58.43	41.57
Potash (K$_2$O)	2.14	1.75	0.62	70.85	29.15
Soda (Na$_2$O)	3.91	3.94	0.50	87.17	12.83
Phosphoric acid (P$_2$O$_5$) .	0.68	0.70	0.08	88.61	11.30
Ignition	2.71	3.74	0.00	100.00	0.00
	100.00%	100.00%	14.93%	....	....

From the figures in column X it appears that there has
been a loss of some 14.93% of all constituents. The increase
in water, as indicated by the ignition, is a natural consequence
of hydration and the presence of a small amount of organic
matter. This increase, it should be stated, is greater than may
be at first apparent, for the reason that the fresh rock contains
a considerable amount of secondary calcite, which is quite lacking in the residual sand. A large part of the ignition in columns I and VIII is therefore to be accredited to carbonic acid,
and not to water of hydration.

From columns XI and XII it appears that of all the essential
constituents, the lime and potash salts have suffered the most,
though the iron oxides have been carried away to the amount
of 18.10%. Magnesia has also proven very susceptible to the
solvent action, disappearing to the amount of 21.70%; and
lastly, silica, to the amount of 18.03%. The small original
amounts of manganese and phosphoric acid render the results

obtained by these calculations of doubtful value, since it is possible they may be due to errors of analysis.

In this case, as in that of the granite from the District of Columbia, we have to do with only the earlier stages of degeneration, with conditions which are as much in the nature of mechanical disintegration as of chemical decomposition. As before, then, it will be instructive to consider cases in which, in rocks of similar nature, the decomposition has proceeded much farther. For this purpose we will select a diabase from Spanish Guiana,[1] and basalts from Bohemia and the Haute Loire as described by Ebelmen;[2] in each instance the actual analysis being recalculated to the basis of 100.

ANALYSES OF FRESH AND DECOMPOSED DIABASE FROM SPANISH GUIANA, VENEZUELA

Constituents	I Fresh	II Decomposed	III Percentage Loss for Entire Rock	IV Percentage of Each Constituent Saved	V Percentage of Each Constituent Lost
Silica (SiO_2)	49.35 %	43.38 %	20.02 %	57.60 %	42.40 %
Alumina (Al_2O_3)	15.30	18.36	3.27	78.62	21.38
Ferric iron (Fe_2O_3)	....	20.39	0.00	100.00	0.00
Ferrous iron (FeO)	12.28	....	....	....	....
Lime (CaO)	9.60	2.37	8.05	16.17	83.23
Magnesia (MgO)	7.38	3.45	5.12	30.63	61.37
Potash (K_2O)	0.85	0.59	0.33	54.12	45.88
Soda (Na_2O)	1.98	0.14	1.82	4.63	95.37
Ignition	3.25	11.34	0.00	0.00[3]	0.00
	100.00 %	100.00 %	39.51 %	....	....

In the case of the diabase, it appears, from a comparison of the figures in columns I and III, that the total loss of material equals 39.51 %, there being the usual gain in volatile matter.

[1] Quar. Jour. Geol. Soc. of London, Vol. XXXV, 1879, p. 586.
[2] Ann. des Mines, Vol. VII, 1845.
[3] Gain.

WEATHERING OF DIABASE AND BASALT

ANALYSES OF FRESH AND DECOMPOSED BASALT FROM KAMMAR BULL, BOHEMIA

Constituents	I Fresh Rock	II Partially decomposed Rock	III Still More decomposed Rock	IV Percentage Loss for Entire Rock	V Percentage of Each Constituent Saved	VI Percentage of Each Constituent Lost
Silica (SiO_2)	43.61%	43.00%	43.27%	15.04 loss	67.01%	32.99%
Alumina (Al_2O_3)	12.26	13.00	18.13	0.00 "	100.00	0.00
Ferric iron (Fe_2O_3)	3.51	5.40 ⎫	11.70	9.10 "	49.83	50.17
Ferrous iron (FeO)	12.16	8.30 ⎭				
Lime (CaO)	11.37	12.10	2.00	9.60 "	54.47	84.53
Magnesia (MgO)	9.14	7.30	3.40	6.83 "	25.90	74.10
Soda (Na_2O)	2.72 ⎫	0.50	0.20	3.39 "	38.31	61.09
Potash (K_2O)	0.81 ⎭					
Water (H_2O)	4.42	9.50	20.70	0.00	100.00	....
	100.00 %	100.00 %	100.00 %	43.96 loss	....	....

ANALYSES OF FRESH AND DECOMPOSED BASALT FROM CROUZET, IN THE HAUTE LOIRE, FRANCE

Constituents	I Fresh Rock	II Decomposed Rock	III Percentage Loss for Entire Rock	IV Percentage of Each Constituent Saved	V Percentage of Each Constituent Lost
Silica (SiO_2)	48.29 %	37.09 %	30.34 % loss	34.44 %	65.56 %
Alumina (Al_2O_3)	13.25	30.75	0.00 "	100.00	0.00
Ferric iron (Fe_2O_3)	0.00	4.31 ⎫	16.04 "	11.16	88.84
Ferrous iron (FeO)	16.06	0.00 ⎭			
Lime (CaO)	7.33	8.97	3.46 "	52.76	47.24
Magnesia (MgO)	7.03	0.61	6.77 "	3.02	96.98
Potash (K_2O)	1.81	0.71	1.51 "	16.66	83.34
Soda (Na_2O)	2.71	1.01	1.40 "	25.59	74.41
Ignition	4.92	16.55	0.00	100.00	0.00
	100.00 %	100.00 %	60.12 % loss	....	....

Of the individual constituents, 83.23 % of the original lime, 61.37 % of the magnesia, 45.88 % of the potash, 95.37 % of the soda, 42.40 % of the silica, and 21.38 % of the alumina have disappeared, the calculations being made on a Fe_2O_3 constant basis.

In the case of the Bohemian basalt, the decomposition commenced with the formation of boulders, which, when the process had not gone too far, still showed fresh, unchanged basalt interiorly, but became more and more altered toward their peripheries. The first stage of decomposition (column II), it will be noted, consists, aside from hydration, in a slight apparent loss of silica, a considerable oxidation of the iron magnesia minerals, accompanied by a slight loss of both constituents, and an almost complete loss of alkalies. In the second stage (column III) lime and magnesia are both lost in considerable amounts, the iron passing over wholly to the condition of sesquioxide, and there is a further slight diminution in the proportional amount of silica. It is evident that here the feldspars were the first of the constituents to yield to the decomposing forces, the augite and olivine proving most refractory. The total loss of material, it will be noted, amounts to 43.96 %, the lime, magnesia, alkalies, iron oxides, and silica disappearing in the order here mentioned.

In the case of the basalt from Crouzet, the analyses show a total of 60.12 % loss, or over one-half of the original material. This loss includes nearly two-thirds of the original silica, 88.84 % of the iron, and 96.38 % of the magnesia. The loss of both iron and magnesia in such proportionally large quantities is quite unusual, and indicates, so far as the iron is concerned, that the decomposition took place under conditions excluding a sufficient supply of oxygen to convert the same into the insoluble sesquioxide, or where subjected to the deoxidizing and solvent action of organic acids. The removal of the magnesia, which must have existed mainly in the mineral olivine, indicates that the decomposition has gone on even to the production of carbonate of magnesia and the separation of free silica and iron oxides.

An analysis by the present writer of a closely related rock, a diorite, and its residual soil, from North Garden, Albemarle County, Virginia, yielded the results given in columns I and II below. The rock here was fine-grained, of an almost coal-black color finely speckled with whitish flecks due to the presence of feldspars. The microscope showed it to be composed mainly of hornblende with interstitial soda-lime feldspars and scattering areas of titanic iron. The clay, or soil, to which it gave rise was deep brownish red in color and highly plastic, though distinctly gritty from the presence of undecomposed minerals. In columns III, IV, and V are given the loss and gain of the various constituents calculated on an alumina constant basis, as before.

ANALYSES OF FRESH AND DECOMPOSED DIORITE FROM ALBEMARLE COUNTY, VIRGINIA

Constituents	I	II	III	IV	V
	Fresh	Decomposed	Calculated Loss for Entire Rock	Per Cent of Each Constituent Saved	Per Cent of Each Constituent Lost
Silica (SiO_2)	46.75 %	42.44 %	17.43 % loss	62.69 %	37.31 %
Alumina (Al_2O_3)	17.61	25.51	0.00 "	100.00	0.00
Iron sesquioxide (Fe_2O_3)[1]	16.79	19.20	3.53 "	78.97	21.03
Lime (CaO)	9.46	0.37	0.20 "	2.70	97.30
Magnesia (MgO)	5.12	0.21	4.07 "	2.83	97.17
Potash (K_2O)	0.55	0.49	0.21 "	61.25	38.75
Soda (Na_2O)	2.56	0.56	2.17 "	15.13	84.87
Phosphoric acid (P_2O_5)	0.25	0.29	0.00	80.11	19.87
Ignition	0.02	10.92	0.00	100.00	0.00
	100.01 %	99.99 %	37.61 % loss	....	....

The ultra basic rocks, — peridotites and pyroxenites, — from the very nature of their composition, must yield on decomposition residues poor in the presence of alkalies and rich in iron or aluminum and magnesian compounds. Owing, further, to their poverty in alkali-bearing silicates, the process of decomposition must be less complex, consisting essentially in hydration, oxidation, and a production of iron, lime, and magnesian carbonates and a liberation of chalcedonic silica.

During the process these rocks as a rule become brownish, and, on the surface, often irregularly checked with a fine network of rifts which become filled with secondary calcite, magnesite, and chalcedony. If the original rock is an olivine-rich

[1] All iron calculated as Fe_2O_3.

peridotite, these clefts may become filled with the silicates of nickel, *noumæite* and *garnœrite*, which may be of sufficient abundance to form valuable ores. This, in brief, is the history of the nickel ores of Riddles, Oregon, and of New Caledonia, though the process is more properly a form of hydrometamorphism than weathering.

The deep green serpentines of Harford County, Maryland, weather slowly down into a gray-brown soil, which consists of 60.17% silica, 10.40% of iron oxides, 14.81% of alumina, and only 7.23% magnesia. The fresh rock, on the other hand, carries nearly 40% of magnesia, 8.50% iron and other metallic oxides, and less than one-half of one per cent of alumina.

Natural joint blocks occur in which the preliminary stages of weathering are manifested by a brown, ferruginous, though tough and hard, vesicular crust of from a millimetre to two or more centimetres' thickness, enclosing the slightly hydrated but otherwise unchanged material.

ANALYSES OF FRESH AND DECOMPOSED SOAPSTONE (ALTERED PYROXENITE)

Constituents	I Fresh Rock	II Residual Soil	III Percentage of Loss for Entire Rock	IV Percentage of Each Constituent Saved	V Percentage of Each Constituent Lost
Silica (SiO_2)	38.85%	38.82%	16.92%	56.42%	43.58%
Alumina (Al_2O_3)	12.77	22.61	0.00	100.00	0.00
Iron sesquioxide (Fe_2O_3) [1]	12.86	13.33	5.33	58.52	41.48
Lime (CaO)	6.12	6.13	2.66	55.55	44.45
Magnesia (MgO)	22.58	9.52	17.20	23.81	76.19
Potash (K_2O)	0.19	0.18	9.03	52.94	47.05
Soda (Na_2O)	0.11	0.20	0.00	100.00	0.00
Ignition	6.52	9.21	1.32	79.74	20.26
	100.00%	100.00%	52.46%	....	....

In columns I and II above are given (I) the composition of an altered pyroxenite (soapstone) from Albemarle County, Virginia, and (II) a residual soil derived from the same, the

[1] All iron calculated as Fe_2O_3.

latter being of a dull, ochreous, brown-red color, somewhat lumpy, but with no appreciable grit when rubbed between the thumb and fingers.

The fresh rock is of a blue-gray color, close texture, and consists, as shown by the microscope, of elongated crystals of colorless tremolite, with folia of talc and chlorite, and occasional opaque granules of chromic iron. The general petrologic features are those of an altered pyroxenite.

Recalculated as before, the analyses give the results shown in columns III, IV, and V.

Total loss of material 52.46%, including water of hydration. The most striking feature brought out is the fact that the magnesia has been carried away in greater proportional quantity than has the lime. A like result was noted by Ebelmen in his analyses of the decomposed basalts of Crouzet, which are given on p. 223.

ANALYSES OF FRESH AND DECOMPOSED SOAPSTONE, FAIRFAX COUNTY, VIRGINIA

Constituents	I Fresh Rock	II Residual Soil	III Percentage of Loss for Entire Rock	IV Percentage of Each Constituent Saved	V Percentage of Each Constituent Lost
Silica (SiO$_2$)	58.40%	54.84%	46.31%	20.70%	79.30%
Alumina (Al$_2$O$_3$) Iron oxides(FeO and Fe$_2$O$_3$)	} 7.44	33.75	0.00	100.00	0.00
Lime (CaO)	0.00	0.00	....	....	....
Magnesia (MgO)	29.19	4.36	28.23	3.29	96.71
Alkalies (K$_2$O and Na$_2$O)	0.00	0.00	....	....	....
Ignition (H$_2$O)	4.97	7.05	3.41	31.28	68.72
	100.00%	100.00%	77.95%		

A varietal form of this same rock occurring near Fostoria in Fairfax County, this state, is thoroughly decomposed throughout nearly the entire area to a depth of twenty or more feet. The fresh rock is composed mainly of a light greenish, almost white talc, with sporadic patches of chlorite some five or more millimetres in diameter, and scattering granules of iron ores. The decomposed material is dull brownish or gray, and when washed and submitted to microscopic examinations is found to consist almost wholly of brown and yellow-brown scales of talcose material, intermingled with an impalpable silt, composed so far as determinable of talcose and chloritic shreds. It is wholly without grit, and with a decided soapy or greasy feeling. Analyses of fresh and decomposed material, and calculations as already given, yielded results as shown in table on p. 227.

The principles involved in the decomposition of fragmental and crystalline stratified rocks are not so different from those we have been discussing as to call for detailed consideration. It is well to note, however, that the materials composing rocks of this type are themselves a product of these very disintegrating and decomposing agencies, but which have become consolidated into rock masses and now, once more in the infinite cycle of change, are undergoing a breaking up. It follows from the very nature of the case that such rocks, with the exception of the purely calcareous varieties, will undergo less chemical change than do those we have been discussing. Their feldspathic and easily decomposable silicate constituents long ago yielded to the decomposing processes, and were largely removed before consolidation took place. Thus, most sandstones are composed largely of quartzose sand, the least soluble and least changeable product, it may be, of many a previous disintegration! Hence, the processes involved in the degeneration of the sandstones, shales, and argillites are largely mechanical, with the exception of those which carry a feldspathic or calcareous cement. In these last-named, the cementing material is gradually leached away, and the rock becomes susceptible to the action of frost, or falls away to loose sand simply through loss of cohesion. Heusser and Claraz [1] described the itacolumites of Brazil as subject to this mechanical degeneration, the process being characterized by fissuration, succeeded by complete disintegration. Among siliceous sandstone it is the binding constituent that yields first, as is naturally to be expected, and as has been shown by the experiments conducted by R. Schutze.[2]

The rocks grouped under the name of argillites, though composed of detrital materials from pre-existing rocks, and of parti-

[1] Ann. des Mines, 5th, Vol. XVII, 1860.
[2] Ueber Verwitterungsvorgänge bei Krystallinischen u. Sedimentärgesteinen, Inaug. Dissertation der Friedrich-Alexanders Universität, Berlin, 1886.

cles reduced to an extreme degree of fineness, are, nevertheless, quite variable in composition, as already noted. As a rule, they are among the most indestructible of rocks, and on breaking down yield only clays which differ from the original argillites mainly in degree of hydration and condition of oxidation of the iron and other metallic constituents. Those argillites which carry appreciable quantities of still undecomposed silicates, particularly alkali-bearing varieties, are, of course, more susceptible, other things being equal, as texture, fissility, etc.

The deep blue-black argillites of Harford County, Maryland, as shown in the analyses given below, do contain very

ANALYSES OF FRESH AND DECOMPOSED ARGILLITE, HARFORD COUNTY, MARYLAND

Constituents	I Fresh Argillite	II Residual Clay	III Percentage of Loss for Entire Rock	IV Percentage of Each Constituent Saved	V Percentage of Each Constituent Lost
Silica [1] (SiO_2)	44.15 %	24.17 %	25.34 %	42.43 %	57.57 %
Alumina (Al_2O_3)	30.84	39.10	0.00	100.00	0.00
Iron oxide (FeO and Fe_2O_3)	14.87	17.61	1.23	91.22	8.78
Lime (CaO)	0.48	None	0.48	0.00	100.00
Magnesia (MgO)	0.27	0.25	0.08	71.84	28.16
Potash (K_2O)	4.36	1.24	3.39	22.04	77.95
Soda (Na_2O)	0.51	0.23	0.33	0.36	99.64
Ignition (C and H_2O)	4.49	16.62	0.00	287.37	None
	99.97 %	100.02 %	40.83 %	....	....

considerable quantities of these undecomposed silicates, and though extremely tough and enduring from a human standpoint, in time decompose in a very interesting manner. In the field these rocks are found standing nearly, if not quite, vertically, that is, with their evident cleavage vertical, and forming steep, high ridges flanked by valleys carved from the softer rocks on either hand. In the fresh cuts made during the work

[1] With traces of TiO_2. Manganese in traces, but not determined. No sulphur; hence no pyrite.

of stripping, to open new quarries, the sound rock is found overlain by a variable thickness of ferruginous residual clay. Joint blocks and splinters of the slate scattered through this clay, in all stages of decomposition leave no doubt as to its origin. Blocks, deep velvety black on the interior, are surrounded by a crust of ochreous brown-red decomposition product, the decay penetrating irregularly like the processes of oxidation into a piece of metal. The first physical indication of decay is shown by a softening of the slate, so that it may be readily scratched by the thumb nail, and an assumption of a soapy or greasy feeling, the entire mass finally passing over to the deep red-brown unctuous clay, sufficiently rich in iron to serve as a low-grade ochre, for paints. The incidental chemical changes are surprisingly large, as shown by the analyses given on p. 229, column I being an average of two analyses of the black, little altered material from the interior of one of these blocks, and II that of the residual clay. In III, IV, and V are given the calculated losses of constituents, as before.

This residual clay, when boiled with hydrochloric acid and sodium carbonate solutions, yielded up nearly 70% of its matter to these solvents, leaving a residue which, when examined under the microscope, shows only faint yellow-brown scale-like particles, rarely over a tenth of a millimetre in diameter, acting very faintly, if at all, on polarized light, and with borders often serrate, through corrosion, though this latter feature may be due, in part, to the action of the solvents used.

Among siliceous rocks poor in alkalies or iron-bearing silicates the degeneration is mainly disintegration, though a small amount of silica, existing in either crystalline or chalcedonic forms, is usually lost through solution. Thus the cherts of southwest Missouri break down into porous friable forms, sometimes passing into the condition of loose powder, or again retaining sufficient tenacity to be utilized for filter discs and tubes, as at Seneca, in Newton County.

Analyses of fresh and altered forms of this material, as given by Dr. E. O. Hovey,[1] show no differences that are of sufficient importance to warrant us in assuming any of them as the direct cause of disintegration. The change is evidently mainly physical, though it is more than probable that a certain amount of interstitial silica has been removed. It is, of course, possible

[1] Appendix A, Vol. VII, Missouri Geological Survey, 1894, pp. 727-739.

that here, as in other forms of decomposition, extensive solution may have taken place, leaving a residue which, so far as composition is concerned, gives no clew to the changes which have occurred. Dr. Penrose, however, describes[1] a process of chert decay, or more properly disintegration, as manifested in the Batesville region of Arkansas, in which the cause of the breaking down is more apparent. There are two stages in the process, as described: (1) A transition into a light, porous, opaque, buff-colored rock of the consistency of ordinary pressed brick, and (2) into an impalpable white or brown powder, locally known as a polishing powder. This second stage is not so conspicuous a feature as the first, since the finer materials thus formed are carried off by surface waters. The white residual powder often contains masses of the porous, semi-decomposed rock, the latter in turn encircling kernels of hard, unaltered chert. Throughout this region, the cherts (of Carboniferous age) are generally decomposed into the condition of a more or less porous mass to all depths up to ten or more feet. In all cases the disintegration may be traced to the removal, by leaching, of a small amount of interstitial carbonate of lime.

When we come to a consideration of the Calcareous rocks, we find, almost invariably, the chemical agencies of degeneration preponderating over those that are purely physical. In arid regions, and with granular crystalline types, physical agencies may for a time prevail, but as a rule the process is largely chemical, and notable for its simplicity. The decomposition is due mainly to the action of meteoric waters trickling over the surface, or filtering through cracks and crevices, under ordinary conditions of atmospheric pressure and atmospheric temperature. Hence the process is one of superficial solution, and the incidental chemical processes set in motion, as in the feldspar-bearing rocks, are almost entirely lacking. It follows that only the lime carbonate is removed in appreciable quantities, while the less soluble impurities are left to accumulate in the form of ferruginous clays, admixed with quartzose particles, chert nodules, etc. Since in many limestones the amount of these constituents is reduced to a minimum, even perhaps to the fraction of one per cent, so it happens that hundreds, or even thousands of feet of strata may

[1] Ann. Rep. Geol. Survey of Arkansas, Vol. I, 1890.

disappear without leaving more than a very thin coating of soil in their place.

An interesting illustration of the changes taking place in the decomposition of an impure Carboniferous limestone is described by Penrose in his treatise on the genesis of manganese deposits.[1] The stone in its least changed condition is of a granular crystalline structure and dark chocolate-brown color. The residual clay from its decomposition is a trifle darker, highly plastic, and quite impervious. Below are given the analyses of (I) the fresh rock and (II) the clay, both being taken from the same pit, the latter being of about fifteen feet in thickness and overlain by a capping of chert, which reduced to a minimum the possibility of any admixture of foreign matter. The materials were dried at a temperature of 110° to 115° C. before analyzing.

ANALYSES OF FRESH LIMESTONE AND ITS RESIDUAL CLAY

CONSTITUENTS	I FRESH LIMESTONE	II RESIDUAL CLAY	III PERCENTAGE OF LOSS FOR ENTIRE ROCK	IV PERCENTAGE OF EACH CONSTITUENT SAVED	V PERCENTAGE OF EACH CONSTITUENT LOST
Silica (SiO_2)	4.13 %	33.69 %	0.00 %	100.00 %	0.00 %
Alumina (Al_2O_3)	4.19	30.30	0.35	88.65	11.35
Ferric iron (Fe_2O_3)	2.35	1.99	2.13	10.44	89.56
Manganic oxide (MnO)	4.33	14.98	2.49	42.41	57.59
Lime (CaO)	44.79	3.01	44.32	1.07	98.93
Magnesia (MgO)	0.30	0.26	6.25	10.62	89.38
Potash (K_2O)	0.35	0.96	0.23	33.63	66.37
Soda (Na_2O)	0.16	0.61	0.085	46.74	53.26
Water (H_2O)	2.26	10.76	0.95	58.37	41.63
Carbonic acid (CO_2)	34.10	0.00	34.10	0.00	100.00
Phosphoric acid (P_2O_5)	3.04	2.54	2.73	10.24	89.76
	100.00 %	100.00 %	97.635%		

These analyses have been recalculated in the same manner as before, excepting that silica, instead of alumina, is taken as the constant factor. This for the reason above suggested. It is believed that one is safe in assuming little or no silica is lost

[1] Ann. Rep. Geol. Survey of Arkansas, 1890, p. 179.

here through the action of alkaline carbonates, since the alkalies are almost wholly lacking in the fresh rock, and a large portion of the silica doubtless exists as free quartz. Recalculating, then, in the same manner as before, but on a silica constant basis, we obtain the matter in columns III, IV, and V.

These columns bring to light some unexpected features, not the least interesting of which is the fact that the residual clay, in spite of its highly hydrated condition, in reality contains scarcely half the amount of water it would, had the small amount (2.26 %) in the original limestone been allowed to accumulate without loss. A more important, though perhaps more to be expected, feature is the entire removal of that portion of the lime which existed as carbonate, as indicated by the absence of carbonic acid in the clay. It will be noted that 97.635 % of the entire rock mass has disappeared through leaching, leaving only 2.365 % to accumulate as an insoluble residue in the form of soil.

This leaching out of the lime carbonates and the accumulation of insoluble residues is a strikingly conspicuous feature in regions abounding in limestone caverns, and to it is due the tenaceous ferruginous clays which cover their floors. So rich indeed are some of these residual deposits in iron oxide that in some instances they are locally used for pigments, under the name of ochre or mineral paint, or again, where occurring in large quantities, as ores of iron. (See p. 267.)

It is possible that loosely consolidated beds of shell limestone may undergo a process of change, perhaps more nearly akin to alteration than decomposition, through agencies quite different from those we have been considering.

Darwin, it will be remembered, found the shells in the raised sea-benches of San Lorenzo, South America, altered to the condition of a white powder without trace of organic structure, and consisting of carbonate, sulphate and chloride of lime with sulphate and chloride of sodium. This alteration he believed to be due to a mutual reaction taking place between the original sodium chloride derived from the sea-water and the lime carbonate of the shells, and he speaks of it as an interesting illustration of the fact that the dry climate of the west coast of South America is much less favorable to the preservation of shell structures than would be a moist one where the salt would be removed too rapidly for the double decomposition to be brought about.

Résumé. — Making all due allowance for possible sources of error in our methods, there are certain general deductions that may be safely drawn. Not, it may be, from our own analyses alone, but from numerous others as found in existing literature.[1]

Let us briefly review the subject and make the deductions accordingly.

In glancing over the columns of our analyses, it is at once apparent that hydration is an important factor, the amount of water increasing rapidly as decomposition advances. In the earlier stages of degeneration it is doubtless *the most* important factor. There is, moreover, among the siliceous crystalline rocks, in every case a loss in silica, a greater proportional loss in lime, magnesia, and the alkalies, and a proportional increase in the amounts of alumina and sometimes of iron oxides, though the apparent gain may in some cases be due to the change in condition from ferrous to ferric oxide. As a whole, however, there is a very decided loss of materials. Among siliceous crystalline rocks, this loss, so far as shown by available analyses and calculations, rarely amounts to more than 50 % of the entire rock mass. Among calcareous rocks, on the other hand, it may, in extreme cases, amount to even 99 %.

Of all the ordinary essential mineral constituents the free quartz is the most refractory toward purely chemical agencies, and the amount of silica lost from this source must be small, though Sorby[2] thinks to have distinguished chemically corroded quartz granules in some of the sands examined by him. It is, however, safe to say that the mineral suffers chiefly from mechanical disruption,—that silica in any rock which is removed during the process of decomposition comes mainly from the silicates, and not from the free quartz. According to Bischof, and as shown by our own work, the silicates in any rock that are most readily decomposed are, as a rule, those containing protoxides of iron and manganese, or lime, and the first indication of decomposition is signalled by a ferruginous discoloration and the appearance of calcite. The evidence bearing upon the relative durability of the various minerals constituting rocks is, however, quite conflicting and unsatisfactory. Doubtless much depends on local conditions.

[1] See especially Roth's Allegemeine u. Chemische Geologie, Vol. III, and Ebelmen's papers in Ann. des Mines, Vols. VII, 1845, and XII, 1847.
[2] Proc. Geol. Soc. of London, 1879.

Dana observed[1] that in the decomposition of the granitic rocks of the Chilean coast the feldspars yielded first, becoming white and opaque and of a friable earthy appearance. But it should be noted that there is nothing in Professor Dana's description to show that this change may not have been a purely physical one, and due to the splitting up of the feldspars along cleavage lines. Fournet, from a study of the processes of kaolinization, was led to state[2] that hornblende yields less readily to decomposing forces than does feldspar, when the two are associated in the same rock. Becker, however, in studying deep-seated decomposition in the Comstock Lode of Nevada, arrived at a precisely opposite conclusion, the feldspars as a whole offering more resistance than the augite, hornblende, or mica.

The present writer has described[3] thick sheets of augite porphyrite in Gallatin County, Montana, in which the feldspathic disintegration has gone on so far that the mass falls away to a coarse sand, from which still perfectly outlined crystals of coal-black augites may be gleaned in profusion. This last is, however, a semi-arid region, and the process thus far one of disintegration more than decomposition. In a moist, or perhaps in *any* climate, minerals consisting essentially of silicates of alumina and magnesia are less liable to decomposition than those containing considerable proportions of iron protoxides or of lime. This for the reason that the first-named are scarcely at all affected by the ordinary atmospheric agents of solution. Bischof goes so far as to say that the silicate of alumina is not at all affected by carbonic acid, but the researches of Müller, to which reference has been made, and our own calculations tend to disprove this. Dana states[4] that in the decomposition of basalt, on the island of Tahiti, the olivine is the earliest to give way, becoming first iridescent and finally falling away to a soft, pulverulent, ochreous yellow or brown powder. The compact base of the rock yielded next, the augites holding out until the last. Those silicates which are least liable to atmospheric decomposition are, as is to be expected, those which have resulted from the alteration of less stable silicates, as serpentine from

[1] Report, Wilkes's Exploring Expedition, Geology, p. 578.
[2] Ann. de Chimie et de Physique, Vol. LV, 1833, p. 240.
[3] Bull, U. S. Geol. Survey, No. 110, 1894.
[4] Op. cit., p. 298.

olivine, epidote from hornblende, or kaolin from feldspar, etc. A few silicates like tourmaline and zircon, or garnet, or oxides like rutile and magnetite, or the salts of rarer earths like monazite, etc., are scarcely at all affected by any of the ordinary agents of decomposition, but remain in the form of residual sands in the beds of streams, from whence the lighter, more decomposed material is removed by erosion.

In the weathering of potash-feldspar rocks carrying black mica, the latter mineral is as a rule the first to give way, and at times almost wholly disappears. With basic rocks, on the other hand, the dark mica is one of the most enduring of the constituents, and in the residual sands may be found in surprisingly large proportions.

In the kaolinized gneisses of northern Delaware, the biotite, as a rule, is in an advanced stage of decomposition, while the small amount of primary muscovite is still fresh and intact, retaining all its original lustre and elasticity.

Among the feldspars the potash varieties are, as a rule, far more refractory than the soda-lime, or plagioclase varieties. This is shown not merely by our own investigations, but by those of others as well. Roth shows[1] from analyses of fresh and weathered phonolites, nepheline basalts, and dolorites, that the loss of soda is almost invariably greater than that of potash.

In the coarse, pegmatitic dikes of Delaware County, Pennsylvania, the microcline masses, as mined for pottery purposes, are beautifully fresh and translucent, while the associated oligoclase is snow-white through a splitting up along cleavage lines and partial decomposition. Where thrown out upon the dumps, this whitened mineral shortly falls away to fine sand, resembling, at first glance, kaolin, but is distinctly gritty.

Max Geldmacher noted[2] that in the weathering of quartz porphyry oligoclase always gave way before the oligoclase.

Indeed, as shown in our analyses, in certain phases of rock degeneration, the potash feldspars may lose very little by decomposition, but be converted into the condition of fine silt merely through a mechanical splitting up. This fact will in part explain the relative scarcity of free potassium salts

[1] Op. cit., 3d ed., 2d Heft.
[2] Beitrage zur Verwitterung der Porphyre, Inaug. Dissertation, Konigl. Freidrich Alexander Universitat, Leipzig, 1889.

(carbonates, sulphates, and nitrates) as compared with those of soda.[1]

The chemical processes involved in this feldspathic decomposition are of sufficient importance to warrant further discussion, even though it may involve a certain amount of repetition of what has gone before.

Berthier, Forschammer, Brogniart,[2] Fournet,[3] and others explained more than fifty years ago the process of feldspathic disintegration through the breaking up of its complex molecule into alkaline silicates soluble in water, and aluminous silicates which are insoluble. The loss in silica, as noted above, was supposed to be due to the removal, by solution, of these alkaline silicates. Ebelmen,[4] however, subsequently showed that silicate minerals poor or quite lacking in alkalies lost a portion of their silica with equal facility, as is also shown in our analyses of pyroxenites on pp. 226 and 227. He accounted for this on the supposition that the silica set free — in a nascent state — was soluble either in pure water, or water containing carbonic acid. Bischof states that when meteoric waters containing carbonic acid filter through rocks containing alkaline silicates, the first

---

[1] An oligoclase occurring in a tourmaline granite on the southern slope of Monte Mulatto, near Predazzo, undergoes, according to Lemberg (Zeit. der Deut. Geol. Gesellschaft, 28, 1876), a much more rapid decomposition than the orthoclase with which it is associated, and gives rise to a green, lustreless, serpentine-like product. The chemical changes incidental to the alteration are as shown in the following tables, I being the fresh oligoclase, and II the decomposition product.

Constituents	I	II
Silica (SiO_2)	59.51 %	45.29 %
Alumina (Al_2O_3)	25.10	25.08
Iron sesquioxide (Fe_2O_3)	1.08	12.49
Lime (CaO)	4.03	0.52
Magnesia (MgO)	Trace	2.88
Potash (K_2O)	2.10	3.00
Soda (Na_2O)	7.26	2.14
Water (H_2O)	0.92	8.00
	100.00 %	100.00 %

[2] Arch du Museum, Vol. I, 1839 (cited by Ebelmen).
[3] Ann. de Chimie et de Physique, Vol. LV, 1833.
[4] Ann. des Mines, Vol. VII, 1845.

action consists in the partial decomposition of these substances by the carbonic acid and the formation of alkaline carbonates, which are dissolved. If the water thus impregnated, on penetrating further below the surface, comes in contact with calcareous silicates, another change will take place consisting of a decomposition and replacement of these calcareous silicates by the alkaline silicates, and a removal of the lime set free, as a carbonate, provided the water still contains a sufficient amount of carbonic acid. This replacing process and the retention of the alkaline silicates is accounted for on the supposition that, in their nascent state, they form new combinations with the other silicates present, while the lime remains as a carbonate to be removed or not, as the case may be. He further states that the alkaline carbonates originating in the manner described are among the most soluble substances known; the carbonate of soda requires for solution only six times its weight of water at ordinary temperatures. Silica, on the other hand, even in its most soluble form, requires ten thousand times its weight of water for solution. If, therefore, the decomposition of feldspar by such carbonated water were ever so energetic, there would be sufficient water for the solution of the carbonate of soda formed. But if the silica separated meanwhile amounted to more than $\frac{1}{10000}$ of the water present, the excess could not be dissolved, but would remain mixed with the kaolin.

The case is very different when the decomposition of feldspar is affected by fresh water containing only the minute quantity of carbonic acid derived from the atmosphere. By the action of such water, only very small quantities of alkaline carbonates are formed; consequently it is possible that the silica separated at the same time, also small in quantity, may find enough water for solution. In such cases the whole of this silica would be removed with the alkaline carbonates, and pure kaolin would be left. Such an action as this does not, however, appear to take place; for the purest of kaolin nearly always contains an admixture of quartz sand, or of free silica in some of its forms.

K. V. Murakozy has shown [1] that in the decomposition of rhyolite from Nagy-Mihaly, the sanidin passes into kaolin and opal, the latter separating out as hyalite in veins or impure concretionary forms.

It follows from this consideration that in the decomposition

[1] Abstract by F. Becke, Neues Jahrbuch, 1894, 1 Band, 2 Heft, p. 201.

of feldspar into kaolin more of the silica separated remains mixed with the kaolin formed, the greater the quantity of carbonic acid in water, and that, perhaps, the amount of carbonic acid is never so small that the whole of the silica separated in the decomposition of feldspar can be removed.[1] The above, however, overlooks the possible presence of nitrates, such as we now know from the researches noted on p. 203 may in many cases exist, even though in extremely small proportions. It is probable that the small amount of nitric acid formed by the bacteria would, if not taken up by plant growth, combine immediately with the alkalies, forming nitrates which, owing to their ready solubility, would be carried away. The larger the proportion of nitric acid, therefore, the greater would be the amount of silica intermingled with the kaolin, since whatever proportion of the alkalies failed to be carried away as nitrates would pretty certainly disappear as carbonate. There is also the possibility, especially in the rocks rich in iron protoxides, that a portion of the silica may combine with the iron, as already noted.

In cases where the decomposition takes place under the influence of a sufficient supply of oxygen, all iron, and presumably the manganese as well, would be converted into the insoluble hydrous sesquioxide form and remain with the residue. Where, however, the supply of oxygen is insufficient, a portion or all of these constituents may be removed in the form of protoxide carbonates, or, in the case of iron, of a ferrous sulphate. These facts well account for the variation in stability of the iron, as indicated in the preceding analyses.

Reference has already been made to the fact that the magnesia from the decomposition of magnesian silicates was sometimes removed in greater relative portions than was the lime. This seeming anomaly is also sometimes met with in calcareous stratified rocks. Roth[2] showed that in the weathering of dolomitic limestones, the magnesia is often removed in greater proportional quantities than the more soluble lime carbonate.

The researches of Hitterman[3] showed, however, that carbonic

---

[1] Chemical and Physical Geology, by Gustav Bischof, Vol. II, pp. 182, 183.
[2] Op. cit., Vol. III.
[3] Die Verwitterungeproducte von Gesteinen der Triasformation Frankers, Inaug. Dissertation, Freidrich-Alexanders Universitat, Munich, 1889.

acid solutions *may* exert a scarcely appreciable effect upon magnesian carbonate, which therefore accumulates in the residual soils.

It is safe to say that while the general process of rock-weathering may be quite simple, as outlined, there are many minor reactions which it is not possible to describe in detail.

It has been shown that even in firm rocks a mutual chemical reaction is not uncommon among minerals lying in close juxtaposition, giving rise to what are known as reaction rims or zones composed of secondary minerals. This is a particularly conspicuous feature in many gabbros, where olivine and feldspar are closely adjacent. In these cases, a mutual interchange of elements may take place, giving rise to garnets, free quartz, or other minerals as the case may be. This is, to be sure, a deep-seated change, to be classed as alteration rather than decomposition, and taking place presumably under conditions of temperature and solution quite at variance with those existing on the immediate surface. It is, nevertheless, self-evident that when elements are set free through any process, they must almost immediately recombine, taking those forms which existing circumstances may dictate and that close contact of particles would be favorable to the more rapid formation of new compounds. In a mass of decomposing rock, circumstances are almost continually changing, and the inference is fair that new combinations are continually being made and unmade, the intricacies of which we are unable to follow.

1

2

3

FIG. 1. Exfoliated granite in the Sierras.
FIG. 2. Talus slopes on Pike's Peak.
FIG. 3. Disintegrated granite, Ute Pass, Colorado.

THE WEATHERING OF ROCKS (Continued)

III. THE PHYSICAL MANIFESTATIONS

Rock-weathering manifests itself in a great variety of ways, much depending upon climate, though naturally the controlling factor is that of mineral composition. The manner of weathering is often sufficiently characteristic to be of great importance in determining surface contours, as well as incidentally affording a means for the identification of rock masses when the outcrops themselves are obscured by decomposition products. Such a means is of only local importance, however, since under varying conditions the resultant forms assumed, even by similar rocks, are themselves quite variable. It is, nevertheless, not without interest to note the varying phases of weathering in different kinds of rocks, the incidental contours assumed, the character of the resultant débris, and, at the same time, the controlling forces that have been instrumental in bringing about the final result.

(1) **Disintegration without Decomposition.** — That in weathering, physical and chemical agencies may go on either singly or conjointly has been noted in previous pages. In the case of single minerals, the preliminary disintegration is beautifully illustrated in the large oligoclase masses associated with microcline in the feldspar mines of Delaware County, Pennsylvania. In the dumps of waste about the mines these are found, in all stages of disintegration, the mineral splitting up along cleavage lines, becoming snow-white, and ultimately falling away to a kaolin-like product, but which, when submitted to microscopic examination, is found to be made up of sharply angular cleavage particles, showing no sign of decomposition other than that indicated by occasional opacity. In the analyses given below are shown (I) the composition of a fresh oligoclase (as given by Dana) from near Wilmington, Delaware, (II) the snow-white

cleaved, but still moderately firm mineral mentioned above, and (III) the flour-like or kaolin-like product.

Constituents	I Fresh Oligoclase	II Opaque White, but still Firm Oligoclase	III Fine Dust from Disintegrated Oligoclase
SiO_2	64.75 %	61.23 %	56.73 %
Al_2O_3	23.56	25.65	28.44
CaO	2.84	2.37	2.95
K_2O	1.11	0.72	1.12
Na_2O	9.04	7.66	5.81
Ignition	....	1.00	5.67
	101.33 %	99.63 %	100.72 %

The fact that granitic and gneissic rocks *may* undergo extensive disintegration with slight decomposition, even in a moist climate, was noted by Nordenskiold[1] in Ceylon. He says: "The boundary between the unweathered granite and that which has been converted into sand is often so sharp that a stroke of the hammer separates the crust of granitic sand from the granite blocks. They have an almost fresh surface, and a couple of millimetres within the boundary the rock is quite unaltered. No formation of clay takes place and the alteration to which the rocks are subjected, therefore, consists in a crumbling or formation of sand, and not, or at least only to a very small extent, in a chemical change. At every road section between Galle, Colombo, and Ratnapoora the granite and gneiss crumbled down to a coarse sand, which was again bound together by newly formed hydrated peroxide of iron to a peculiar porous sandstone, called by the natives *cabook*.[2] This sandstone forms the layer lying next the rock in nearly all the hills on that part of the island which we visited. It evidently belongs to an earlier geological period than the Quaternary, for it is older than the recent formation of valleys and rivers. The cabook often contains large, rounded, unweathered granite blocks, quite resembling the rolled stone blocks in Sweden. In this way

[1] Voyage of the *Vega*, Vol. II, 1881, p. 420.
[2] Laterite? It seems so regarded by H. F. Alexander, Trans. Edinburgh Geol. Society, Vol. II, 1869-74, p. 113.

there arises at places where the cabook stratum has again been broken up and washed away by currents of water, formations which are so bewildering, like the ridges (osars) and hills with erratic blocks in Sweden and Finland, that I was astonished when I saw them."

The same features are brought out in the previous descriptions relative to the weathering of the granite of the District of Columbia, the diabase of Medford, Massachusetts, and other localities mentioned in these pages. (See pp. 206 and 218.) This tendency toward disintegration without decomposition is exaggerated among coarsely crystalline rocks, as is abundantly exemplified in the rocks of the Pike's Peak (Colorado) area. (See Pl. 18.) Among those of finer grain, particularly the quartz-free varieties, as the Fourche Mountain (Arkansas) syenites, decomposition may follow so closely on disintegration that little or no sand is formed, sound fresh rock passing within the space of a few millimetres into the condition of residual clay.[1]

(2) **Weathering influenced by Crystalline Structure.** — It is elsewhere observed that, other things being equal, a coarsely granular rock will disintegrate more rapidly than one of finer grain.

Lone Mountain, one of the high eruptive peaks on the west side of the Madison valley in Montana, presents in its upper portions all the features of a volcanic crater broken down on one side by the lava flow. The facts of the case are, however, that the coarser grained central portion has been disintegrated, and swept by wind and rain into the valleys, while the fine-grained, more compact outer portions, those which solidified near the line of contact with adjacent rocks, remain intact. Professor Bell[2] describes an interesting case of this kind where the coarsely crystalline central portion of a "greenstone" dike has yielded more readily to erosion than at the sides and afforded channel-way for the Mattagami River, north of Lake Huron, in Canada. The gneiss adjoining the dike having been shattered, yielded also to decomposing agencies and forms now a second parallel channel on each side of the central one. "Between them the finer grained, hard, and undecayed 'greenstone' con-

[1] Dr. Max Fesca has noted that the granitic rocks of Kai province, Japan, yield on decomposing gravel, sand, and clayey loams, while those rocks poor in quartz, such as the syenites, give rise only to clays (Abhandlungen und Erlauterungen zur Agronomischen Karte de Prov. Kai, Kaiserlich Japanischen Geologischen Reichsanstalt, 1887).

[2] Bull. Geol. Soc. of America, Vol. V, 1894, p. 364.

stituting the outer portions of the dike rises up in the shape of ridges and chains of islands, so that the river flows as a main, central channel, more or less separated from the smaller lateral ones." The same writer describes several instances in which long straight valleys in the Archæan regions of Canada, now occupied by straight river stretches, long narrow lakes or inlets of the larger lakes, are due to the decay and removal of the wide " greenstone " dikes, or of parallel dikes with narrow belts of rock between. Long Lake, north of Lake Superior, some 52 miles in length, is mentioned as typical of lakes of this class.

(3) **Weathering influenced by Structure of Rock Masses.** — In any rock mass weathering is greatly augmented by lines of weakness, such as joint and bedding planes, since these furnish so many additional points of attack. In homogeneous massive rocks the rate of disintegration is retarded by a lack of vulnerable points, and the resultant form is that of rounded bosses such as are shown in plate 1.

As a rule, however, the most massive of rocks are traversed by one or more series of joints (see Pl. 14) whereby they are

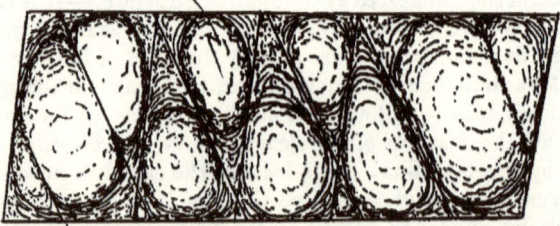

FIG. 17. — Showing the influence of joints in the production of boulders.

divided up into rhomboidal blocks of varying sizes. Even when not sufficiently developed to be conspicuous, such joints not infrequently exist as lines of weakness along which moisture and the accompanying agents of disintegration make their way, gradually rounding the corners until there is left an oval mass of which the so-called "niggerheads" of the gabbro area about Baltimore are typical examples. In nearly all such rocks the exfoliation and decomposition take place in the form of concentric layers, like the coatings on an onion. This holds true with the huge granitic bosses, as well as with the smaller joint blocks, and has been argued by some of the earlier geologists as indicative of an original concretionary structure. Such an assumption seems, however, wholly uncalled for. If the block or mass is reasonably homogeneous, the agencies of decomposition will penetrate nearly uniformly from all exposed surfaces, producing an exfoliation nearly parallel to that surface, and the concentric structure is inevitable, as was long ago pointed out by Werner.

In some cases the tendency to assume the boss-like form is accentuated through the presence of joints running approximately parallel to the exposed surface, such joints as give rise to the step-like arrangement of the stone so frequently seen in granite quarries. Stone Mountain, Georgia, an immense boss of light gray granite some 2 miles long by 1½ wide and 650 feet

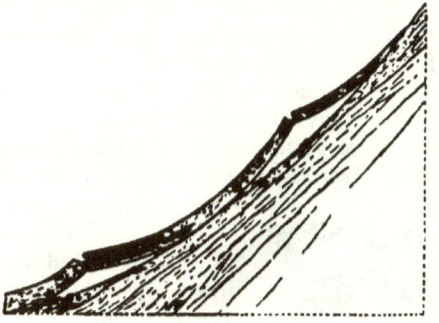

FIG. 18.—Exfoliation of granite.

high, owes its form, apparently, wholly to exfoliation parallel to pre-existing lines of weakness. The entire mass, so far as exposed by quarrying operations, is made up of imbricated sheets of granite, which, of unknown thickness beneath the surface, thin out to mere knife edges above, like shingles on a roof. Through prolonged exposure the superficial layers have become detached from the parent mass, and doubtless hundreds of feet in vertical thickness completely disintegrated and swept away. With many geologists these joints, in themselves, would be accepted as due to atmospheric action. In the writer's present opinion they are, however, the result of torsional strains and once existing are lines of weakness which become more and more pronounced as weathering progresses. The boss-like form is therefore incidental and consequent. The process of exfoliation has, in the case mentioned, been productive of some peculiar results which may be described in detail.

As above mentioned, the sheets of granite, varying from a few inches to several feet in thickness, conform in a general way to the present surface of the hill. Constant expansion and contraction from temperature changes have, in the manner already described, so expanded these sheets that, bound at the sides, they have found relief in an upward direction where resistance was least, and risen in dome or roof shaped forms, as shown in the sketch. The weight of the sheets higher up the slopes, impinging upon the edges of those below, has in some cases undoubtedly aided in the work, but the larger part is due to simple expansion, such as was referred to on p. 180.

These ruptured sheets are rarely more than 10 inches thick, but 10 or 20 feet in diameter. The material, though quite fresh appearing, is loosely granular and friable, easily reduced to sand. In a few instances small avalanches have been caused by the giving way of the sheets below and the consequent sliding down of those above through lack of support. (See Fig. 18.)

This same mass of granite sometimes shows upon its surface peculiar circular depressions, one within another, separated by intervening ridges of low relief, such as have been described in a much more perfect stage of development by Dr. Robert Bell[1] in the Huronian rocks of Canada. These, as shown in Fig. 19 from Bell's paper, are some 3 or 4 feet in diameter and 3 or 4 inches high. The cause of this form of weathering at Stone Mountain is not apparent, though Bell, in the case figured, regards it as induced by an original concretionary structure.

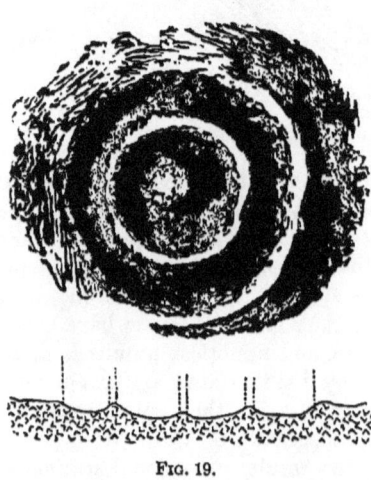

Fig. 19.

The spheroidal structure so frequently seen in basaltic rocks, and as typified in the *sphaeroidische absonderung* of German writers, may perhaps be due to an original spheroidal ten-

[1] Bull. Geol. Soc. of America, Vol. V, 1894, p. 362.

dency caused by cooling,[1] but a very large proportion of the spheroidal masses so typical of the decomposition of massive rock is, as already suggested, due wholly to external causes. W. P. Blake in 1855 called attention to this form of disintegration in the massive sandstones near San Francisco (California) and pointed out the true explanation.[2]

This sandstone is described as occurring in the form of layers from a few inches to 6 and 8 feet in thickness, alternating with beds of slate and shale. Down to a depth of 10 or 20 feet, or to the limits of atmospheric action, all the beds have turned from gray to rusty brown or drab. "There are, however, parts of the upper beds that have not yet been reached and changed by decomposition; these parts are found in the condition of spherical or ellipsoidal masses, from which the weathered parts scale off in successive crusts. These nuclei have the appearance of great rounded boulders, and have accumulated in great numbers at the base of the cliff." In this case the sandstone is composed mainly of grains of quartz and a little feldspar cemented by calcite, the disintegration being due mainly to the removal of this cement by percolating water, while the change in color is doubtless due to oxidizing pyrite or ferrous carbonate.

The effect of percolating waters is not, however, always immediately destructive. Though in themselves carrying cementing materials, or causing an oxidation of the iron carbonates or sulphides, a local induration may be induced along the joint lines such as becomes conspicuous only through the weathering away of the non-indurated portions. Resultant forms may be extremely regular or again irregular, according to the character of the lines along which percolation takes place, and that of the rock itself. An interesting illustration of this form of weathering is that given by Wyville Thompson[3] as occurring on the islands of Bermuda.

"This dissolving and hardening process," he writes, "takes place irregularly, the water apparently following certain courses in its percolations, which it keeps open, and the walls of which it hardens; and in consequence of this, the rock weathers most unequally, leaving extraordinary rugged fissures and pinnacles,

---

[1] T. G. Bonney, Quar. Jour. Geol. Soc. of London, Vol. XXXII, 1876, p. 153.
[2] Expl. and Survey for a Railroad from the Mississippi to the Pacific Ocean; Report on the Geology of the Route, near the 32d Parallel, by W. P. Blake.
[3] See The Atlantic, Vol. I. Also Bull. 25, U. S. National Museum.

and piled up boulders, the cores of masses which have been eaten away, more like slags or cinders than blocks of limestone. The ridges between Harrington Sound and Castle Harbor are a good example of this. It is like a rockery of the most irregular and fantastic style, and there seems to be something specially productive in the soil; for every crack and crevice is filled with the most luxuriant vegetation, mossing over the stones and training up as tier upon tier of climbers, clinging to the trees and rocks. Frequently the percolation of hardening matter, from some cause or other, only affects certain parts of a mass of rock, leaving spaces occupied by free sand. There seems to be little doubt that it is by the clearing out of the sand from such spaces, either by the action of running fresh water or by that of the sea, that those remarkable caves are formed which add so much to the interest of the Bermudas."

A form of weathering due to similar causes, but productive of results much more regular in arrangement, is shown in Fig. 4, Pl. 20, from a block of weathered sandstone in the National Museum. The original joints through which the waters filtered are easily recognized in the sharp straight lines running diagonally across the specimen. Blocks of fine shale and argillite, in their incipient stages of weathering, often show concentric bands of varying color, due to the oxidizing effect of water percolating inward from all sides of the natural joints as shown in Fig. 3, Pl. 20.

In stratified rocks there is, as a rule, a lack of homogeneity, certain layers being more porous than others, or containing mineral constituents more susceptible to the attacking forces. Such rocks, therefore, weather unevenly, and give rise to exceedingly ragged contours. The finely fissile schists standing nearly on edge along the coast of Casco Bay, in Maine, under the combined influence of wave and atmospheric action, weather into peculiarly fantastic forms resembling nothing more than piles of old lumber in which the multitudinous channels formed by boring coleopterous larvæ have become irregularly enlarged by decay. (See Fig. 1, Pl. 19.) The numerous quartz veins by which these schists are traversed stand out in bold relief until no longer supported by the matrix, when they fall to the beach, where, together with fragments of the schist, they are gradually reduced to pebbles and fine sand.

(4) **Weathering influenced by Mineral Composition.**—Although

PLATE 19

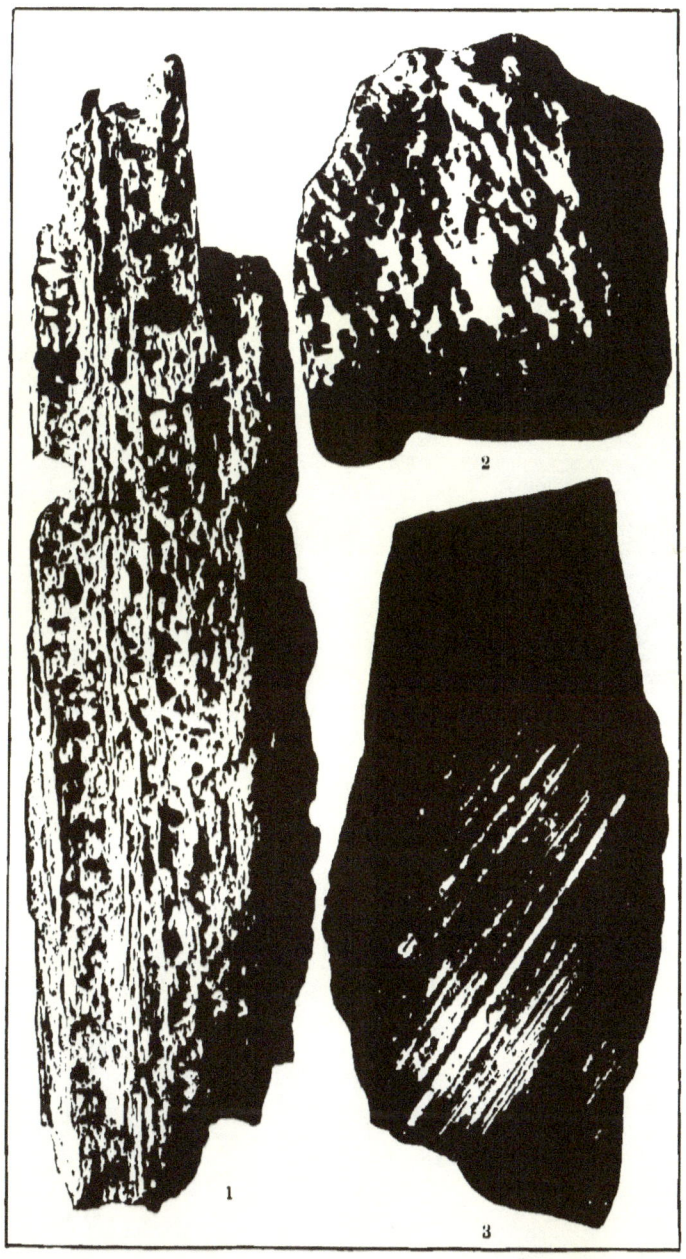

the soda-lime feldspars yield to the decomposing agencies more readily than the potash varieties, basic eruptives do not in all cases decompose more rapidly than the granitic rocks into which they are intruded, as is well illustrated in some of the glaciated areas about Boston, where small, compact dikes form low ridges a few inches above the surface of the enclosing granite. Much seems to depend upon the character of the secondary minerals which have been generated in a rock at some period prior to its decomposition proper. Thus those dikes containing so large a proportion of secondary epidote as to be of a dull greenish hue are almost invariably more enduring than the granites, while those, on the other hand, in which the secondary minerals are largely chlorite, calcite, and zeolitic compounds, yield to the decomposing agencies more readily. Even when the dike as a whole gives way, the presence of epidotic aggregates frequently manifests itself in protruding knots and bunches above the corroded surface. Knots caused by segregations of black tourmalines stand out in the same way from the surface of the granite boss called Stone Mountain, near Altanta, Georgia. Garnets, staurolites, quartz veins, and other of the less easily decomposed minerals may stand out in like manner from the surface of the rocks of which they form a part.

Granitic and other complex crystalline granular rocks will, on exposure, sometimes take on a pitted surface, owing to the removal of the more easily decomposed materials. The boulders of nepheline syenite in the glacial drift about Portland, Maine, are thus corroded to the depth of several millimetres through the removal of the granular nepheline, while the feldspars and hornblendes project irregularly.

Calcareous rocks containing silicates, like the amphiboles or pyroxenes, show like roughened surfaces due to the dissolving away of the calcareous matter, leaving the silicates projecting, or, as is the case with some of the tremolite-bearing dolomites used for building, may become pitted by the dropping out of the tremolite as the calcareous cement gives way.[1]

Many sandstones become likewise roughened through the removal of a portion of the cementing constituent, leaving the siliceous granules projecting. In the coarsely crystalline limestones and dolomites the solution and weathering effects are often first manifested along cleavage lines and the contacts of

[1] As in the U. S. Capitol Building at Washington.

the individual granules, as may be observed in many an old tombstone or polished column.

Even where the decomposition is almost purely chemical, the corroded surfaces are peculiarly irregular, as shown in Pl. 17. This feature is doubtless due to some imperceptible difference in the texture of the stone, or to the presence of joints and flaws which give direction to the solvent fluids. Calcareous rocks consisting of an admixture of calcite and dolomite crystals may undergo disintegration through a complete or partial removal of the calcite granules by solution, the dolomite remaining almost untouched. Certain dolomitic limestones near Stockton, Minnesota, have been described[1] as peculiarly subject to this form of disintegration. The mass of the rock consists of dolomitic crystals and granules, but often interlaminated with narrow bands of calcite. Through the removal of the latter, the stone becomes porous and its degeneration so complete that "shovelfuls of loose sand consisting of dolomitic rhombohedra can be taken up."

Fine-grained, compact, and seemingly homogeneous rocks may, on account of imperceptible differences in composition and structure, weather out in strikingly irregular and peculiar forms. Figure 2 on Pl. 15 is that of a limestone fragment from Harrisonburg, Virginia. The resemblance to cuneiform characters is so close that it is not surprising that such were at first supposed to be of human origin.

Massive granitic rocks seemingly of quite uniform composition will sometimes weather very irregularly, giving rise to oven-like cavities, in general shape resembling the pot-holes in the beds of streams. Reusch has described[2] such in exposed faces of granite ledges on the island of Corsica, the holes extending inward horizontally, or sometimes with a slight upward tendency. The cause of this is not apparent from the description given, but it is presumably due to slight textural differences such as are not readily discernible in the decomposed rock.

In any rock consisting of a variety of minerals, disintegration is likely to constitute a more prominent feature of weathering than in one of less complexity of composition, owing to the unequally refractory properties of its constituents. Thus a

[1] Hall and Sardeson, Bull. Geol. Soc. of America, Vol. VI, 1895, p. 184.
[2] Forhandlinger i Videnskabs-Selskabet i Christiania, 1878, No. 7, pp. 24–27.

granite must yield a sand, while a purely feldspathic, pyroxenic, or calcareous rock may yield only clays.

Beds of feldspathic quartzite, through the decomposition of the feldspar, undergo disintegration, giving rise to beds of friable siliceous sand interlaminated with kaolin, as described by Dana.[1] The same author also describes an interesting pseudobreccia formed by a quartzite divided up by a succession of cracks into which limonite from decomposing pyrite has filtered and acted as a colored cement. He says: "Many of the pieces lie in place barely separated from one another, and appear to be undergoing new divisions. But in the lower part, large pieces look as if there had been wide displacements; yet the hardly disturbed condition of the upper half proves that the apparent displacement is due to the extension of the coloring and penetrating limonite. The cracks are made in part by the extremely slow, wedge-like action of the depositing limonite."

Heusser and Claraz[2] describe somewhat similar breccias formed in Brazil through the weathering of crystalline schists rich in iron. These breccias consist of angular fragments of schist, more or less decomposed, firmly cemented by limonite.

The boulders of Oriskany quartzite in the Cretaceous gravel about Washington, District of Columbia, are composed of rounded and angular quartz fragments tightly bound together by a fine granular crystalline aggregate of quartz and feldspar. Disintegration first manifests itself on the exterior of the boulders in the form of an irregular network of grooves or channels, which gradually become more and more conspicuous until the boulder falls into bluntly pyramidal fragments and finally into sand. The microscope shows that the disintegration is due wholly to the disaggregation and partial kaolinization of the binding constituents whereby all cohesion is lost, and disintegration follows from necessity. (Fig. 1, Pl. 20.)

This form of disintegration seems to take place only in boulders exposed at or near the surface, and is believed to be due primarily to expansion and contraction from alternations of temperature.

Many rocks, owing to a lack of homogeneity, weather with extreme irregularity and give rise to odd and sometimes fan-

[1] Am. Jour. of Science, Vol. XXVIII, 1884.
[2] Ann. des Mines, 5th Series, Vol. XVII, 1860, p. 290.

tastic forms. In the case of a friable sand, or limestone, subject to wind or rain erosion or to solution, certain portions may be protected by a capping of other rock while the intervening material is carried away. There thus arise spindle-shaped forms of varying proportions, each capped by the roof or hat-like block to which it owes its origin. Such have been noted in many regions, and have been described by Hayden as occurring on a colossal scale in Colorado. In the case of strata lying nearly horizontal, it rarely happens that all possess the same power of resistance, the more friable weathering away with the greatest rapidity, leaving the harder layers for a time projecting in rib-like masses, to ultimately break down in large angular blocks as the support below is gradually removed. Friable beds of sedimentary rock are thus not infrequently protected by a capping of impervious lava. Continual percolation of water through existing joints and fractures in time, however, erode away, in part, the underlying material, causing the landscape to assume the Table Mountain appearance, where each flat-topped hill represents residual masses of a once continuous plateau, now isolated in the manner described.

(5) **Results due to Position.** — In very many instances loose blocks of stone lying exposed upon the ground, will undergo a more rapid disintegration from the lower surface, a feature evidently due to the fact that this portion of the rock is kept in a state of continual moisture. This form of disintegration results in the production of oval, flattened, scale-like masses, quite independent of the original jointing. In other cases decomposition going on from all exposed sides of a joint block may in time produce the so-called rocking-stones or "logans" and "tors" of English writers, though some of these are undoubtedly nicely balanced boulders from the glacial drift.

A mass of rock may be prevented from undergoing disintegration, even though partially decomposed, by its surroundings. Thus, in driving the tunnel for the waterworks extension, in Washington, natural joint blocks of hard and apparently firm rock brought to the surface would fall away to loose sand in course of a few days, or months, as the case might be, much depending on the conditions of the weather and the state of decay. This characteristic was sufficiently pronounced to attract even the attention of the workmen, who described the rock as "slaking" and believed it to contain quicklime.

The fact was that percolating waters had brought about a partial kaolinization of the feldspar, and hydration, without great oxidation of the iron-magnesian constituent. The original pressure, coupled with that incidental to expansion from hydration, had, however, been sufficient to hold the mass intact until exposed briefly to atmospheric influences.

The protective action of water, as sometimes shown in the beds of streams and in deep ravines, may be only apparent, and due to the fact that erosion exceeds decomposition, the stream having cut its way down to fresh bed-rock. Professor Dana, to be sure, writing more than half a century ago,[1] described the basaltic rocks of Kiama, Australia, as in a condition of advanced decomposition except where protected by sea-water. He says : " It is a general and important fact that a rock which alters rapidly when exposed to the united action of air and water, is wholly unchanged when immersed in water, or exposed to a constant wetting by the surf." While no exception can be taken to the conclusion regarding those rocks wholly immersed, the question naturally arises in one's mind, if the absence of decomposition products in those rocks constantly wetted by the surf and in many stream beds may not be due, in part at least, to erosion. That rocks so situated are in a condition far from fresh, is well known to any petrologist who has attempted to gather specimens.

It is obvious that where a large series of sedimentary rocks composed, it may be, of interbedded limestones, sandstones, and argillites are turned up on edge and exposed alike to atmospheric agencies, they will become eroded very unequally. If chemical agencies alone prevail, the limestone will dwindle away and perhaps give rise to long valleys or depressions walled in by the more enduring sands and shales, and carrying upon its bottom a fertile clayey soil representing not merely the insoluble impurities contained by the original limestone, but also the mechanically disintegrated particles washed in from the hills on either hand. This indeed may be considered the history of the fertile Shenandoah valley of Virginia, famous alike for soft contours, beautiful scenery, and the exuberant fertility of its soils.

When stratified rocks lie nearly or quite horizontally, much must depend upon the character as regards permeability, etc.,

[1] Reports of Wilkes's Exploring Expedition, Geology, p. 514.

of the upper layers, since these may so protect the lower lying as to retard or quite stop further disintegration. Further than this, an easy and rapidly disintegrating superficial layer may yield a residual clay so impervious as to protect the underlying rocks as securely as a mass of rock itself, or so hard and tough as to put a stop to purely mechanical erosion, as in the case of the laterite beds of central India.

In cases where thinly bedded rocks lie sharply inclined, it nearly always happens that certain layers decompose more readily than others. There may thus arise strikingly ragged saw-tooth contours, the more enduring layers standing out in sharply serrate or wall-like masses, while the softer give way and become obscured by their own débris.

(6) **Induration on Exposure.** — Many rocks, instead of becoming disintegrated on exposure, undergo a kind of induration upon the exposed surfaces. This is particularly the case with some siliceous sandstones. The water with which the stone is permeated holds in solution certain constituents, as silica, carbonate of lime, or iron oxides. When the rock is so situated that this "quarry water," as it is popularly called, is brought to the surface and evaporated, it binds together the granules composing the stone, forming thus a more or less superficial coating of a more enduring nature. The induration sometimes takes place so rapidly that even an exposure of but a few months is sufficient to produce very marked results on freshly broken surfaces. This peculiarity of certain classes of rocks has long been known to quarrymen and stone workers, who recognize the fact that a well-seasoned stone yields much less readily under the chisel than one that is newly quarried.[1]

A somewhat similar induration, due to purely superficial causes, has been described by Dr. M. E. Wadsworth as taking place on the surface of exposed blocks of siliceous sandstone in Wisconsin. "The St. Peters Sandstone" he writes,[2] "is composed almost wholly of a pure quartz sand, and in the outliers of it found on the hilltops south of the town, the parts covered by the soil were more or less friable, and the grains distinct; while the exposed portions of the same blocks and slabs were greatly indurated, the grains almost obliterated, and the rock possessed the conchoidal fracture and other characteristics of a

[1] See Stones for Building and Decoration, p. 415.
[2] Proc. Boston Soc. of Natural History, Vol. XXII, 1883, p. 202.

quartzite." In this and other cases cited by Dr. Wadsworth, the cementing matter is silica.

The explanation given (in letter to the present writer) is to the effect that all water, including that of rains, as well as terrestrial, dissolves silica, which is again deposited under suitable conditions. Part of the silica apparently comes from the solution of the quartz, chalcedony, and opal, and a part from the alteration and destruction of the silicates. Both solution and deposition seem at times to take place on the immediate surface, the interior waters in such cases playing no part.

P. Choffat regards it as possible that silica set free through feldspathic decomposition in granitic rocks may, on evaporation, be redeposited in an insoluble form in the interstices of the fresh rock in the immediate vicinity, thus retarding if not wholly preventing further decay in that direction.[1]

Professor W. O. Crosby, in a personal memorandum to the writer, calls attention to the fact that in the disintegrated granites of the Pike's Peak, Colorado, area, the rock is almost invariably exceptionally firm and impervious along the joints, indicating a local induration due perhaps to infiltration of iron oxides or silica. Where a joint face bounds a ledge of rock, it often maintains its integrity, weathering out in relief like a quartz vein, while the granite is in a condition of advanced degeneration all around. A slight break in the face of a joint plane, in such cases, may lead to extensive disintegration behind it, until it finally falls away from the disintegrating mass, a slab of relatively sound rock.

Andesitic rocks in regions of limited rainfall have been noted by Professor G. Vom Rath as having become covered on the upper surface with a thin layer of brown iron oxide, which protected them from further disintegration. Such crumbled away only from the under surfaces, where they absorbed moisture from the ground, and gave rise thus to peculiar tent-like and mushroom-shaped forms.

The present writer has noted in the Madison valley, north of the Yellowstone Park, rounded masses of a vesicular rhyolite which have, through the same causes, been reduced to the condition of mere shells with openings on the under sides and that

[1] Sur quelques cas d'erosion atmospherique dans les garnites du Minho, Communicações da Direcção Dos Trabalhos Geologicos de Portugal, Tome 3, Fasc. 1, 1895-96, p. 17.

facing the direction of the prevailing winds. In these cases, however, the wind seemed to have aided their formation, not merely through transporting the disintegrated material, but by catching up and whirling about the loosened granules within the gradually enlarging cavity, where, by force of impact, as already described, they became themselves agents of abrasion. Some of the cavities observed were of sufficient size to afford shelter for a human being and had in some instances served as temporary dens for wild animals.

Roth mentions [1] an induration evidently somewhat similar to that described by Vom Rath above, as having taken place, on the surface of a reddish yellow sandstone in Fezzan, North Africa. The crust thus formed was so dense and hard as to break with a shell-like fracture resembling basalt. A similar incrustation on sandstone from the Lydian desert was found to consist of: manganese oxide, 30.57%; iron oxide, 36.86%; alumina, 8.91%; silica, 8.44%; barium oxide, 4.89%; sulphuric acid, 4.06%; phosphoric acid, 0.25%; and water, 5.90%.

W. P. Blake has described boulders from the Colorado desert colored exteriorly by what he regarded as organic matter received from water during a period of submergence. Similarly discolored quartzitic boulders brought by G. K. Gilbert from the Sevier desert in Utah, and examined by the present writer, show a thin dark varnish-like coating, not inaptly named by Mr. Gilbert "desert varnish," and which consists largely of oxides of iron and manganese, though a slight amount of organic matter is present. In this case the rock is composed not wholly of quartz granules, but carries interstitial calcite and feldspathic granules. Near the discolored surface of the boulders these interstitial calcites are found quite dissolved away, leaving cavities stained by a dark deposit which reacts for iron and manganese. Inasmuch as acid solutions obtained from fresh and uncolored portions of the boulders give faint reactions of the same nature, it seems very probable that the crust is due to a concentration of these metals in a condition of higher oxidation on the surface, whither they have been brought by capillarity, while the more soluble lime carbonate was removed.[2]

[1] Allegemeine u. Chemische Geologie, 2d ed., Vol. III, p. 215.
[2] Although such discolorations seem to have been noted principally in desert regions, they are by no means limited thereto. The quartzitic boulders in the superficial deposits of the District of Columbia show at times a like discoloration, due to a very thin coating of iron and manganese oxide.

The Potsdam quartzites of Minnesota have had, in many instances, an almost glass-like polish imparted to their exposed surfaces through no other apparent agency than that of wind-blown sand. Unlike a polish produced by artificial methods, this wind polish extends to the bottoms of every little groove and cavity, or over every protruding knob alike. In softer rocks, or rocks of less homogeneous structure, the same agencies carve out the softer portions, leaving the more resisting protruding, as already described on p. 186. This polish is so perfect, even on rough surfaces, as to suggest a partial solution of the granules, and a redeposition of the dissolved matter in the form of a glaze, but the microscope proves to the contrary. The gloss is due wholly to superficial smoothing and has no thickness whatever, nor has any new matter been deposited either on the surface or between the granules.

(7) **Changes in Color incidental to Weathering.** — That in nearly every rock a change in color, the assumption of a brownish or reddish hue, is an early indication of decomposition has been made sufficiently apparent in the chapter devoted to a discussion of the chemical changes involved. This discoloration is, however, merely incidental, and not essential, and is found to diminish, if not wholly disappear, as the distance from the surface increases, as was noted in the case of the granites of the District of Columbia (p. 207) and the diorites of the Sierra Nevadas (p. 274. See further under Color of Soils, p. 385).

Granitic and other highly feldspathic rocks carrying proportionately small amounts of iron become almost invariably bleached or whitened on the immediate surface, owing in part to kaolinization and in part to the splitting up of the feldspars along cleavage lines.

In extreme cases rocks consisting of an admixture of feldspars and iron-bearing silicates, but in which the first-named, owing to its glassy nature, is in the fresh rock quite inconspicuous, become almost snow-white in the earlier stages of weathering. This, as in the case above mentioned, is due to the change in the feldspars and the consequent obscuring of the darker silicates by the white product of kaolinization. Continued decomposition must, however, attack the ferruginous constituent and the usual staining ensue, unless, as in some cases possible, sufficient carbonic acid may exist to convert the iron immediately into carbonate and permit of its removal in solution.

Allusion has been already made to the fact that oxidation or other chemical action, with the possible exception of hydration, practically ceases below the permanent water level. Hunt and Le Conte have both called attention to the fact that the hornblendic and feldspathic rock fragments occurring in the Pliocene auriferous gravels of California are firm and intact in those portions below the drainage level (the blue gravel layer), but more or less completely oxidized, kaolinized, and otherwise altered in the red or upper gravel.

Van den Broeck has called attention[1] to the possibility that the so-called red and gray diluvium of the Quaternary deposits near Paris may be but portions of one and the same geological body, the "*diluvium rouge*" being but an upper member of the "*diluvium gres*," oxidized and impoverished in lime by the action of meteoric waters.

The same feature is noticeable in many of our quarries for building stone, as those in the Berea sandstones of Ohio. These below the drainage level, are of a gray or blue-gray color, while above, where they have been subjected to the oxidizing influence of meteoric waters, they are buff. The Jurassic oölites of England, are blue-gray at some depths below the surface, but white above.

In cases where natural joint blocks are exposed to the percolation of meteoric waters, the weathering may for a time manifest itself only in differential oxidation and zonal segregation of the iron whereby are produced concentric bands of varying hues. Figure 3, Pl. 20, is a slab from a natural joint block of argillite in the collections of the National Museum, in which the bands, due to this cause, vary from yellow-brown, drab, to ochreous yellow and red, while the rock as a whole still retains its compact structure and susceptibility to polish, forming an ornamental stone of no mean order.[2]

(8) **Relative Amount of Material removed in Solution.** — Among siliceous rocks, chemical action proceeds but slowly, and the amount of material actually removed in solution is rarely over 50 %, and may be so small that, as the writer has shown,[3] the residue in extreme cases occupies some 80 % more space than the rock from whence it was derived. Carbonate

---

[1] Bull. Soc. Geologique de France, 5, 1876-77, p. 298.
[2] Stones for Building and Decoration, p. 169.
[3] Bull. Geol. Soc. of America, Vol. VI, 1895, pp. 321-332.

PLATE 20

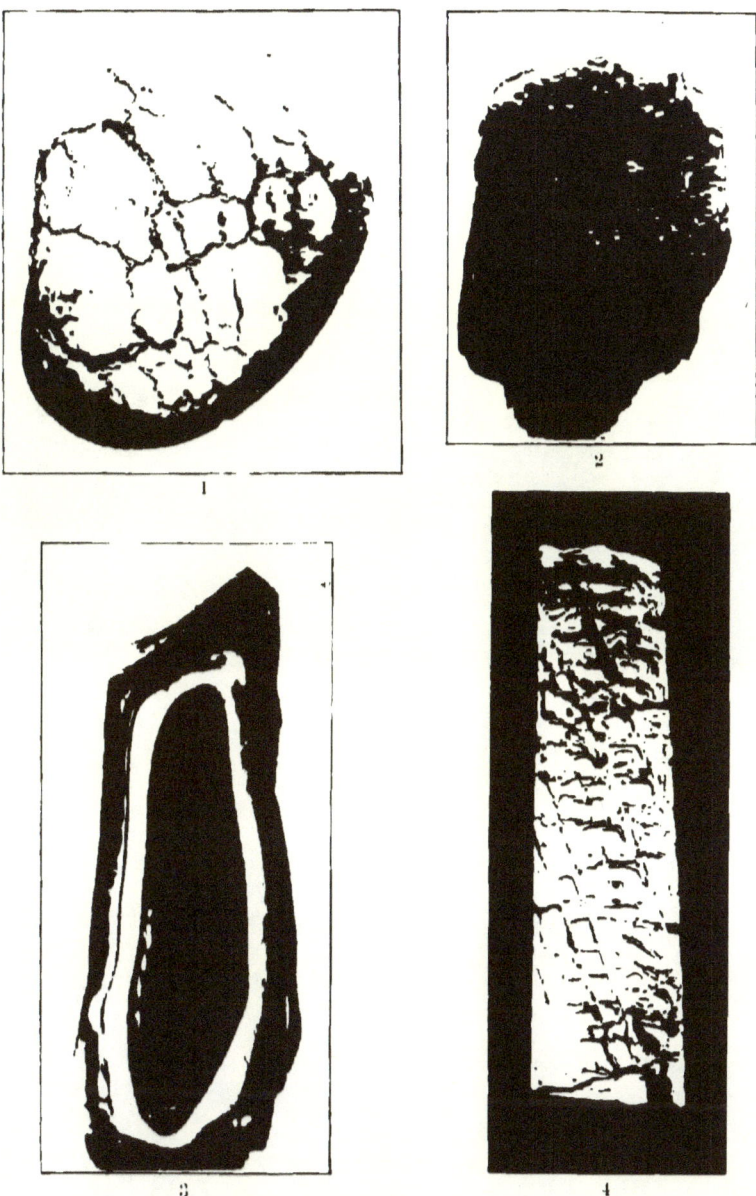

FIG. 1. Weathered boulder of Oriskany sandstone.
FIG. 2. Concentric weathering in diabase.
FIG. 3. Zonal structure in weathered argillite.
FIG. 4. Weathered sandstone, showing induration along joint planes.

of lime, the essential constituent of ordinary limestone, is, however, as has been observed, soluble in the carbonated water of rainfalls, and, in time, may undergo complete removal, leaving but the insoluble impurities behind. This is, indeed, the almost universal history of limestone soils. They are not infrequently so siliceous or ferruginous as to be quite barren and of a nature to be benefited by the application of lime as a manure.

Throughout the areas occupied by the Trenton limestones, in Maryland, nearly every farm has, in years past, had its quarry and lime-kiln where the stone was fitted for supplying lime once more to soils from which it had been so thoroughly leached as to render them lean and poor. It is almost wholly to this solvent action that is due the formation of the multitudinous caverns, large and small, of the limestone regions. Even where caverns are not apparent, the corrosive action is evident to the practised eye. In the quarry regions of Tennessee surface blocks of limestone are often grooved to a depth of an inch or more with wonderful sharpness, simply from the water of rainfalls with its acids absorbed from the atmosphere and surface soils, while in the quarry bed the stone is found no longer in continuous layers, but in disconnected boulder-like masses. (Figs. 3 and 2, Pls. 16 and 21.) In such cases casual examinations give very little clew to the rapidity of the destruction going steadily on, since all is removed in solution excepting the comparatively small amount of insoluble matter (usually clay or silica) existing as an impurity.

(9) **Incidental Surface Contours.** — In limestone regions the solvent action of water has frequently gone on so extensively as to leave its imprint upon the topographic features of the landscape. The drainage is no longer wholly superficial, but by subterranean streams sinking entirely into the ground to reappear again at lower levels, it may be miles away, having traversed the intervening distance in some of the numerous passages (fissures enlarged by solution) with which the rocks abound. Entire landscapes are undulating through the abundance of sink-holes — shallow depressions down through which the water has percolated and escaped into the underground passages.

The writer recalls a beautiful illustration of this nature seen in the limestone regions of southern Indiana, some years ago.

The season was that of the wheat harvest. On every side, far as the eye could reach, were undulating fields of waving grain, of that charming golden hue of which poets sing, with intervening patches of woodland. From every farm was heard the click of the reaper, and from every fence the whistle of the "Bob White." Owing to the fact that the ridges between these depressions were drier than the bottoms, the wheat here ripened earlier, and field after field showed long reaches of saucer-shaped depressions green in the centre, with intervening ridges of golden brown, making, with that charming hazy atmosphere, a picture long to be remembered. Through accident or design, the opening in the bottom of these sink-holes sometimes becomes closed, giving rise thus to temporary pools, or ponds, as shown in the accompanying plate. It is this same action that has given rise to the so-called "sandpipes" of the English geologists. These are slender funnel- or tube-shaped cavities found in chalk and calcareous sandstone, sometimes filled with drift gravels, sands, brick-earths, or again with fragmental materials fallen into them from the overlying beds as the support beneath was gradually removed. In all these cases it is assumed that direction was given the percolating water by pre-existing fissures or lines of weakness.[1] (Fig. 1, Pl. 21.)

In regions underlaid by massive siliceous crystalline rocks, and where mechanical erosion is reduced to a minimum, landscapes are softly undulating, with few abrupt escarpments or precipitous ledges, owing to the uniform rotting away of the materials, and the gradual accumulation of the débris. It is to this form of weathering that is due the beautiful rolling hills of southwestern Maryland. The prevailing rock is granite or gneiss. Decomposition follows out each line of weakness. Streams erode through the softened material down to hard bed-rock, while the relatively large proportion of insoluble débris is left to accumulate on the gentle slopes which form such an enchanting feature of these landscapes.

In regions of gneissic or granitoid rocks traversed by large veins of quartz, as in the northwestern part of the District of Columbia, the superior resisting power of the quartz causes it to stand out in relief from the gradually dwindling rock masses on either hand, giving rise thus to prominent knolls, or ridges,

[1] See Prestwich's paper, Quarterly Journal Geological Society of London, 1855, p. 62.

the occasion for which is a mystery until we come to examine their foundation materials. Belt, in describing the auriferous quartz lodes at San Domingo,[1] states that the prevailing trend of the main ranges is nearly east and west, and is probably due to the direction of the outcrops of the lodes which have resisted the action of the elements better than the soft dolerites.

So striking a feature of the landscape as the Devil's Tower or Bear Lodge on Little Sun Dance River, Wyoming, is due to the weathering away and erosion of sedimentary beds from around a dense crystalline core or plug of eruptive rock intruded into them in some past period of volcanic activity. Through its greater powers of resistance, this still stands, towering over 1000 feet above the level of the river, though in time this, too, must go. Quite similar forms have resulted, within a comparatively brief geological period through the erosion of tufaceous cones from around the compact, crystalline plug of lava which solidified within the crater when volcanic activity ceased. Beautiful examples of these are to be seen in Arizona and New Mexico, where they are known as "volcanic necks." The formation of bosses through the influence of joint planes has been described elsewhere (p. 244).

In regions abounding in intrusive olivine or pyroxene rocks which have undergone alteration into serpentine and talc or "soapstone," one frequently finds these materials forming the main mass of the hills, while the valleys are carved out of the softer, more readily decomposed granite, or whatever the country rocks may be. The same feature is prominently developed in the slate regions of Harford County, Maryland, where the slate is the more enduring rock, and forms steep ridges, flanked by valleys, carved out from less resisting materials. Regions of trappean dikes in siliceous schists or gneisses, particularly along sea-shores where swept by incoming tides, are often characterized by narrow, straight-walled chasms, or cañons due to the weathering out of the basic rocks, while the more refractory schists on either hand remain.

In cases where trappean dikes have cut through friable sandstones, they have in some instances so indurated these rocks along either contact as to cause them to be more durable than the original rock or than even the trappean rock itself. There may thus arise long parallel ridges of indurated sandstone sepa-

[1] The Naturalist in Nicaragua.

rated by an intervening depression due to the weathering out of the dike material.

In regions where climatic conditions or the nature of the rock are more favorable to mechanical disintegration than chemical decomposition, contours may be ragged in the extreme. Entire crests may be but successions of jagged peaks and intervening narrow valleys which are gradually becoming choked up by the débris fallen from the cliffs above.

(10) **Effacement of Original Characteristics through Weathering.** — In cases of extreme decomposition, in place, the residual products may so slightly resemble the parent rock as to give rise to very conflicting opinions concerning their origin. This was for a long time the case with the laterite of India, already described, and the terra rossa of Europe.

Dana describes[1] an interesting case of basaltic decomposition which, on account of the peculiar nature of the residual product, is worthy of mention here. He writes: "The process of decomposition is finely exhibited on the second cliff north of Kiama (Australia) towards the north end. At first sight, a distinct argillaceous deposit was supposed to overlie the columnar basalt; for it was twenty feet thick, and of a whitish color, resembling a soft crumbling marl, thus wholly unlike the basalt, and the common results of basaltic decomposition. Still it had proceeded from the alteration of a regular columnar variety, having a dull grayish blue color. The original rock is exceedingly compact, showing no trace of crystallization, excepting an occasional minute crystal of feldspar; and within the reach of the swell, it was still compact and solid.

"The rock has a concentric structure, and to this it owes in part its rapid decomposition. The alteration commences between the concentric layers, rendering them apparent, although not so before. At first a thin ochreous line appears, arising from iron; either magnetic iron disseminated in the rock, or from that of the constituent mineral augite. This ochreous color afterwards mostly disappears, and the concentric coats become separated by thin clayey layers of a white color, more or less striped with ochreous lines. In a more advanced stage of the process large ovoidal masses of basalt (but little changed in appearance excepting the development of a slaty concentric structure) lie in the cliff separated by a considerable thickness

[1] Reports Wilkes's Exploring Expedition, Geology.

of the whitish clayey layers, which are stained by irregular ochreous lines. At last the centres of the spheroidal masses yield, and finally the change is so complete that the concentric arrangement is entirely lost, and a soft whitish or yellowish-white argillaceous deposit, with few ochreous spots or lines, takes the place of the compact basalt.

"In basalts of more compact structure these changes take place more slowly. The grayish blue basalt in the Illawarra range, near Broughton's Head, when long exposed, is discolored exteriorly to a depth of an inch and a half. The colors, beginning within, are dirt-brown, grayish yellow, ochre-yellow, brownish red; and they are evidently dependent mostly on changes in the condition of the iron which the rock or its minerals contain.

"When the rock includes much chrysolite, the results of decomposition in some instances give a fissile or micaceous appearance to the rock. At Prospect Hill, five miles west of Paramatta, this change is in progress. The rock is a black ferruginous basalt of homogeneous aspect, breaking with a smooth fracture and no appearance of crystallization. It contains chrysolite; but the grains are small and not apparent except on very close examination. . . .

"Were we unable to trace the transitions, and distinguish the columnar structure through the whole, we should scarcely suspect its basaltic origin. Indeed, it was pointed out to me as an instance of mica slate overlying basalt. Particles of rusted mica, as they seemed, were distinct, and it much resembled a decomposing variety of that rock. On close inspection and an examination of the rock in different stages of change, it became evident that the pseudo-mica was nothing but altered chrysolite, which had rusted from partial decomposition, and split into thin cleavage scales.

"The crystals of chrysolite have evidently a parallel position in the rock, and hence the plane of easiest cleavage lies in the same direction, or, as the cleavage shows, parallel with the upper surface, that is, at right angles with the vertical axis of the columns. The passage from the compact to the decomposed rock is, in this case, unusually abrupt. Alteration takes place (through the elimination of oxide of iron as before suggested) slowly at the surface, which therefore chips off as soon as decomposed and exposes a new portion. This sudden transition may, in part, proceed from the absence of any natural planes of fracture (which are brought out when there is a concentric structure), and perhaps in part also from the presence of chrysolite. The layer of pseudo-mica schist is in some places five feet thick and has a rusty brownish color. Above it passes into three feet of earth of the same origin, having a brownish black color, and this is covered again by four feet of brownish red soil."

Such an effacement is not, however, an invariable accompaniment of decomposition, since where the amount of residuary material is relatively large, and allowed to accumulate in place, the mass may for a long period retain its original structural characteristics. Indeed, the original features are sometimes so perfectly preserved that casual inspection alone quite fails to reveal the havoc that has gone on. Every detail of bedding, jointing, or foliation, or even of internal structure, as brought about by the arrangement or size of the individual particles, may be retained with perhaps only a slight change of color due to oxidation. This feature is often strikingly conspicuous in the newer railway cuts of the southern Appalachian regions, particularly where the country rock is of the nature of gneisses or schists. In the work of grading the streets, in the extensions of the city of Washington, masses of strongly foliated granites, so soft as to be readily removed with pick and shovel, would be cut through, and which yet showed every vein or other structural detail as plainly marked as in the original rock, and it was only when by thrusting one's cane or other implement into it that its thoroughly decomposed condition became apparent. Russell describes[1] a similar condition of affairs prevailing in the coarse Triassic conglomerate near Wadesborough, North Carolina. This conglomerate is here composed of rounded and angular pebbles of talcose schist and other crystalline rocks. In the fresh cuts along the line of the North Carolina railroad, every detail of the original rock is brought out almost as sharply as in the so-called "Potomac marble" phase of the same formations as used in the Capitol building at Washington. "On examining more closely, however, one is surprised to find that it is completely decomposed, and that when moist it can be cut with a pocket knife through pebbles and matrix alike, as easily as so much potter's clay. The full depth of the alteration in this

[1] Bull. 52, U. S. Geol. Survey, 1889.

instance is not revealed, but it extends more than 30 feet below the surface without change in character."

W. B. Potter described [1] the feldspar porphyry of Iron Mountain, Missouri, as decomposed to the extent that it can be easily whittled away with a penknife or scratched with the thumb nail. "At the same time," he writes, "the original porphyritic structure of the individual crystals scattered through the mass is beautifully preserved, and is even frequently more distinctly visible than in the original rock, owing to stronger contrasts of color in the kaolinized material." In many dense massive rocks, indeed, such features as flow structure and inequalities of texture are frequently rendered evident only on weathered surfaces. The same is often true of fossiliferous limestones, a weathered surface revealing the presence of organic forms wholly imperceptible on one freshly broken.

The crude kaolin as removed from the pits near Brandywine Summit, Pennsylvania, and at Hockessin, Delaware, still retains more or less distinctly the structure of the original gneiss or conglomerate from whence it was derived. The quartz granules of the gneiss are, in these cases, almost invariably shattered, as though crushed by dynamic agencies, and show distinctly corroded surfaces, presumably caused by the alkaline carbonates formed during the kaolinizing of the feldspars. The black mica makes its former presence known by rust-colored spots which, in those cases where the mineral was sufficiently abundant, have ruined the material for the purposes of the potter.

(11) **Simplification of Chemical Compounds, incidental to Weathering.** — It has been noted on p. 172 that the process of weathering is but an attempt on the part of the elements in their various combinations to adjust themselves to existing conditions. This adjustment consists in the formation of new compounds which are characterized by a less complex structure than those first formed.

Indeed, one of the most striking features of chemical geology is the tendency toward simplification in composition as manifested all over the superficial portions of the earth. During the process of decomposition there is almost invariably a constant breaking down of complex molecules of mixed silicates of alumina, iron, lime, magnesia, and the alkalies, and a recombi-

[1] Jour. U. S. Assoc. Charcoal Iron Workers, Vol. VI, p. 25.

nation of their various elements as simpler silicates, carbonates, sulphates, and oxides.

(12) **Other Results incidental to Decomposition and Erosion.** — That all the minerals of a rock mass are not equally acted upon by atmospheric agencies has been sufficiently noted in previous pages. The more refractory, freed by the breaking down of their host, remain to gradually accumulate in vastly greater proportions than they existed in the original rock. If, in addition to their refractory qualities, such possess, as is usually the case, greater density, decomposition and erosion may act but as agents of concentration, and in such residues minerals like xenotime and monazite have been found in abundance, although occurring so sparingly in the fresh rock that their existence was scarcely suspected.

It is in this manner that has originated the gem sand of Ceylon. Precious stones have been found disseminated in limited numbers in the granite converted into the cabook described on p. 242. In weathering, the difficultly decomposable precious stones have not been attacked, or attacked only to a limited extent. They have therefore retained their original form and hardness. When in the course of thousands of years streams of water have flowed over the layers of cabook, their soft, already half-weathered constituents have been for the most part changed into a fine mud, and as such washed away, while the hard gems have only been inconsiderably rounded and little diminished in size. The current of water therefore has not been able to wash them far away from the place where they were originally embedded in the rock, and we now find them collected in the gravel bed, resting for the most part on the fundamental rock which the stream has left behind, and which afterwards, when the water has changed its course, has been again covered by new layers of mud, clay, and sand. It is this gravel bed which the natives call *nellan*, and from which they chiefly get their treasures of precious stones.[1] The same process in states bordering along the Appalachian Mountain system in North America has given rise to auriferous sands, as well as to sands bearing monazite, zircons, and other valuable minerals, which become segregated merely through their greater density and power to resist decomposition. The stream tin ores of the Malayan Peninsula, the

---

[1] Nordenskiold, Voyage of the Vega. See also Judd, On the Rubies of Burma, etc., Philos. Trans. Royal Soc. of London, Vol. CLXXXVII, 1896, p. 151.

Fig. 1. Sink-hole near Knoxville, Tennessee.
Fig. 2. Beds of marble corroded by meteoric waters, Pickens County, Georgia.

diamond-bearing gravels of Brazil, and indeed placer deposits in general are illustrative of this same principle. The very soil itself, although so indispensable to human existence, is but an incidental and transitory phase of rock-weathering, as has been made sufficiently apparent in previous pages. The deposits of kaolin in western Pennsylvania and northern Delaware, as elsewhere noted, are but decomposed highly feldspathic gneisses and conglomerates, while the phosphate deposits of middle Tennessee are insoluble residue left by the leaching out of the calcium carbonate from phosphatic limestones.[1]

According to Russell,[2] the Clinton iron ore of Alabama is but the insoluble residue from ferruginous Silurian limestones. On the immediate surface the ore is quite pure, containing, it may be, but a trace of lime. When followed downward, the amount of lime is found to gradually increase, until the ores may become so poor in iron as to be valueless. The following figures show this gradual increase in lime carbonate, and necessary decrease in iron, from the surface downward.[3]

PERCENTAGE OF CALCIUM CARBONATE IN CLINTON IRON ORE

Depth	Per Cent	Depth	Per Cent
Surface	Trace	70 feet below surface	25.61
10 feet below surface	Trace	80 feet below surface	29.92
20 feet below surface	Trace	90 feet below surface	20.80
30 feet below surface	Trace	100 feet below surface	23.37
40 feet below surface	21.06	110 feet below surface	28.82
50 feet below surface	23.00	120 feet below surface	21.32
60 feet below surface	27.01	130 feet below surface	30.55

William Whitaker in 1864[4] noted the decomposition of the English chalk beds, in Middlesex, and the gradual accumulation of a stiff brown-red residual clay interspersed with many flint nodules. It is by this same leaching action on aluminous limestones that is formed the so-called "rottenstone" so commonly used in polishing brasses and other metals.

[1] J. M. Safford, American Geologist, October, 1890, p. 201.
[2] Op. cit., p. 22.
[3] Trans. Am. Ins. of Mining Engineers, Vol. XV, 1886, p. 189.
[4] Mem. Geological Society of Great Britain, 1864, p. 04.

THE WEATHERING OF ROCKS (*Continued*)

IV. TIME CONSIDERATIONS

Concerning the rate of decomposition of rocks of various kinds, only very general rules can be laid down, since much depends upon climatic conditions and the position of rock masses relative to the action of frost, moisture, and the various growing organisms.

(1) **Rate of Weathering influenced by Texture.** — From the study of building materials it has become apparent that a coarsely crystalline rock will, all other conditions being the same, disintegrate more rapidly than one of finer grain. This is doubtless owing in part to expansion and contraction from ordinary temperature variations, which act the more energetically the larger the crystalline particles.[1]

It has already been remarked (*ante*, p. 44) that crystalline rocks have a greater density than do glassy forms of the same chemical composition. This indicates a contraction during the processes of crystallization, which manifests itself, according to at least one authority, in the development of minute interspaces between the individual crystals. The coarser the crystallization, then, the greater the amount of interstitial space, and hence the greater the absorptive power.

These coarser rocks, owing to their tendency to undergo a mechanical disintegration, or disaggregation, may also yield to

[1] The coefficient of cubical expansion for several of the more common rock-forming minerals has been determined as follows: —

Quartz	0.0000360	Tourmaline	0.000022
Orthoclase	0.0000170	Garnet	0.000025
Hornblende	0.0000284	Calcite	0.000020
Beryl	0.0000010	Dolomite	0.000035

The strain brought to bear upon a mass of rock through the unequal rate of expansion of its various constituents is further complicated through the unequal expansion of the individual minerals along the direction of their various axes. Thus quartz gives a coefficient of 0.00000769 parallel to the major axis, and of 0.000001385 at right angles thereto. Adularia gives 0.0000156, 0.000000659, and 0.00000204 for its three axes, and hornblende 0.0000081, 0.00000084, and 0.0000095 (Stones for Building and Decoration, p. 419).

the decomposing agencies more readily than those of finer grain, though from the fact that they first fall away to coarse sand, whereby the rock-like character is lost, one might, on casual inspection, be led to the opposite conclusion. It need scarcely be said that, among rocks having the same composition, whether fragmental or crystalline, those of a granular structure will undergo disintegration more quickly than will those in which the individual minerals are closely compacted or interknit, as in many quartzites or diabases.

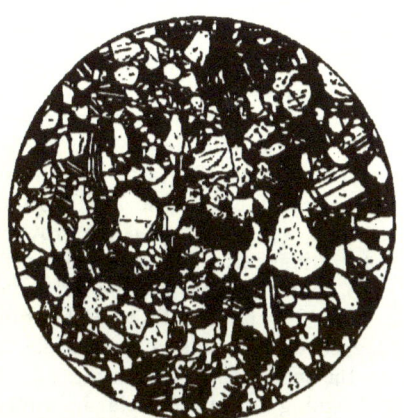

Fig. 20.

(2) **Rate of Weathering influenced by Composition.** — Among rocks of the same structure as regards crystallization and size of particles, the basic varieties, such as the diabases and gabbros, as a rule succumb more readily than do the more acid varieties like the granites. This for the reason that the iron-magnesian as well as the soda-lime minerals are more susceptible than are the potash silicates and other essential constituents of

Fig. 21.

Microstructure of sandstone (Fig. 20), showing relatively large amount of interstitial space and absorptive power, and (Fig. 21) of diabase, with relatively little.

the rocks of the granitic group. It is possible also that these dark colors cause them to become more highly heated, where exposed to direct sunlight, and hence subject to mechanical disintegration. The fact that many of our trappean rocks, as seen in dikes cutting other rocks, do not in all cases succumb with greater comparative rapidity is due to their very compact structure, whereby percolating waters are so largely excluded.

(3) **Rate of Weathering influenced by Humidity.** — The rapidity of rock weathering and soil formation is, even among rocks of the same nature, widely variable, being dependent upon climatic conditions of any particular locality. In the arid regions north of Flagstaff, Arizona, are wide areas of country covered with coal-black lapilli ejected from volcanoes whose craters are now occupied by growing pines upwards of two feet in diameter. Yet these fields are, with the exception of the pines, as bare of vegetation as though but yesterday scorched by fire. The fine lapilli, resembling nothing more than crushed coke, cover everywhere the undulating plains, greedily absorbing the moisture from melting snows and scanty rainfalls, but undergoing no appreciable decomposition and affording foothold for only a few desert shrubs and grasses. Yet in a moister clime, and one more adapted for luxuriant vegetation, we might expect that these lapilli should long ago have succumbed and given fairly fertile soils.

(4) **Rate of Weathering influenced by Position.** — Among the siliceous crystalline rocks superficial disintegration is undoubtedly greatly aided by temperature variations, which, by rendering the rocks porous, facilitate chemical decomposition. Such action must, however, be merely superficial, and at considerable depths below the surface the change must be purely chemical. The chief conditions favoring chemical action are those of continual percolation by waters carrying the organic acids already described. It naturally follows, therefore, that a purely chemical decay will progress more rapidly where the rock mass is covered by such a layer of vegetable soil as shall give rise to the decomposing solutions. Hence, that such an accumulation having begun, decomposition will keep on at an ever-increasing rate to a depth concerning which we have at present no data for calculation. It must not be too hastily assumed from this that rocks thus protected do in reality break down more rapidly than those exposed on bare hillsides, since here, where physical causes predominate, the loosened particles are removed as fast as formed, and, besides leaving no measure of the destruction going steadily on, new surfaces for attack are being continually

exposed. Moreover, in assuming that rocks decay rapidly where covered by vegetation, we must not overlook the fact that the character of the overlying soil may be such as to be protective rather than otherwise. Thus in glaciated regions it is a well-known fact that the striæ on rock surfaces are found best preserved where they have been protected from heat and frost by a mantle of drift, or the compact turf so characteristic of the Northern states. (See further under Influence of Forests, p. 280.)

(5) **Relative Rapidity of Weathering among Eruptive and Sedimentary Rocks.** — As to the relative rapidity of chemical decomposition among eruptive and sedimentary rocks, there can — with the exception of the calcareous varieties — be no question, the eruptives being far the more susceptible. This for reasons which will be at once apparent when we consider their origin. The eruptive rocks result from the comparatively sudden cooling of magmas originating far below the action of atmospheric agencies, and which are pushed up and allowed to solidify under conditions which are not at all conducive to chemical equilibrium. They are compounds of elements which have combined according to the conditions under which they temporarily existed, but which undergo continual changes as they become exposed by erosion and other causes. They become, in short, out of harmony with their surroundings, and there are at once set up a series of physical and chemical changes such as shall result in products more in harmony with existing conditions, and hence more stable. These changes, briefly put, are those involved in the weathering processes we have described. Indeed, we may well say that rock weathering and all the seemingly endless processes of rock decay and rock consolidation are but stages in the continual efforts being made by these inorganic particles to adjust themselves to existing conditions. But the sedimentary rocks (exclusive of the calcareous varieties) are themselves the actual products of these adjustments. The conglomerates, sandstones, shales, and argillites are but the detrital remains of eruptive rocks which under the various weathering influences have become disintegrated and decomposed, their more soluble constituents quite or in part removed, and the residues laid down and consolidated under conditions such as to-day exist upon or near the surface of the earth. They have, it is true, been laid down under water; they are subaqueous, but

their decomposition and disintegration was subaerial. Hence, when elevated above the ocean's level to become a part of the dry land, they are for the most part comparatively stable, subject to only such chemical changes as oxidation, and it may be dehydration. All other things being equal, then, those siliceous rocks which are the product of mechanical sedimentation will be found far less susceptible to the chemical action of the atmosphere and meteoric waters than are the eruptives. While they may undergo a transformation into soils, it is mainly through the disintegrating effects of heat and frost. Sedentary soils resulting from such disintegration resemble, therefore, their parent rock more than those of any other class.

Turning now to calcareous rocks, we shall find a quite different state of affairs prevailing, owing to the different chemical nature of the material and its ready solubility. These rocks represent, in fact, the soluble portions of the eruptive rocks which have been leached out during the process of decomposition. They are themselves solution products, although their immediate deposition may have been brought about through mechanical agencies, as in the laying down of beds of shell marl upon a sea-bottom. The lime leached out of terrestrial rocks is carried in solution into the sea, where, taken up by molluscs and corals as a carbonate, it becomes precipitated to the bottom on their death, and may reappear as a limestone, or, if mixed with sufficient quantities of other constituents, as a marl, calcareous sandstone, or shale. Such on their re-elevation are still subject to chemical change, owing to the ready solubility of lime carbonate in terrestrial waters, and so the endless round begins once more. Reference has already been made to the amounts of lime carbonate that may thus be annually removed from the earth's surface, but one may add here, that, according to J. G. Goodchild, certain English limestones waste away, superficially, at the rate of one inch in 300 years.[1]

(6) **Time Limit of Decay.** — We are sometimes enabled to put a time limit on the beginnings of decomposition such as shall enable us to gain at least a geological measure of the rapidity of the process. This is the case with the disintegrated granite of the District of Columbia described on p. 206. The residual material is here now overlaid by clastic deposits of such a nature as to force the conclusion that they were laid down by

[1] Geological Magazine, 1890, p. 403.

water under such conditions as would have thoroughly eroded away all underlying pre-existing decomposed material. It is therefore inferred that this decomposition has taken place since the clastic material was deposited, or, since these are of Cretaceous age, that it has taken place since the close of Cretaceous times. In the same way, since glaciation must have carried away the pre-existing disintegrated matter from the dike of diabase at Medford, leaving the surface smooth and hard, so here it is inferred that the decomposition is post-glacial. It is but rarely that the rate of decomposition of any rock has been sufficiently rapid since the beginning of human history, to be of geological significance, though weathered surfaces in old quarries, or the walls of old buildings, not infrequently offer abundant illustration of what we might expect, could observation be extended over whole geological periods instead of at most a few years. We must not forget, however, that, in the latter case, the conditions are quite different from those existing in nature, and the rate of weathering may be accelerated or retarded, as the case may be.

Stone implements, made by prehistoric man, as now found in graves, or dug from the soil, sometimes show incipient signs of decomposition, as indicated, when broken across, by a change in color and texture from without inward. Flint arrow and spear-heads from prehistoric caves or mounds in Europe, England, or America, often present on the outer surface a thin crust or *patine* of a gray or white color extending inward, it may be, for the distance of two or more millimeters. A grooved stone axe of diorite found in eastern Massachusetts and now in the collections of the National Museum at Washington,[1] shows concentric exfoliation in every way comparable to that on the diabase boulder figured on Pl. 20, extending inward to a depth of from three to six millimetres. It is of course possible that the axe was made from a boulder, itself not quite fresh, but this seems scarcely probable, and the inference is fair that both the patine and the exfoliation are due wholly to weathering subsequent to the manufacture of the implements on which they occur.

Mills[2] regards the extreme condition of decomposition existing in the Archæan rocks of Brazil as having taken place prior

---

[1] Specimen No. 172,794, Archæological Series.
[2] American Geologist, June, 1889, p. 345.

to the deposition of the loess, that is, in the long interval between the elevation of the Archæan rocks and the beginning of Quaternary times. Inasmuch, however, as the Quaternary gravels and loess are all readily permeable by water and not of a nature to be themselves readily affected, it would seem possible that at least a portion of the decomposition might have been brought about since their deposition and, indeed, to be still in progress.

The writer is informed by Mr. W. Lindgren that the granitic diorites of the Sierra Nevadas of California, and which are of

FIG. 22. — Flint implement showing weathered surface.

late Jurassic or early Cretaceous age, are often decomposed and disintegrated to a maximum depth of 200 feet, the extreme upper, more superficial portions being reduced to the condition of a red clay, while the lower are merely rendered soft and friable, with little if any change in color. This disintegration has gone on to such an extent that where the rock is traversed, as is sometimes the case, by numerous gold-bearing quartz veins, the entire mass of material is washed down by water — hydraulicked — as in the ordinary process of placer mining. The Pliocene andesites are also in places decomposed to a depth of

20 feet. The region is one of heavy annual precipitation, but the rainfall is limited almost wholly to the winter season.

Rock disintegration and decomposition, after the manner already described, has been by no means limited to the present era, but has been going on since the first land appeared above the surface of the primeval ocean. The results of the recent decomposition are more apparent, since the derived materials are still recognizable as rock débris, while that formed in past ages may have been so changed by the solvent and assorting power of water, the chemical action of the atmosphere, and the general agents of metamorphism, as to have quite lost its identity.

Dr. R. Bell, of the Canadian Geological Survey, has described [1] an interesting illustration of pre-Palæozoic decay in the crystalline rocks north of Lake Huron. The red granite, where it has been protected from glacial action, is found to be eaten into hollows in the form of round and sack-like pits and small caverns, the last-named generally occurring on steep slopes or perpendicular faces of the rock. These pits are, in places, of sufficient size to allow two men to crouch within. The sack-like ovens, such as are shown in Fig. 23, are most usually on sloping surfaces. The granite around these pits shows no indications of decay. That they are of pre-Palæozoic origin is demonstrated by the presence in them of residual patches, *in situ*, of the fossiliferous Black River limestone and which Professor Bell regards as veritable inliers of the Black River formation, which once filled all the inequalities and still overlies the granite at lower levels, though elsewhere almost wholly removed by erosion. Figure 23, after Bell, shows diagrammatically the old granitic corroded floor up on which the calcareous sediments were laid down, with pits still containing residual masses of the limestone, and the intact beds passing under the waters of Lake Huron at the lower right.

Pumpelly, too, has shown [2] that the diabase dike at Stamford, Massachusetts, had undergone extensive decomposition prior to the deposition of the Cambrian conglomerates. Of equal interest and still greater economic importance was the suggestion by this same authority, subsequently abundantly confirmed by W. B. Potter,[3] that beds of iron ore lying on the western

[1] Bull. Geol. Soc. of America, Vol. V, 1894, pp. 35–37.
[2] Ibid., Vol. II, 1891, p. 209.
[3] Jour. U. S. Assoc. Charcoal Iron Workers, Vol. VI, p. 23.

flank of Iron Mountain, Missouri, and covered by Silurian limestones, were true detrital deposits resulting from the pre-Silurian breaking down of the ore-bearing porphyry forming the mass of the mountain. These and other [1] illustrations that might be given point unmistakably to the identity of geological processes and correspondence in results since the earliest times, even did not analogy and the thousands of feet of secondary rocks furnish us safe criteria upon which to base our inferences.

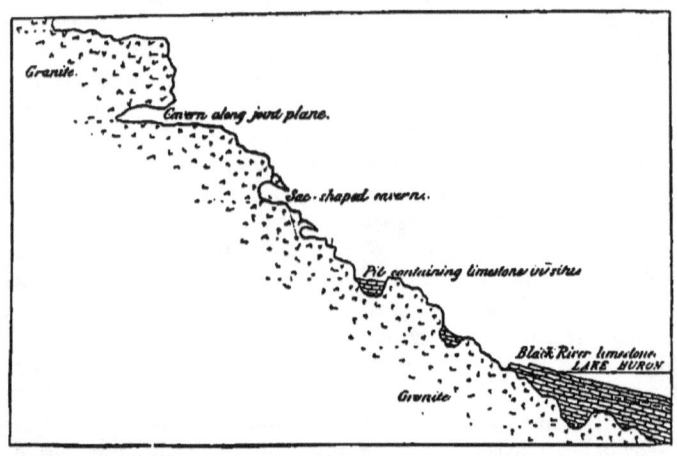

Fig. 23.

(7) **Extent of Weathering.** — The depth to which weathering has penetrated necessarily varies greatly. In cases where the detrital material is removed nearly or quite as rapidly as formed, it may go on indefinitely, until, it may be, thousands of feet of material have melted away; where, however, remaining in place, decomposition must be gradually retarded until a time comes when it practically ceases. In the region about Washington, District of Columbia, the writer has observed the granitic rock so disintegrated at a depth of 80 feet from the present surface as to be readily removed by pick and shovel. Even greater depths have been noted by writers on the geology of our own Southern states and Central and South America.

[1] See also T. Sterry Hunt, The Decay of Rocks Geologically Considered, Am. Jour. of Science, Vol. XXVI, 1883, p. 190.

Spencer states[1] that in the region about Atlanta, Georgia, the rocks are "completely rotted" to a depth of 95 feet, while "incipient decay" may reach to a depth of 300 feet. W. B. Potter describes[2] the feldspar porphyry of Iron Mountain in Missouri as decomposed to a visible extent as far into the hill as mining operations had been carried, while to depths varying from 10 to 80 feet the kaolinization is complete.

The coarse granite of Pike's Peak, Colorado, is reported as disintegrated to a depth of from 20 to 30 feet. Belt[3] describes dolerites in Nicaragua as decomposed, as shown by deep cuttings in mines, to a depth of 200 feet. "Next the surface," he says, "they were often as soft as alluvial clay, and might be cut with a spade."

Derby describes[4] certain shales in Rio Grande do Sul, Brazil, as decomposed into the condition of reddish, drab, greenish, black, and umber-colored clays to the depth of 120 metres (394 feet).

W. H. Furlonge has described[5] the granite of the Dekaap gold fields, in the Transvaal, South Africa, as decomposed to a depth of 200 feet. Rain erosion has carved out from this decomposed mass deep "dongas," as they are locally called, and which sometimes present most striking and picturesque appearances.

The apparent depth to which weathering has gone on is often greater among siliceous than calcareous rocks. This is, however, due merely to the facts that (1) the siliceous rocks are composed largely of insoluble materials, and hence leave a proportionately large amount of débris, and (2) that among calcareous rocks the change is mainly chemical and takes place only from the immediate surface. As a result of this, residuary nodules of limestone may be found perfectly fresh and unaltered at a depth of but a few millimetres below the surface, while granites and allied rocks may show signs of disintegration and incipient decay for many inches, or even feet.

Pumpelly states[6] that in the Ozark Mountains of Missouri the secular dissolving away of limestones containing from 2 to

[1] Geol. Survey of Georgia, 1893.
[2] Jour. U. S. Assoc. Charcoal Iron Workers, Vol. VI, p. 25.
[3] The Naturalist in Nicaragua, p. 86.
[4] Am. Jour. of Science, February, 1884, p. 138.
[5] Trans. Am. Inst. of Mining Engineers, Vol. XVIII, 1890, p. 337.
[6] Am. Jour. of Science, 1879, p. 136.

9 % of insoluble matter has left residual clays from 20 to 120 feet in thickness, indicating a removal of not less than 1200 vertical feet by solution. According to Whitney, the dark, reddish brown, residual clays of southern Wisconsin, of an average depth of perhaps 10 feet over the entire area, represent the insoluble accumulations from the decomposition of from 350 to 400 vertical feet of dolomite, limestone and calcareous shale.

(8) **Relative Rapidity of Weathering in Warm and Cold Climates.** — For many years an impression has prevailed to the effect that rocks decomposed more rapidly in warm and moist than in cold climates. While, owing to abundance of vegetation and other supposed favorable conditions, a more rapid decomposition might be expected, such has not as yet been proven to actually take place, and indeed many facts tend to prove the impression quite erroneous. Lack of decomposition products in high latitudes is not infrequently due to glaciation or erosion by other means. Whitney,[1] Irving,[2] Chamberlain, and Salisbury[3] have shown the presence of residual clays of all thicknesses up to 25 feet in the driftless area of Wisconsin, and Chamberlain has described[4] limited areas of strongly decomposed gneiss in the non-glacial areas of Greenland.

Moreover, we have no actual proof that the action of frost is, on the whole, protective, as is stated by Branner.[5] It must be remembered that frost, excepting in the extreme north, penetrates to but a slight depth, and while it undoubtedly puts a temporary stop to chemical action on the immediate surface, it remains yet to be shown that the mechanical disruption that ensues, and as described in previous pages, is not as efficacious as would have been the chemical agencies alone, had they been permitted to continue their work. Through bringing about a finely fissile or pulverulent structure, whereby a vastly greater amount of surface becomes exposed, frost prepares the way for chemical action at a thousand-fold more rapid rate than could otherwise have been possible. If, further, as the writer has elsewhere at least suggested,[6] hydration is the most potent

[1] Rep. Geol. Survey of Wisconsin, 1861.
[2] Trans. Wisconsin Acad. of Science, Vol. III, 1875.
[3] Ann. Rep. U. S. Geol. Survey, 1884–85, p. 254.
[4] Bull. Geol. Soc. of America, Vol. VI, 1895, p. 218.
[5] Bull. Geol. Soc. of America, Vol. VII, 1896, p. 282.
[6] Bull. Geol. Soc. of America, Vol. VI, p. 331.

factor in rock disintegration, the process can go on uninterruptedly below the level of freezing.

Professor H. P. Cushing has described[1] the argillites in the vicinity of Glacial Bay, Alaska, as in a condition of great disintegration, wholly through the action of frost. "Disintegration," he says, "takes place with amazing rapidity, as shown by the enormous piles of morainic matter furnished to the tributaries of Muir Glacier, whose valleys are adjoined by mountains of argillite, and by the massive talus heaps that are rapidly accumulating at the bases of other mountains made up of the same material." In a private communication to the present writer, he further states that the diabases of the region are fully as much decomposed as are those in the Adirondacks of New York, and that the blocks of eruptive rocks occurring in the moraines of Muir Glacier are far gone in decomposition.

Mr. C. W. Purrington has made similar observations, and states[2] that on the south side of Silver Bow Basin, some three miles west of Juneau, at an elevation of 2000 feet above sea-level, he found schistose diorites disintegrated over a considerable area to a depth of 20 feet. The particular locality cited was on a mountain slope, where landslides were frequent, and other conditions prevailed such as would prevent the accumulation of the débris throughout a prolonged geological period or to a very great depth. There could be, however, no doubt as to the residuary character of the material observed, and the inference drawn was to the effect that the disintegration had taken place within a comparatively brief space of time. G. E. Culver has also described[3] a diabase dike in Minnehaha County, South Dakota, an arid region lying within the glacial area, as decomposed throughout the whole exposures from its upper surface down to a depth of 20 or 25 feet, the limit of disintegration being the drainage level of the region as marked by the bed of a stream cutting through it.

On the other hand, Professor I. C. Russell, who has devoted much attention to the subject of rock-weathering in both high and low latitudes, is of the opinion that rock *decay* is a direct result of existing climatic conditions. He states further that decay goes on most rapidly in warm regions where there is an

[1] Trans. N. Y. Academy of Science, Vol. XV, 1895.
[2] Personal Memoranda to the writer.
[3] Wisconsin Academy of Sciences, Art, and Literature, 1886-91, p. 200.

abundant rainfall, and is scarcely at all manifest in arid and frigid regions.[1] Professor Russell's observations are of more than ordinary value, since he has discriminated between decay and disintegration, which most writers have failed to do.

Relative to the subject of rock degeneration in temperate regions, we have further to consider the possible increased amounts of atmospheric gases brought down by snowfalls, over those brought by rain. The snowflakes, in falling, so completely fill the air as to rob it of a larger proportion of its impurities than would a corresponding amount of precipitation in the form of rain. Further, the snow in melting slowly away affords the water better facilities for soaking into the ground than though it was poured down during the comparatively brief period of a shower. How far these agencies may go toward counterbalancing the effects of the continued higher temperatures of the tropics, we have no means of judging.[2]

It is even questionable if decomposition has actually gone on to greater depths in regions covered by forests, as contended by Hartt[3] and Belt[4] than elsewhere. The accumulation of a large amount of organic matter is undoubtedly favorable to decomposition, but the growing vegetation constantly robs the soil beneath of moisture and other elements necessary for its growth, storing it away in the form of woody fibre or sending it off into the atmosphere once more. The amount of moisture that a full-grown tree evaporates daily through its leaves is simply enormous, and is often made conspicuously apparent by the dry knolls that may be seen surrounding isolated trees or groups of trees in swampy areas. Indeed, Mr. R. L. Fulton, in discussing[5] the influence of forests in the mountain regions of the West, states it as his belief that the local springs and streams are "more diminished by the water used by the trees than by evaporation in their absence."

It has been shown[6] that the total amount of moisture returned

[1] Surface Geology of Alaska, Bull. Geol. Soc. of America, Vol. I, 1890.
[2] There is an old saying among Eastern farmers to the effect that a late spring snowstorm is as good as a dressing of manure. It undoubtedly arose from an appreciation by the farmers of the fact that the snow was more beneficial than rain for the reasons above mentioned.
[3] Physical Geography and Geology of Brazil.
[4] The Naturalist in Nicaragua, p. 86.
[5] Science, April 10, 1890.
[6] See Bull. No. 7, Forestry Division, U. S. Dept. of Agriculture, 1893.

into the atmosphere from a forest by transpiration and evaporation from the trees and underlying soil, is about 75 % of the total precipitation. For other forms of vegetation it varies between 70 % and 90 %, the forest as a rule being surpassed by the cereals, while the evaporation from a bare soil is but 80 % of the precipitation. To this should be added the fact that the activity of evaporation from forested areas is continued throughout a longer period of each year, as a rule, than in non-forested, for the simple reason that the grasses and cereals early ripen, and practically cease to exhale altogether. This is particularly the case in cultivated areas and prairie regions. Hence, while the daily evaporation from given areas might for a time be nearly equal, the annual amount is likely to be greatest for that which is forested.

Further, it has been shown that only 70 % as much rainfall reaches the soil in the woods as in the open fields, the rest being caught in the leaves, branches, and trunks, whence it is returned directly to the atmosphere by evaporation. These percentages naturally vary with the character of the forest growth. In this connection the following table, showing the measured amounts of water at varying depths in a loamy soil under forests of spruce, twenty-five, sixty, and one hundred and twenty years old, and one base of all vegetation, is instructive. It will be observed that the average amount is appreciably greater in the bare soil, and that the least amount is found under forests 60 years old, when we may assume the trees are in their prime.

WATER CONTENTS OF A LOAMY SAND; RESULTS BY SEASONS EXPRESSED IN PERCENTAGES OF THE WEIGHT OF THE SOIL

	SPRUCE					
SEASON	25 Years Old			60 Years Old		
	16 inch	32 inch	Average	16 inch	32 inch	Average
Winter (January and February) .	20.23 %	17.00 %	18.61 %	18.06 %	17.76 %	17.91 %
Spring (March to May)	18.62	18.02	18.32	15.29	16.28	15.78
Summer (June to August) . . .	15.10	16.22	15.90	14.42	17.03	15.72
Fall (September to November) . .	16.57	17.57	17.07	13.40	16.52	15.00

Season	Spruce 120 Years Old			Naked Soil		
	16 inch	32 inch	Average	16 inch	32 inch	Average
Winter (January and February)	19.75%	22.44%	21.09%	19.96%	24.73%	22.35%
Spring (March to May)	17.47	20.83	19.15	20.66	20.51	20.58
Summer (June to August)	17.78	20.90	19.97	19.77	19.98	19.97
Fall (September to November)	14.88	19.46	17.17	20.04	20.20	20.12

Other experiments have shown a marked difference in the distribution of the water in the forest-covered and naked soils, in the first-named a much larger proportion being held in the extreme upper portion than in that which was unprotected. This is a natural consequence of the absorptive properties of the accumulated humus. The following table, as compiled by Fernow[1] from the work of Ebermayer, illustrates this point.

AVERAGE OF WATER CAPACITY, EXPRESSED IN PERCENTAGES OF THE WEIGHT OF THE SOIL

Depth	Spruce			Unshaded Soil
	25 Years Old	60 Years Old	120 Years Old	
0 to 2 inches	30.93%	29.48%	40.32%	22.33%
6 to 8 inches	19.19	18.99	19.30	20.62
12 to 14 inches	19.19	16.07	18.28	20.54
19 to 20 inches	18.40	16.26	20.16	20.14
30 to 32 inches	17.91	17.88	21.11	20.54

It is obvious that it is only that portion of the water which passes through this superficial blanket of mould that can be instrumental in promoting rock decomposition. Hence the presence of such a blanket *may* exert a protective, or at least conservative, rather than destructive action. Further than this, we have to remember that plant growth tends to reduce the extremes of temperature and, even more, to diminish evapora-

[1] Bull. No. 7, Forestry Division, U. S. Dept. of Agriculture, 1893.

tion from the immediate surface. The constant action of gravity and capillarity in pumping the water down and up through the soil is therefore largely diminished. Since it is by temperature changes and water action that decomposition is so largely brought about, it is apparent that we must not be too hasty in assuming that forest action is actually destructive; it may be largely conservative. It is possible that the apparent amount of decomposition in wooded areas is due to protection from erosion, and the consequent accumulation of the residuary material. More facts are necessary before this question can be decided.

(9) **Difference in Kind of Weathering in Cold and Warm Climates.**—That, however, there may be a difference *in kind* in the degeneration in warm and cold climates, or at least in moist and dry climates, is possible and even probable.[1] In cold and in dry climates subject to extremes of temperature, as in the arctic regions or in the arid regions of lower latitudes, the weathering is at first almost wholly in the nature of disintegration, a process of disaggregation whereby the rock is resolved into, first, a gravel and ultimately a sand composed of the isolated mineral particles which have suffered scarcely at all from decomposition. The writer has elsewhere referred to this form of degeneration as manifested in the desert regions of the Lower Californian peninsula.[2] In a warm, moist climate chemical decomposition may or may not keep pace with the disintegration, according to local conditions, so that the resultant material may be in the form of an arkose sand, as in the District of Columbia, or a residual clay, as in the more superficial portions of the residual deposits to the southward. In certain cases, or among certain classes of rocks, the decomposition proceeds at so rapid a rate that there is scarcely any apparent preliminary disintegration. Local circumstances and character of rock masses being the same, we are, however, apparently safe in assuming that in warm and moist climates decomposition follows so closely upon disintegration as to form the more conspicuous feature of the phenomenon, while in dry regions, or those subject to energetic frost action, mechanical processes prevail and disintegration exceeds decomposition.

[1] The majority of writers have failed to discriminate between decomposition and disintegration. That there may be a very marked difference, due mainly to climatic conditions, is the point I wish to emphasize here.
[2] Bull. Geol. Soc. of America, Vol. V, 1894, p. 400.

Accepting these facts, there is at once suggested the idea that the lithological nature of sedimentary rocks, as well as their fossil contents, may be regarded as indicative of prevalent climatic conditions.

The possibility of estimating these conditions by the character of the débris resulting from the degeneration of feldspathic rocks was first suggested by the geologists of the Indian Survey,[1] the undecomposed feldspars in the Panchet (Mesozoic) sandstones being regarded as indicating a recurrence of a cold period during which mechanical forces preponderated over those purely chemical. The same idea was subsequently put forth, quite independently, by the present writer.[2] That rocks in arid regions do actually undergo less decomposition during the weathering processes is shown not only by the fresh character of the residuary material. Judd has shown[3] that rivers like the Nile, draining regions of great aridity, though in a condition of high concentration from prolonged evaporation, carry, in solution, smaller proportional amounts of derived salts than do those of humid regions.

Russell has noted that in the Yukon River region of Alaska disintegration so far exceeds decomposition that the talus from the mountains, composed of loose, angular masses of rock quite free from vegetation, forms what he calls *débris* streams, which actually creep slowly down the slopes, the movement taking place principally in the winter time and being due apparently to the slow settling, or creep, of deep snows. He states it as his opinion that the mountains of the region have suffered more through this form of disintegration than have those of Colorado or the southern Appalachians, but less than those of the Great Basin area. The range of limestone mountains along the Yukon is pictured as presenting a crest of sharp, blade-like crags, flanked by vast slopes of loose, angular stones on either side, the rock being everywhere fresh and undecomposed, but badly shattered and fissured.

(10) **Relative Amount of Material lost.**— Other things being equal, it is also safe to infer that more material has actually been lost through disintegration and decomposition in moun-

[1] Geol. of India, 2d ed., Vol. I, p. 201.
[2] Bull. Geol. Soc. of America, Vol. VII, p. 302.
[3] Report on Deposits of the Nile Delta, Proc. Royal Society of London, Vol. XXXIX, 1885.

PLATE 22

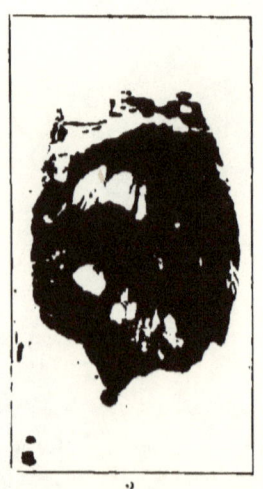

FIG. 1. Forest destroyed by wind-blown sand.
FIG. 2. Calcareous conglomerate carved and polished by wind-blown sand.
FIG. 3. Rock being undermined by wind-blown sand.

tainous and hilly countries than from the level plains. This for the reasons that (1) through the upturning of the beds there were exposed, it may be, friable and soluble strata that might otherwise have been protected, and (2) that through the shattering incident to this upturning the rocks were rendered more susceptible to the weathering forces. Further, (3) the steeper slopes in mountain regions promote more rapid removal of the resultant débris, whereby fresh surfaces are continually exposed, such as might otherwise shortly become protected through its accumulation, as above noted.

PART IV

THE TRANSPORTATION AND REDEPOSITION OF ROCK DÉBRIS

It rarely happens that more than a comparatively small proportion of the products of disintegration and decomposition are left to accumulate on the site of the parent rock. In most instances a very considerable proportion, in some instances *all*, the débris is removed immediately, or soon after its formation, and deposited elsewhere. A portion of this material is removed in solution, as has already been described (*ante*, p. 194). A still larger portion is transported mechanically, and it is to a discussion of the method of this transportation that a few pages may now be devoted with profit.

The chief agencies involved in this transportation are gravity, water, in either a solid or liquid form, and the wind. Undoubtedly the major part of the work is done by water, but as the wind's action is so frequently overlooked, and as, moreover, the results thus produced are of more than ordinary interest from our present standpoint, it may perhaps be well to dwell upon this branch of the subject with considerable detail.

(1) **Action of Gravity.**—Gravity, especially when aided by the lifting power of frost, may locally exert no insignificant influence. The tremendous power of landslides, or avalanches, have, owing to their devastating effects, been impressed upon us from the beginnings of written history. There are, however, other results, due to similar causes, but which, going on on an almost microscopic scale, are wholly overlooked by the ordinary observer, and the full meaning of which can be discovered only when the results of years are taken into account. Professor W. C. Kerr, in 1881, described [1] the manner in which the superficial cap of soil from the decomposition of micaceous

[1] Am. Jour. of Science, 3d Series, Vol. XXI, p. 345.

and hornblendic gneisses near Philadelphia had crept down the inclined surface on which it rested, and the gradual attenuation of the bands of variously colored débris of which it was composed. This creeping process he ascribed wholly to the expansive action of included water passing into the condition of ice, the expansion taking place laterally and the material being pushed down the slope along the line of least resistance. Mr. C. Davidson has since taken up the subject experimentally

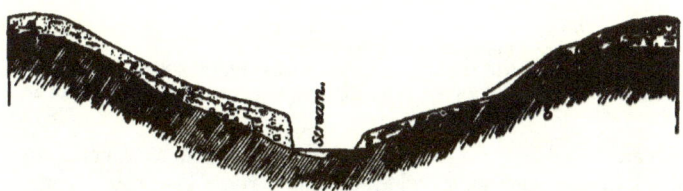

FIG. 24. — Showing direction and rate of motion of soil; the arrows showing, by their relative lengths, the rate of movement at various points. *a*, soil; *b*, bedrock.

and shown that the amount of the creeping could be accounted for by the ordinary laws of gravity, the frost, by its expansion, raising the individual particles a slight distance, and, on thawing, allowing them to drop back again a greater or less distance down the slope, according to the angle of inclination. Dr. Milton Whitney has, however, shown[1] that there is an almost continual movement among soil particles, dependent upon meteorological conditions quite aside from those involved in freezing and thawing. The creeping appears therefore to be but the manifestation, in mass, of the inclination of each individual particle to slide down the slope.

The accumulations of talus at the foot of every cliff and on the slopes of hills and mountains are matters of such every-day observation as to need no mention in detail.

(2) **The Action of Water and Ice.**[2] — The power of a stream to transport rock débris depends naturally upon its volume and the rapidity of its current. This, on the supposition that the character of the sediment to be transported remains the

[1] Some Physical Properties of Soils, Bull. No. 4, U. S. Weather Bureau, 1892.
[2] Students are referred to Professor R. D. Salisbury's article on Agencies which transport Material on the Earth's Surface, Journal of Geology, Vol. III, 1895, p. 70.

same. According to the calculations of Hopkins, as quoted by Geikie,[1] the capacity of transport increases as the sixth power of the velocity of the current; that is to say, the motor power is increased sixty-four times, by doubling the velocity. The following table is taken from the work quoted as showing the power of transport of river currents of varying velocities: —

INCHES PER SECOND	MILES PER HOUR	
3	0.170	: will just move fine clay.
6	0.240	: will lift fine sand.
8	0.4545	: will lift sand as coarse as linseed.
12	0.6819	: will sweep along fine gravel.
24	1.3638	: will roll along rounded pebbles 1 inch in diameter.
36	2.045	: will sweep along slippery, angular stones of the size of an egg.

There are, of course, other factors that should be taken into consideration, such as the character of a river bed, the density of the water, etc., but which lack of space prevents our touching upon here, and which are, moreover, sufficiently enlarged upon in other works.

The writer has stood at the head waters of the Missouri, and seen the Jefferson, Madison, and Gallatin rivers uniting their floods to form one grand rushing stream of clear green water, full of trout and grayling. He has seen it again at Mandan, Dakota, a sluggish stream actually yellow with suspended silt. At St. Louis, one beholds it a mighty torrent, whirling along trunks and stumps of trees, twigs, and all manner of organic débris and inorganic detritus picked up from its banks, or washed in by rains and tributary streams, till, one vast sea of liquid mud, it pours every year into the Gulf of Mexico a mass of sediment equal to 812,500,000,000,000 pounds (7,468,694,400 cubic feet), or enough to cover a square mile of territory to a depth of 268 feet. But only a portion of the detritus carried by running streams reaches the ocean; otherwise we need devote little time here to its consideration. Nearly all streams, in some part of their courses, flow through level plains with low banks which are subject to inundation during seasons of high water. Picture a muddy stream such as is shown in cross-section in Fig. 25, and which at ordinary periods is confined within the narrow channel near the centre. In time of freshet, however, the volume of water is so greatly augmented as to

[1] Text-book of Geology, 3d ed.

cause it to overflow its banks and spread out over the plains on either hand. But no sooner does the water leave the channel than the force of its currents becomes checked, its carrying power lessened, and it therefore begins to deposit its load of silt upon this *flood plain*, as it is called, where it remains to permanently enrich the land when the waters subside. It is to such processes of formation that we owe some of the most fertile lands in existence, as the valley of the Mississippi, that of the Red River of the North, the Nile, and scores of others that might be mentioned readily attest.[1]

FIG. 25.

To the same process, coupled with the accumulation of organic matter, we owe the filling in and gradual extinction of thousands of glacial lakes throughout New England and the North, and the formation of rich, flat-bottomed valleys known locally as meadows, swales, and bogs.

Ice in the form of glaciers is an efficient agent for transportation as well as for erosion, as already noted. While the work being done by existing glaciers may seem comparatively insignificant, that done by the ice sheet of the glacial epoch was by no means so, and deserves a more than passing notice. The manner in which the ice carries and deposits its load has already received attention in speaking of its erosive power, and but little more need be said on the subject. That material which existed in a loose, unconsolidated condition, on the surfaces on

[1] The Arkansas River is stated by Owen (Geol. of Arkansas, 2d Rep., 1860, p. 52) to be at certain seasons of the year almost blood-red from the quantity of suspended fine ferruginous clay and saliferous silt brought down from the regions of ferruginous shales, which prevail in the Cherokee County, through which the river flows. This material, deposited along the banks and in the eddies of still water, produces the celebrated red buckshot land. Material washed from the bluffs of argillaceous shell marl, near the confines of Jefferson and Pulaski counties, is deposited again farther down the stream as a fine silt, imparting, like the red silt, extraordinary fertilizing properties to the soil.

U

which the glacier formed, was pushed and dragged along by the onward movement of the ice, which in extreme cases may have exerted a pressure of 200,000 pounds to the square foot. On the final retreat of the glacier, this was left in the form of a compact structureless mass of almost stony hardness, commonly known as *till* or *ground moraine*. Materials falling upon the surface from greater heights were likewise transported, so long as the ice sheet continued to advance, and finally deposited in the form of *terminal* or *frontal* moraines.

Inasmuch as the ice sheet was almost continually melting upon its surface, it is practically impossible to consider its action wholly independent of that of water also. Thus, streams resulting from such melting would gradually wear channels in the ice, as on the land. In these channels would accumulate sand and boulders of such size and weight as to resist the current, and such accumulations would, on the final melting of the sheet, be deposited on the surface of the ground in the form of ridges known as *eskers*, or *osars*. Other forms of water action on the materials of the ice sheet, are hillocks of stratified sand and gravel deposited near the terminal moraines, and known as *kames*. Since during the advancing of this ice sheet existing rivers flowing eastward must have been dammed, we can safely imagine the formation of large temporary lakes, on the bottom of which would be deposited the glacial silt, like the so-called *loess* of the Mississippi valley. Lake Agassiz, a glacial lake of this type, is supposed to have occupied an area of more than 100,000 square miles in northwestern Minnesota, northeastern Dakota, and a considerable portion of Manitoba. On the bottom of this lake there was deposited during the comparatively brief time of its existence, silt to a depth as yet undetermined, but known to be at least 100 feet.[1]

Waters issuing from the melting ice sheet tend to reassort the material of the terminal moraine, redepositing it in approximately concentric zones beyond its margin. These deposits are naturally thicker and coarser near the moraine and thinner and finer at increasing distances. Their form and mode of occurrence is such as to have suggested for them the name of glacio-fluvial aprons, or frontal aprons. Their materials are nearly always loose sands and gravels, the lithological nature

[1] Ice Age in North America, by G. F. Wright, p. 355.

of the individual particles being of course dependent upon that of the moraines from which they are derived.

The effects upon the landscapes of this ice sheet have been lasting and peculiar. We may safely imagine that, before the ice invasion, the surface was covered with decayed and softened materials like the residual soils of our Southern states, and which had been cut up into valleys and intervening ridges by the stream of that time. The ice sheet stripped from these surfaces their mantle of decomposed materials, and in addition cut, in many cases, into the fresh rock, actually planing the entire country so deeply that in most cases the preglacial surface is no longer recognizable. The hills were thus lowered and the valleys in some cases deepened or again filled by sand and gravel. Since a protruding rock mass would, from necessity, be most eroded on the side from whence the ice sheet approached, and since, moreover, such would serve to catch and hold back a part of the loose earth and stony matter brought from the north, a peculiar feature in the topography of glaciated hills has been brought about as shown in Fig. 2, Pl. 25.

The direction taken by this drift material was quite variable. It was, as a rule, from the north toward the south, with many minor deflections. Boulders of Laurentian rocks north of Lake Huron are abundant in the drift about Oberlin, Ohio, and even further south. Boulders of native copper from the Lake Superior region are found even as far south as Kankakee, Illinois, and a large boulder of a peculiar conglomerate known in place only near Ontario, has been found a few miles south of the Ohio River in Kentucky. Dawson states "that boulders from the Laurentian axis of the continent, which stretches from Lake Superior northward to the west of Hudson Bay, have been transported westward a distance of 700 miles, and left upon the flanks of the Rocky Mountains at an elevation of something over 4000 feet."[1]

All over the states once occupied by this ice sheet the material originating from the decomposition of rocks *in situ*, or deposited on alluvial plains, was, with a few minor exceptions, carried away to the southward and in part dumped into the Atlantic, while its place was supplied by mongrel hordes from the north. In process of digging for the foundations of the

[1] Ice Age in North America, p. 171.

Maine Experiment Station at Orono, the fresh and highly polished slaty rock was found but a few feet below the surface, proving incontestably that, with the exception of the small amount of organic matter that had since been added, not an ounce of the soil was truly native, but all of foreign birth, and a mongrel creature of compulsory migration. We shall dwell more fully upon the character and distribution of these soils later. The single illustration above given will answer present purposes.

In a less degree the ice along the shores of lakes and rivers may exert a transporting influence. Thus the ice first formed along the shores encloses sundry pebbles, boulders, and sand. Through the expansion force of the freezing water as the entire surface becomes frozen over, this shore ice, together with its enclosures, may be pushed up some distance beyond the water line, where the included débris is deposited on melting. Or, on the breaking up of the ice in the spring, the shore ice may be drifted to other parts of the lake, or down the stream, perhaps for miles before melting sufficiently to cause it to deposit its load.

(3) **Action of Wind.**[1] — While abrasion by the wind is impossible without transportation, the converse is by no means true; indeed it is as an agent of transportation for rock detritus, without appreciable abrasion, that the wind accomplishes its greatest work, though in like manner this phase is most manifest in arid regions.

It is stated by Darwin that for several months of the year large quantities of dust are blown from the northwestern shores of Africa into the Atlantic over a space some 1600 miles in width and for a distance of from 300 to 600 and even 1000 miles from the coast. During a stay of three weeks at St. Jago in the Cape Verde Archipelago, this authority found the atmosphere almost always hazy from the extremely fine dust coming from Africa and falling upon the land and water for miles around. So abundant was this dust that a distance of between 300 and 400 miles from the coast the water was distinctly colored by it. In the arid lands of Central Asia the air is also reported as often laden with fine detritus which drifts like snow around conspicuous objects and tends to bury them in a dust

[1] See article on Erosion performed by the Wind, by Professor J. A. Udden, Journal of Geology, Vol. II, 1894, p. 318.

drift. Even when there is no apparent wind, the air is described as often thick with fine dust, and a yellow sediment covers everything. In Khotan this dust sometimes so obscures the sun that even at midday one cannot see to read fine print without the aid of a lamp. The tales of the overwhelming of travellers and entire caravans by sand storms in the Great Desert of Sahara are familiar to every schoolboy. Greatly exaggerated though these may be, the accounts of Layard and of Loftus show us that the sand storms which are of frequent occurrence during the early part of summer throughout Mesopotamia, Babylonia, and Susiana are by no means of insignificant proportions. Layard states that during the progress of the excavations at Nimrud, whirlwinds of short duration but almost inconceivable violence would suddenly arise and sweep across the face of the country, carrying along with them clouds of dust and sand. Almost utter darkness prevailed during their passage, and nothing could resist their force; the Arabs would cease their work and crouch in the trenches almost suffocated and blinded by the dense cloud of fine dust and sand which nothing could exclude.

The accounts of Loftus are equally impressive. Describing their departure from Warka to Sinkara, he says: "A furious squall arose from the southeast and completely enveloped us in a tornado of sand, rendering it impossible to see within a few paces. Tellig and his camels were as invisible as though they were miles distant. A continuous stream of the finest sand drove directly in our faces, filling the eyes, ears, nose, and mouth with its penetrating particles, drying up the moisture of the tongue, and choking the action of the lungs." With such descriptions before one it is not difficult to believe that these ruined cities have in the course of centuries been completely hidden and their sites obscured by mounds of wind-drifted sand and dust.

We need not, however, confine ourselves wholly to the Old World for illustrations. Not longer ago than May of 1889 a dry southwesterly wind which for several days had prevailed in various parts of the Northwest, particularly in Dakota, culminated in a storm peculiarly suggestive from a geological standpoint. It is stated[1] that during the prevalence of this wind, on the 6th and 7th of the month mentioned, the air be-

[1] American Geologist, June, 1889, p. 308.

came filled with flying particles caught up from the ploughed fields, fire-blackened prairies, public roads, and sandy plains. The particles formed dense clouds and rendered it as impossible to withstand the blast as it is to resist the blizzard which carries snow in winter over the same region. The soil to a depth of 4 or 5 inches in some places was torn up and scattered in all directions. Drifts of sand were formed in favorable places, several feet deep, packed precisely as snowdrifts are packed by a blizzard. It seemed as if there were great sheets of dust and dirt blown recklessly in mid air, and when the wind died down for a few moments, the dirt, fine and white, appeared to lie in layers in the atmosphere, clouding the sun and hiding it entirely from sight for an hour or more at a time. (See also on p. 184.)

Over the wide, dry, and bare flat-topped terraces of the upper Madison valley the wind sweeps in a strong steady current for days together, or during the heated portion of the year, when the sun pours from a cloudless sky its hottest rays upon the parched soil, starts up spasmodically here and there in the form of small whirlwinds made visible by the dust they carry, and which wander spectre-like across the plain to noiselessly disappear in the distant mid air.

Dust columns of this nature are common in all arid regions, and doubtless have been observed by the many who have crossed the Humboldt desert in Nevada. Seated comfortably in a Pullman car on the Union Pacific, one may not infrequently see at a single view not less than a half dozen of these geological spectres, each in the distance doing its apportioned task and silently disappearing, laying down its load of sand as its strength gives out and leaving it for its successor.[1]

Under proper conditions such of these wind-blown sands as are too heavy to be carried into the air as dust accumulate upon the surface in the form of drifts, or dunes, all lying with their longer axes approximately at right angles with the prevailing currents. Excepting during periods of calm, such are in a state of almost constant, though it may be imperceptible, motion, ever changing their shapes and moving onward like long parallel drifts of snow. The rate of motion of a dune

---

[1] Professor J. A. Udden estimates that the dust in a cubic mile of lower air during a dry storm weighs not less than 225 tons, while in severe storms it may reach 126,000 tons (Popular Science Monthly, September, 1886).

from necessity is governed by the strength and constancy of the winds, and the fineness and dryness of the sand. Urged into temporary activity, each little grain goes scurrying up the slope, across the crest, and tumbles to rest in the steeper declivity upon the leeward side, to be slowly buried by those which follow. This is the sum total of the movement taking place in the march of a dune, whatever its pace and however great its bulk. Yet in this very faculty of moving itself forward by but a ten billionth part of its bulk at a time lies the whole secret of its power. Silently, imperceptibly it may be except when measured by months and perhaps years of time, retarded by no walls nor ordinary declivities, it relentlessly performs its task.[1]

A writer in one of the recent popular magazines estimates the dunes of Hatteras and Henlopen as in some cases upwards of 70 feet in height and moving at least 50 feet a year. Swamps have thus been filled, forests and houses buried, and it is stated that but a few years can elapse before the entire island lying north of Cape Hatteras will be rendered uninhabitable. The sand dunes on the coast of Prussia commenced not over a century ago, and already fields and villages have been buried and valuable forests laid waste by them. In one instance a tall pine forest covering many hundred acres was destroyed during the brief period intervening between 1804 and 1827. Loftus, writing of Niliyga, an old Arab town a few miles east of the ruins of Babylon, says that in 1848 the sand began to accumulate about it, and in six years the desert within a radius of six miles was covered with little undulating domes, while the ruins of the city were so buried that it is now impossible to trace their original form and extent. A still more striking illustration of the rapidity of these sand accumulations is offered by the same authority in describing the burial customs of some of these ancient people, it being stated that the earthen coffins were merely stacked in layers one on top of another, and left thus to be covered by the finer sand sifted over them by the winds from the desert. Even Nineveh, founded some twenty centuries before Christ and destroyed 1400 years later, became so covered by drifted sands that at the time of the Greek Xenophon (about 400 B.C.) the very site of the once famous

---

[1] The Wind as a Factor in Geology, Engineering Magazine, 1892, p. 506.

city was unknown. Marsh[1] gives the rate of movement of dunes along the western coast of Jutland and Schleswig-Holstein as averaging 13½ feet a year, while Anderson estimates the average depth of the sand over the entire area as about 30 feet, equalling therefore about 1½ cubic miles for the total quantity.

It is not in all cases possible to trace the drifted sands to their various sources. Dunes along the sea-coasts are in nearly all cases composed of materials thrown up by the waves on the beaches in the immediate vicinity. This is the case with those of Hatteras, Cape Cod, Gascony, Algeria, and Schleswig-Holstein. But the origin of the large inland dunes, like those of Nevada, is not always so clear. It has been suggested that these last are formed of beach sand driven in by the prevailing westerly winds from the Pacific coast. This is, however, a matter of very grave doubt, and it seems more probable, as stated by geologist Russell,[2] that they were derived from the disintegrating granites of the Sierras. They certainly have travelled far, and are not a product of disintegration of rocks in the immediate vicinity.[3]

By wind action, accompanied by the carrying power of spasmodic or perennial streams, were formed the wide stretches of adobe in the western United States, and according to many authorities the deposits of loess which cover, as in Europe and Asia, areas aggregating many square miles and which have a depth, in extreme cases, of 2000 feet.[4]

The tendency of the wind is not, however, in all cases toward

---

[1] The Earth as modified by Human Action, p. 562.

[2] Quaternary History of Lake Lahonton, Nevada, Monograph, U. S. Geol. Survey, 1885.

[3] The sands covering the Egyptian Sphinx and Pyramids are stated to have come mainly from the sea on the north, and not from the desert, as is popularly supposed. Sand showers having their origin in the desert of Sahara extend across the Mediterranean, and as far as northern Italy (Nature, July 18, 1889, p. 286).

[4] The wind plays an important part in the transportation of soils in Wyoming, owing to the incoherent state of the soils, due to the lack of clay. The arid regions of this state, which are chiefly Tertiary and Cretaceous plains and tablelands, receive very little rain. Consequently the soils become loosened by great earth cracks, and during the dry and windy winter weather are transported in dense clouds, which almost suffocate travellers, to the broken country and distant hills and mountains. In a single season it is not an uncommon sight to see banks of earth, like huge banks of snow, behind a reef of rock, or in the lee of large bunches of sage brushes (U. S. Dept. of Agriculture, Office of Experiment Stations, Vol. V, No. 6, 1894, p. 567).

forming drifts and ridges, but at times rather to reduce the land to one general level. Thus J. Flinders Petrie[1] states that near the ancient cemetery of Tell Nebesheh, on the Isthmus of Suez, the surface of the country has been cut down at the rate of 4 inches a century until some 8 feet have been removed from the dry areas and deposited in the intervening depressions, slowly converting the existing lakes into marshes, and the marshes into dry land. An even more rapid change of contours is that described by Dwight[2] as having taken place on Cape Cod, Massachusetts. The entire country here is composed of sand so susceptible to the drifting action of the wind that it has for years been the custom of the people to sow pines and coarse beach grass to hold it in place. In the instance described by Dwight, however, reckless pasturage had so far destroyed the grass as to lessen its protecting power, and under the strong breezes from the open Atlantic it began to drift rapidly. Over an area of about 1000 acres the sand was blown away to a depth, in many places, of 10 feet. "Nothing," says Dwight, "could exceed the dreariness of this scene. Not a living creature was visible; not a house, nor even a green thing except the whortleberries which tufted a few lonely hillocks rising to the height of the original surface, and prevented by this defence from being blown away also. The impression made by this landscape cannot be realized without experience. It was a compound of wildness, gloom, and solitude. I felt myself transported to the borders of Nubia, and was well prepared to meet the sand columns so forcibly described by Bruce, and after him by Darwin. A troup of Bedouins would have finished the picture, banished every thought of my own country, and set us down in an African waste."

One more instance of contour changes of this sort must suffice. It is stated[3] that in Pipestone and Rock counties in Minnesota, the bluffs facing to the westward are, as a rule, more precipitous and more rocky than those facing in the opposite direction. This fact is regarded by Professor Winchell as due to the action of the prevailing westerly winds, combined with the drying effects of the southwestern sun in summer. Such winds would uncover and keep bare the coarser materials of

[1] Proc. Royal Geographic Soc., November, 1889, p. 648.
[2] Travels in New England and New York, Vol. III, p. 101.
[3] Geol. of Minnesota, Vol. I, p. 575.

the western surface by blowing away the sand and clay, while the bluffs on the east are not only protected from the winds, but collect upon their slopes all the flying particles from the prairies above.

The finely comminuted rock dust blown from volcanic vents is often drifted for long distances by atmospheric currents, and ultimately deposited in beds of no insignificant proportions. Dense clouds of such dust were blown from Icelandic volcanoes to the coast of Norway in 1875, and subsequent to the eruption of Krakatoa (in 1883) the ship *Beaconsfield* of Philadelphia, while at a distance of 831 miles from the source, sailed for three days through clouds of dust which fell upon her decks at the rate of an inch an hour. That such are not or have not in the past been unusual instances is shown by results obtained by the *Challenger* Expedition, volcanic ashes and sand being repeatedly dredged up from almost abysmal depths at points in the central Pacific far remote from land areas. The day following the explosive eruption of St. Vincent, in 1812, the Barbadoes Island, 80 miles to the windward, was completely shrouded in darkness for many hours, the light of the sun being almost wholly obscured by the cloud of impalpable dust which in the form of a slow, silent rain fell over the whole island. "The trade wind had fallen dead; the everlasting roar of the surf was gone; and the only noise was the crushing of the branches snapped by the weight of the clammy dust. About one o'clock the veil began to lift, a lurid sunlight stared in from the horizon, but all was black overhead. Gradually the dust cloud drifted away; the island saw the sun once more, and saw itself inches deep in black, and in this case fertilizing, dust."[1]

[1] Kingsley, as quoted by Belt, in The Naturalist in Nicaragua, p. 354.

PART V

THE REGOLITH

THROUGHOUT the millions of years which have elapsed since the earth assumed its present form and essentially solid condition, the rocks composing its more superficial portions have been constantly undergoing degeneration in the manner described, and, in so doing, have given rise to the immense masses of materials which constitute the thousands of feet of secondary rocks, and the still unconsolidated sands, gravels, and other products which will be considered in detail later. With those products which have undergone lithification, which are now in the state of consolidation commonly ascribed to rocks by the popular mind, we shall have little more to say. These have already been sufficiently described as rocks in Part II of this work. It is to the most superficial and unconsolidated portion of the earth's crust that we will now devote our attention.

Let the reader for a moment picture to himself the present condition of this crust, with particular reference to the land areas. Everywhere, with the exception of the comparatively limited portions laid bare by ice or stream erosion, or on the steepest mountain slopes, the underlying rocks are covered by an incoherent mass of varying thickness composed of materials essentially the same as those which make up the rocks themselves, but in greatly varying conditions of mechanical aggregation and chemical combination.

In places this covering is made up of material originating through rock-weathering or plant growth *in situ*. In other instances it is of fragmental and more or less decomposed matter drifted by wind, water, or ice from other sources. This entire mantle of unconsolidated material, whatever its nature or origin, it is proposed to call the regolith, from the Greek words ῥῆγος, meaning *a blanket*, and λίθος, *a stone*. Within certain limits it varies widely in composition and structure, and many names have, on one ground and another, been applied to its local phases, the more important of which are given in tabular form below, and described in detail in the pages following. According to its origin, whether the product of transporting agencies as noted above, or derived from the degeneration of rocks *in situ*, the regolith is found lying upon a rocky floor of little changed material, or becomes less and less decomposed from the surface downward until it passes by imperceptible gradations into solid rock.

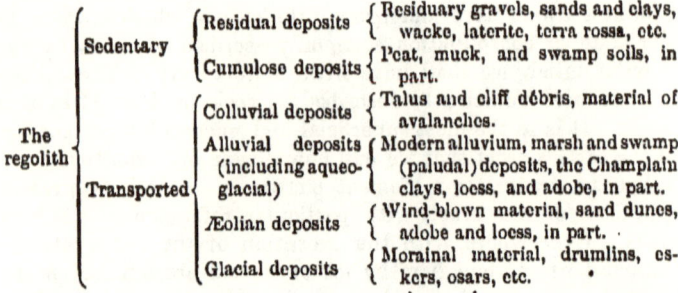

The extreme upper, most superficial portion of this regolith, that which affords food and foothold for plant life, is commonly designated as soil; that immediately underlying the soil, and passing into it by insensible gradations, is known as the sub-soil. This last differs from the soil proper only in degree of compactness and in such chemical changes as may have been induced in the soil through growing organisms and more extensive weathering. Indeed, the soil is but derived from the sub-soil, and were it entirely removed, would shortly be replaced through the same agencies as first gave it birth.

The characteristics of individual soils can be best discussed when speaking of those local phases of the regolith of which they form a part, and with this understanding we will proceed.

1. SEDENTARY MATERIALS

Here are to be considered those deposits which, resulting from chemical decomposition or disintegration, from any or all of the processes involved in rock-weathering, or from organic accumulation, are found to-day occupying their original sites. They are, in fact, the primeval types of nearly all soils and secondary rocks, since those of drift origin are but derived from sedentary materials through the transporting agencies of air and water. They may be conveniently divided into two classes, (1) residual [1] and (2) cumulose.

(1) **Residuary Deposits.** Under this name, then, are included all those products of rock degeneration which are to-day found occupying the sites of the rock masses from which they were derived, and immediately overlying such portions as have as yet escaped destruction. The name is peculiarly appropriate, since, they are actually residues, left behind while the more soluble portions have been leached away by meteoric waters.

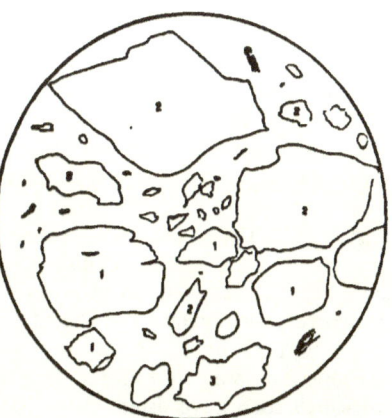

FIG. 26. — Showing angular outlines of residuary particles from decomposed gneiss. 1, mica; 2, feldspar; 3, quartz.

The residual deposits of North America reach their maximum development in the portion of the United States east of the Mississippi and south of the southern margin of the ice sheet of the Glacial epoch. Their mode of accumulation and general characteristics have been very thoroughly discussed by Professors I. C. Russell, Chamberlain, and Salisbury,[2] on whose papers we shall draw for some of the facts given here.

---

[1] Various names have from time to time been proposed for deposits of this nature, but obviously it is impossible to include under a single lithological term materials so widely variable. The term *saprolite* (from the Greek σαπρος, rotten, recently suggested by G. F. Becker, 16th Ann. Rep. U. S. Geol. Survey, Part III, p. 289) is objectionable as conveying the idea of putridity. The old provincial term *geest* adopted by De Luc, and recently endorsed by McGee (11th Ann. Rep. U. S. Geol. Survey, 1889-90, p. 279), has lost whatever precise meaning it may have had, being defined in both the Standard and Century dictionaries as (1) a bed derived from rock decay *in situ*, (2) high gravelly land, and (3) gravel or drift. The term *gruss*, although advocated by some American authorities, is of old German origin and open to the same objection.

[2] Bull. 52, U. S. Geol. Survey and Ann. Rep. U. S. Geol. Survey, 1884-85.

The prevailing characteristic of an *old* residual deposit, from whatever rock it may be derived, is a ferruginous clay. Examined by a microscope, its mineral particles, when not too thoroughly decomposed, are found to be sharply angular in outline. With the exception of the quartz, the various mineral constituents are often in an advanced stage of decay, and the more soluble constituents are wholly or partially lacking, having been leached out, in the manner already described.

Owing to the prevalence of the aluminous constituents, these deposits, when thoroughly decomposed, as on the immediate surface, are very tenacious, and may well be termed clays. Their colors are dull, or some shade of brown or red, owing to the higher oxidation and perhaps dehydration of the ferruginous matter set free by the decomposition of the iron-bearing silicate constituents. Such in general are the residual soils of the southern Appalachian regions of the United States and which are apparently in every way comparable with the *terra rossa* of Europe, but only in a slight degree with the laterite of India, to which they have often unfortunately been referred.[1] From a chemical standpoint the soils forming the upper portion of the residuary deposits, though of a prevailing aluminous character, vary widely from the rock masses from whence they were derived, much depending upon their age and the amount of actual decomposition and leaching that has taken place. On p. 306 are given a few typical but widely varying analyses which will serve to illustrate this point.

Deposits of this nature are never truly stratified, excepting where, through having remained wholly undisturbed, they retain the original structure of the parent rock. (See under Effacement of Original Characteristics, p. 262.)

The residuary differ from the drift deposits in that they contain no materials foreign to their vicinity, but only such more enduring matter as has been handed down to them from the

---

[1] The term *terra rossa*, according to Neumayer (Erdgeschichte, Vol. I, p. 405) was first applied to the red residual deposits in the Karst maritime lands of the Adriatic Sea. The material is described as a highly ferruginous clay resulting from the leaching out, by meteoric waters, of the soluble portions of the prevailing limestones. Its distribution is by no means limited to the maritime provinces of the Karst, but it is found also on the Grecian coasts and in the Schwabia-Frankonia Jura Plateaus of Bavaria. In fact it is to be found anywhere in these regions where the prevailing country rock is a marine limestone and erosion not sufficiently active to remove the residuary material.

parent rock. In the case of limestones such matter consists mainly of aluminous and ferruginous matter, grains of sand, and nodular masses of chert which existed as mechanically admixed impurities.

The inherited characteristics of deposits of this nature may be illustrated by the accompanying exaggerated section across central Kentucky where, it is easy to see, the regolithic material overlying the Lower Silurian and Cambrian limestones may

FIG. 27.

contain a portion of all the insoluble residues from the hundreds of feet of Upper Silurian, Devonian, Lower and Upper Carboniferous beds which formerly stretched above them. Upon the nature of this inheritance must depend the adaptability of the regolith to soil purposes and its consequent fertility.

The transition from a regolith of this type to fresh rock is usually quite sharp, owing to the fact that limestones decompose mainly through solution from the immediate surface. Nevertheless there is a gradual change in the character of such a deposit from above downwards, owing to the oxidizing influence of the air and percolating waters. (See p. 307.)

As above noted, the mineral particles in the older residuary deposits are, with the exception of the quartz, found to be as a rule in a state of advanced decomposition. Nevertheless the ultimate individual constituents of even the darkest clays of the driftless regions of Wisconsin, as examined by Messrs. Chamberlain and Salisbury, are transparent, although stained by iron oxides.

Concerning the physical properties of limestone residues as occurring in this driftless area, the following statements are made by Messrs. Chamberlain and Salisbury. "Above, the clay graduates into soil which, outside the valleys, is uniformly shallow. Beneath the soil, the clay loses the dark color of the latter, due to the presence of organic matter, but is for a certain distance downward not unlike the superior portion in texture. The deeper lying clay, where limestone is the subjacent rock, is the most characteristic member of the residuary earth series.

It is not like that above, structureless, although, like that, it is without trace of stratification. It generally shows a tendency to cleave, breaking up into little pieces which are roughly cubical. This is often conspicuous, and especially so on the faces of sections which are thoroughly dry. In such situations large quantities of the clay in small angular blocks may be removed by slight friction. The size of the cuboids varies, within somewhat narrow limits, from a small fraction of an inch to one or two inches in diameter. This cleavage is probably a phenomenon of shrinkage due to drying, as it partially disappears when the clay becomes wet. This structure has given rise to the local name of 'joint' clay, an appellation not altogether inappropriate.

FIG. 28. — Showing angular character of quartz particles in decomposed gneiss.

"Upon drying, this variety becomes very hard and rock-like. It only becomes adapted to serve as soil by surface amelioration, as is shown by the fact that, from the thousands of mineral holes scattered over the southern part of the mining district, the material ejected still lies beside the excavations as heaps of clay, without covering of vegetation, although it has been exposed in most cases for many years. Notwithstanding this fact, the clay, even in its deepest parts, wherever examined, is found to abound in minute perforations. These, in many cases at least, indicate the penetration of rootlets, for the rootlets themselves may sometimes be found. In some cases, too, the perforations have been seen to undergo a gradual variation in size, and to branch now and then, much as rootlets do. On the other hand, it is probable that some of the perforations have had a different origin, for in one case a small insect was found in one of the little canal-ways. The clay is exceedingly tenacious, and hence the perforations, once formed, would endure for long periods of time.

"Another characteristic of certain portions of the clay is its power of retaining moisture. It can rarely be found, even in the driest season, unless exposed to the direct rays of the sun, without visible moisture a few inches from the surface. The regions where it is present are conspicuously less affected by drought than adjacent localities where it is wanting. For this reason it is a valuable sub-soil.

"Fragments of residuary rock are not uncommon in the deeper portions of this earth. Of these, chert fragments are most abundant, and occur scattered sparingly throughout the clay or sometimes arranged in more or less distinct layers in it. Even where they appear to be entirely wanting, the microscope often reveals minute flakes scattered sparsely throughout the clay. The larger pieces are more numerous near the basal portion of the clay than higher up.

"It is natural to suppose that the residuary earths derived from the decomposition of limestone would differ very notably from those which take their origin from sandstones or from shales or mixed crystalline rocks. Yet the difference is far less than might be anticipated. There usually overlies the sandstone strata a loamy earth not very far removed in character from that which mantles limestones. It is somewhat more sandy, and consequently less cohesive, and presents the opposite variations in vertical sections, becoming less cohesive below, instead of more so. In the limestone region the toughest clay lies next to the rock. In the sandstone regions the soil graduates below into sand. The difference is most conspicuous where the mantle has been washed and redeposited and mingled with mechanically derived sand and secondary products, as occurs in some of the valleys."[1]

The following analyses, in part from this same report, will answer, in connection with those already given, to show the prevailing type of the residuary deposits throughout widely separated areas. It will be noted that silica exceeds as a rule all other constituents, while alumina, iron oxides, and moisture make up the main bulk of the residue. This generalization holds good of nearly all sedentary soils, whatever the character of the rocks from which they were derived, and is the more pronounced the more advanced the decomposition.

[1] 6th Ann. Rep. U. S. Geol. Survey, 1884–85, pp. 240–242.

ANALYSES OF RESIDUARY DEPOSITS OF VARIOUS KINDS

Constituents	I	II	III	IV	V	VI	VII	VIII	IX	X
Silica (SiO$_2$)	71.13 %	49.90 %	53.09 %	49.13 %	55.42 %	40.127 %	39.55 %	66.27 %	77.24 %	55.39 %
Alumina (Al$_2$O$_3$)	12.50	18.64	21.43	20.08	22.17	13.75	28.76	15.25	26.17	20.16
Ferric oxide (Fe$_2$O$_3$)	5.52	17.19	8.63	11.04	} 8.30	12.315	10.80	6.97	7.76	8.79
Ferrous oxide (FeO)	0.45	0.27	0.86	0.93						
Titanium oxide (TiO$_2$)	0.45	0.28	0.16	0.13	....	....	0.64	....	....	....
Phosphoric acid (P$_2$O$_5$)	0.02	0.03	0.03	0.04	....	0.620	0.10	0.07	0.14	0.04
Manganese (MnO)	0.04	0.01	0.03	0.06	....	....	....	....	....	....
Lime (CaO)	0.85	0.93	0.95	1.22	0.15	3.518	0.37	0.24	0.18	0.51
Magnesia (MgO)	0.38	0.73	1.43	1.92	1.45	0.479	0.59	0.43	0.38	1.27
Soda (Na$_2$O)	2.19	0.80	1.45	1.33	0.17	0.006	Trace	0.40	0.29	0.79
Potash (K$_2$O)	1.61	0.93	0.83	1.60	2.32	0.118	Trace	0.86	4.41	4.03
Water (H$_2$O)	(a)4.03	(a)10.46	(a)10.79	(a)11.72	(a) 9.86	27.441(b)	13.26	8.20	....	7.38
Carbonic acid (CO$_2$)	0.43	0.30	0.29	0.30	....	2.251	....	....	....	....
Carbon (C)	0.19	0.34	0.22	1.09	....	....	....	....	....	....
	100.39 %	100.50 %	100.09 %	100.68 %	99.84 %		100.07 %	98.69 %	99.27 %	98.36 %

(a) Contains hydrogen of organic matter. Dried at 100° C.
(b) Contains 11.21 % of organic matter.

Columns I, II, III, and IV of this table (see opposite page) are limestone residuals from southern Wisconsin. Columns I and II are from the same vertical section, I being $4\frac{1}{2}$ feet from the surface, and II $8\frac{1}{4}$, and in contact with the underlying limestone. Columns III and IV are similarly related, III being 3 feet from the surface, and IV $4\frac{1}{2}$ feet, the lower sample lying on the unchanged rock. The larger percentages of silica in the samples from nearest the surface indicate a higher state of decomposition, the soluble portions having been more largely removed. The presence of larger percentages of alkalies in these same samples indicates that these salts existed in the form of silicates which have resisted the decomposing influences, and remain mechanically included in the residues. Column V is a clay from the decomposition of the Knox dolomite at Morrisville, Alabama; VI the characteristic red earth from the decomposition of coralline limestone on the islands of Bermuda; VII a product of the decay of a diabase dike at Wadesboro, North Carolina; VIII a gabbro sub-soil from Maryland; IX a sub-soil from the decomposition of Trenton limestone near Hagerstown, Maryland; and X a residual soil from the decomposition of a Triassic sandstone, Maryland.

A microscopic examination of the material represented by analyses I and IV, as given by the authorities quoted, showed it to consist of particles in an extreme condition of comminution. An actual measurement of over 700,000 of these particles yielded results as below: —

Particles less than .0025 mm. in diameter . . . . 721.866 %
Particles between .0025 mm. and .005 mm. in diameter 9.812
Particles over .005 mm. in diameter 0.034
                                                               732.312 %

Of those over .005 millimetre in diameter, particles reaching 0.06 millimetre were not rare. Nearly all those above 0.1 millimetre were found to be of flints and cherts which graded up into chips and flakes of notable sizes. Particles much coarser than those above enumerated do indeed occur, but their actual number is comparatively small, though their comparative bulk may be considerable.

Work of a like nature, but done under somewhat different conditions, by Dr. Milton Whitney, showed the residues from the Trenton limestones near Hagerstown, Maryland, to contain on an average some 45% of finely comminuted material, the individual particles of which vary in size between .005 and .0001 millimetre in diameter, and which may appropriately be termed *clay*. As Dr. Whitney has calculated, there are approximately 22,000,000,000 grains of sand and clay in each gramme of such a sub-soil, presenting in every cubic foot not less than 158,000 square feet of surface to the action of water and air, as well as to the roots of growing plants.

The results of mechanical analyses of (I and II) residues from the Trenton limestone, (III) Triassic sandstone, (IV) gabbro, and (V) gneiss are presented in tabular form below.[1]

Diameter of Particles mm.	Conventional Names	I	II	III	IV	V
		%	%	%	%	%
2-1	Fine gravel	0.54	0.17	0.00	0.00	0.19
1-.5	Coarse sand	0.32	0.00	0.23	0.26	1.80
.5-.25	Medium sand	0.72	0.15	1.20	0.18	3.12
.25-.1	Fine sand	0.62	0.25	4.03	0.66	6.96
.1-.05	Very fine sand	4.03	2.34	11.57	6.73	8.76
.05-.01	Silt	36.02	19.04	38.97	47.32	34.92
.01-.005	Fine silt	14.99	20.88	8.84	10.04	12.14
.005-.0001	Clay	41.24	51.77	32.70	34.90	28.82
	Total mineral matter	88.48	94.60	97.63	94.44	96.71
	Organic matter, water, and loss	1.52	5.40	2.37	5.56	3.29
		100.00	100.00	100.00	100.00	100.00

Many of the products of weathering of siliceous crystalline and calcareous rocks are of economic importance as soils, clays, and iron ores, as elsewhere noted. The kaolin beds of northern Delaware and southwestern Pennsylvania are mainly decomposed, highly feldspathic, gneissic rocks, and which as dug from the pits still retain their gneissic structure, but which are now plastic clays full of angular quartz fragments, mica scales and feldspar particles in various stages of decomposition. The change that has taken place consists in a kaolinization of the feldspars, whereby the alkalies are largely removed, and a residue consisting essentially of a hydrous silicate of

[1] Bull. No. 21, Maryland Agricultural Exp. Station, by Milton Whitney, 1893.

alumina left in their place. The quartz granules are disaggregated, and their surfaces sometimes slightly etched by the action of the alkaline carbonates; the black mica, where such existed, decomposed, giving rise to rust-colored spots. The material is dug from the pits and washed with water to separate the impurities, the "kaolin" or clay remaining in suspension, and being ultimately saved by filtration through canvas. This finest material, as seen under the microscope, still contains particles of undecomposed feldspars and shreds of white mica, together with other extremely irregularly outlined, sometimes almost amœba-shaped forms, as shown in Fig. 29. An average of two mechanical analyses of this clay, made under Dr. Whitney's direction, yielded the results given below: —

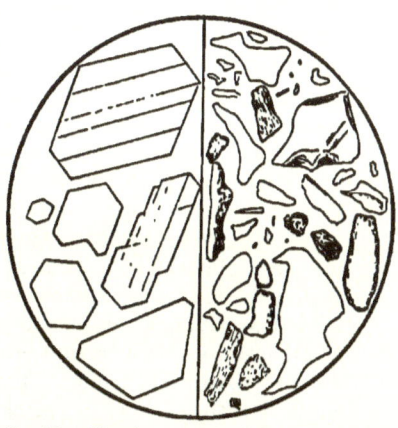

Fig. 29.—Showing, on the left, the mineral kaolinite as seen under the microscope, and on the right, washed kaolin.

Moisture in Air-Dry Material at 100° C.	Moisture on Ignition	Silt .05-.01 mm.	Fine Silt .01-.005 mm.	Clay .005-.0001 mm.
0.41 %	11.41 %	31.70 %	7.31 %	47.78 %

Chemical analyses of the same material, made in the laboratories of the United States Geological Survey, yielded: —

Silica (SiO_2)	48.73 %
Titanic oxide (TiO_2)	0.17
Alumina (Al_2O_3)	37.02
Ferric iron (Fe_2O_3)	0.79
Lime (CaO)	0.16
Magnesia (MgO)	0.11
Potash (K_2O)	0.41
Soda (Na_2O)	0.04
Water at 100°	0.52
Ignition	12.83
Phosphoric acid (P_2O_5)	0.03
	100.81 %

Among the special names that have from time to time been given to local phases of residuary accumulations, there remain two, the *laterite* and *wacke*, which are sufficiently common to merit some attention. The first mentioned of these, *laterite*, like loess and several other terms that might be mentioned, has to a considerable extent lost its true lithological significance through careless usage. Originally the name was applied to a vesicular highly ferruginous clay, soft in the mass, but hardening on exposure to the weather, and which has a wide distribution throughout India and Ceylon. Two forms are commonly recognized, — the one capping the summits of hills and plateaux on the highlands of central and western India, and underlain by the Deccan traps; and the second occurring on the lowlands, in part overlying gneisses and granites. The prevailing colors of the laterite, when freshly broken, are various tints of brown, red and yellow mottled, or whitish ; after exposure it is usually covered with a brown or blackish brown coating of limonite. When first dug out, the material is sufficiently soft to be cut with a pick or shovel, but becomes greatly indurated on exposure. In some instances the material is of so compact a texture and so hard as to resemble jasper. In many forms of laterite the material is traversed by "small irregular tortuous tubes from a quarter of an inch to upwards of an inch in diameter." These penetrate the mass in all directions, though most commonly nearly vertical, and are often lined with a coating of limonite. On weathering, these give rise to extremely irregularly pitted or scoriaceous surfaces, which, together with the dense, often botryoidal structure, cause it to resemble certain types of igneous rocks, for which it has more than once been mistaken. The more massive forms show usually a horizontal banding. Some forms of laterite show a brecciated structure, due to its detrital fragments becoming recemented into masses closely resembling the original rock. The high level form, that which occurs capping the hills and plateaux on the highlands of central and western India, is fine grained and compact and of a fairly homogeneous structure, although the iron oxide may be somewhat irregularly distributed and sometimes segregated in pisolitic nodules sufficiently abundant to form an ore. The lower level form, that which covers large areas of both east and west coasts, frequently contains grains of sand and pebbles embedded in a ferruginous matrix. It is, as a rule, less homogeneous than the high level form, but nevertheless passes into it by insensible gradations.

The origin of both high and low level forms of the laterite has been the subject of much speculation. It is probable that all of it is of a residual nature, *i.e.* represents the less soluble portions of pre-existing rock masses. That which is found on the high levels occurs overlying the Deccan trap sheets, into which it can in many instances be traced, proving conclusively its origin from this rock by the ordinary processes of weathering. The low-lying variety can, in many instances, in like manner be traced back to its origin from more siliceous, gneissic, and granitic rocks. A part of the material, however, has the appearance and structure of a clastic rock of sedimentary origin, and so it is considered by the best authorities to be.

The chemical composition of a very ferruginous laterite from Rangoon is as below: —

Constituents	Insoluble	Soluble	Bulk
Silica (SiO_2)	30.728 %	0.848 %	37.576 %
Alumina (Al_2O_3)		5.783	
Iron sesquioxide (Fe_2O_3)	2.728	46.279	52.802
Lime (CaO)		0.742	
Magnesia (MgO)			
Alkalies	6.802	0.000	6.892
Water and loss			
	40.258	59.742	100.00 %
	100.00 %		

"The surface of the country composed of the more solid forms of laterite is usually very barren, the trees and shrubs growing upon it being thinly scattered and of small size. This infertility is due, in great part, to the rock being so porous that all the water sinks into it, and sufficient moisture is not retained to support vegetation. The result is that laterite plateaux are usually bare of soil, and frequently almost bare of vegetation."[1]

Wacke is an old German name now but little used, designating the gray, brown to black earthy residue or clay resulting from the decomposition in place of basic eruptive rocks, as

[1] Manual of the Geology of India, by R. D. Oldham, 2d ed., 1893, pp. 369–390.

basalt, melaphyr, etc. In composition the material naturally varies with the character of the rock from which it was derived, and the amount of decomposition and leaching it may have undergone.

It seems advisable to call attention here, a little more emphatically, to the fact that the same processes which in ages past have been instrumental in the formation of sandstones, shales, slates, or marls are to-day, and have in late Tertiary and in Quaternary times, given us soils; in other words, many of our soils are but secondary rocks in a state of loose consolidation, and many of the accumulations classed as residual were derived by disintegration, *in situ*, of alluvial materials; materials brought down years ago and deposited in shallow seas. The amount of consolidation undergone by the more recent of these sediments has in many instances been so slight that on elevation above the water level they are ready almost at once to assume the rôle of soil with little if any preparatory disintegration. Nevertheless consistency demands that such be here grouped as residuary.

Over what is known as the coastal plain of the middle Atlantic slope, a narrow belt bordering on the Atlantic and extending from the Hudson River on the north to the Roanoke on the south, have been deposited in late Mesozoic and Tertiary times a series of gravels, sands, and clays which constitute the well-known Potomac, Appomattox, and Columbian formations of Darton, McGee, and others. These are all detrital deposits from the eastern Appalachian regions, brought down by streams and deposited in the shallow estuaries and deltas of these periods, but which have remained in a condition of slight consolidation, and through subsequent elevation and weathering form the soils. Such vary widely and abruptly. In the region northeast of Washington, the Potomac formation consists of feldspathic sands, gravels, and clays irregularly bedded and often enclosing notable accumulations of rounded pebbles of quartzite brought down from the Appalachian and Piedmont regions. The Appomattox formation, from which was derived surface soil in the vicinity of the Rappahannock and Appomattox in Virginia, is a yellowish or orange-colored clay and sand with sometimes interbedded gravel. The Columbian formation, which yields the surface soil of the main portion of Washington City and the immediate valley of the Potomac and contributary streams southward, is a delta and littoral deposit made up of materials worked over from the older Potomac and Lafayette formations and also of granitic sands and clays from the decomposed rocks of the Piedmont plateau.

The clays of the Potomac formation above mentioned are not infrequently sufficiently homogeneous and plastic to be utilized in the manufacture of brick, tiles, and pottery. The following table shows the finely comminuted condition of the materials which go to make up these clays in Maryland, as determined by Whitney.[1]

Diameter mm.	Conventional Names	Red Clay, Tile	Red Clay, Puddling	Blue Clay, Stoneware
2–1	Fine gravel	0.00 %	0.31 %	0.00 %
1–.5	Coarse sand	0.00	0.82	0.00
.5–.25	Medium sand	0.50	2.69	0.29
.25–.1	Fine sand	2.03	3.23	1.27
.1–.05	Very fine sand	9.02	8.89	8.03
.05–.01	Silt	25.13	26.17	20.16
.01–.005	Fine silt	13.44	11.18	10.72
.005–.0001	Clay	42.34	42.36	50.02
	Total	93.76 %	95.65 %	97.39 %
	Organic matter, water loss	6.24	4.35	2.61

(2) **Cumulose Deposits.** — To be classed with the sedentary deposits, in that they result from the gradual accumulation of material *in situ*, but differing radically in both composition and origin from those just described, are those portions of the regolith which result from the gradual accumulation of organic matter with only small amounts of foreign detritus; which are made up almost wholly of the combined accumulations, organic and inorganic, of growing plants. Such may not infrequently be found in all stages of formation, in enclosed ponds or lakes, without appreciable inlet or outlet, being merely due to standing water in low places. "Such pools, when not exposed to periodical drying up, are invaded by a peculiar vegetation, first mostly composed of confervæ, simple thread-like plants of various color and of prodigious activity of growth, mixed with a mass of infusoria, animalcules, and microscopic plants, which,

[1] Bull. 4, U. S. Dept. of Agriculture, 1892.

partly decomposed, partly containing the floating vegetation, soon fill the basins and cover the bottom with a coating of clay-like mould. So rapid is the work of these minute beings, that in some cases from 6 to 10 inches of this mud is deposited in one year. Some artificial basins in the large ornamental parks of Europe have to be cleaned of such muddy deposits of floating plants, mixed with small shells, every three or four years.

"When left undisturbed, this mud becomes gradually thick and solid; in some cases, of great thickness; affording a kind of soil for marsh plants, which root at the bottom of the basins or swamps and send off their stems and leaves to the surface of the water or above it; where their substance becomes in the sunshine hard and woody.

"As these plants periodically decay, their remains of course drop to the bottom of the water; and each year the process is repeated, with a more or less marked variation in the species of the plants. After a time the basins become filled by these successive accumulations of years or even centuries, and the top surface of the decayed matter, being exposed to atmospheric action, is transformed into humus and is gradually covered by other kinds of plants, making meadows and forests. In other cases when basins of stagnant water are too deep for vegetation

FIG. 30. — Section across a small lake. *a*, bed rock; *bb*, drift; *cc*, growing peat; *dd*, decaying peat; *ee*, climbing bog.

of aquatic plants, nature attains the same result by a different special process; namely, by the prolonged vegetation of certain kinds of floating mosses, especially the species known as sphagna. These grow with prodigious speed, and expanding their branches in every direction over the surface of ponds or small lakes, soon cover it entirely. They thus form a thin floating carpet, which as it gradually increases in thickness serves as a solid soil for another kind of vegetation, — that of the rushes, the sedges, and some kinds of grasses, which grow abundantly mixed with the mosses, and which by their water-absorbing structure furnish a persistent humidity sufficient for the preservation of their remains against aerial decay. The floating carpet of moss becomes still more solid, and is then overspread by many species of larger swamp plants, and small arborescent shrubs, especially those of the heath family; and so, in the lapse of years, by the continual vegetation of the mosses, which is never interrupted, and by the yearly deposits of plant remains, the carpet at last becomes strong enough to support trees, and is changed into a floating forest, until, becoming too heavy, it either breaks and sinks suddenly to the bottom of the basin, or is slowly and gradually lowered into it and covered with water."[1]

It is to such processes that are due, in large part, the inland swamp soils of many localities. Beginning at and near the shore and upon a soil of wet sand, the organic matter has accumulated year by year till now several feet in thickness and in some cases covering miles of territory. The proportion of organic matter in such a deposit naturally increases from the shore outward until in the upper and central layers it may comprise 90% of the total weight.

This feature is well brought out in the following analyses of material from an open ground prairie swamp in Carteret County, North Carolina.

Constituents	I	II
Silica (insoluble) (SiO_2)	80.84 %	1.52 %
Silica (soluble) (SiO_2)	3.70	0.00
Alumina (Al_2O_3)	2.60	0.30
Oxide of iron (Fe_2O_3)	1.18	0.15
Lime (CaO)	0.44	0.36
Magnesia (MgO)	0.22	0.14
Potash (K_2O)	0.07	0.06
Soda (Na_2O)	0.02	0.13
Phosphoric acid (P_2O_5)	0.08	0.06
Sulphuric acid (SO_3)	0.06	0.00
Chlorine (Cl)	Trace	0.02
Organic matter (C)	7.70	87.25
Water (H_2O)	2.50	9.60

Column I of the above is from the margin — the oak fringe — of this great swamp, near North River, about 8 miles north of

[1] Geol. Survey of Pennsylvania, 1885, p. 100.

Beaufort; it is light gray to ash-colored with a growth of white oak, gum, maple, pine, and palmetto trees; the situation is low and flat. "This margin belt of semi-swamp is from a half mile or less in width to above a mile. The surface rises towards the interior and is covered by a soil, if it may be called such, represented by column II, which is 2 to 3 feet deep and upwards, and lies on a bed of white sea-sand. It consists of a loose open mass of half-decayed woody matter, of a brown color, and is in fact a superficial, uncompressed lignite; for it will be observed that the analysis includes nearly 10 % of water, so that the dry substance would give but $3\frac{1}{2}$ % of inorganic matter, not more than would be accounted for by the ash of the woody matter. The growth is a dense thicket of spindling shrubs with small scattered maples and bays."[1]

Wiley has described[2] deposits of a somewhat similar nature as covering 1,000,000 acres in the Kissimmee valley of Florida. These, which are of a dark brown to deep black color, contain in some cases as much as 96.16 % of volatile matter, and vary from 3 to 20 feet in depth. Such, when properly drained, may be made extremely fertile, though in periods of drought endangered by fire which, once started, may burn for months, doing immense damage. The partially reclaimed areas of the Great Dismal Swamp of Virginia are fair representative types of swamp soils.

The formation of cumulose deposits is not, however, limited to lakes, stagnant ponds, or even to swamps as the word is ordinarily used, excepting as the swamp itself may be incidental and consequent. Regions of poor drainage, particularly in moist and cool climates, may give rise to growths of sphagnous mosses and subsequently to plants of a higher type, which in course of years assume no insignificant proportions.

In accounting for such accumulations, we have but to remember that ordinarily when a plant dies, its organic constituents are returned to the atmosphere once more in a comparatively brief period of time through the usual processes of decay. It needs only such conditions of moisture as shall prevent the complete decay and hence favor the accumulation of the organic matter, to give us beds of peat and ultimately of coal. Plants of the type of sphagnous mosses, growing continuously above

[1] Geology of North Carolina, Vol. I, 1875.
[2] Agricultural Science, Vol. VII, No. 3, 1893, pp. 106–120.

and dying beneath, hold in their mass sufficient moisture to exclude atmospheric air, and thus themselves bring about the proper conditions for bog making. In virtue of this property such may gradually rise above the level of the surrounding country, as is the case with the Great Dismal Swamp of Virginia and numerous others that need not be mentioned here. Instances are on record where bogs of this nature have grown so far above the natural level, that during seasons of unusual rainfall they have burst, and flooded adjacent regions, with disastrous results. The rate of growth of such accumulations is naturally quite variable. H. S. Gesner, as quoted by T. Rupert Jones,[1] states that in Bavarian moors the observed increase in peat, in forty-five years, amounted to from 2 to 3 feet in thickness; in Oldenberg, in one hundred years, to 4 feet; in Hammelsmoor, Denmark, to $2\frac{1}{2}$ feet; and in Alpine districts to 4 and 5 feet in from thirty to fifty years.

The peat bogs, so characteristic of Ireland, Scotland, and other northern latitudes, are of this type. A section of the well-known Bog of Allen, made in county Kildare, is given below.[2]

	THICKNESS
(1) Dark reddish brown; mass compact; no fibres of moss visible; surface decomposed by atmosphere	2 feet
(2) Light reddish brown; fibres of moss very perfect	3 "
(3) Pale yellowish brown; fibres of moss very perceptible	5 "
(4) Deep reddish brown; fibres of moss perceptible	$8\frac{1}{4}$ "
(5) Blackish brown; fibres of moss scarcely perceptible, contains numerous twigs and small branches of birch, elder, and fir	3 "
(6) Dull yellow-brown; fibres not visible; contains much empyreumatic oil; mass compact	3 "
(7) Blackish brown; mass compact; fibres not visible; contains much empyreumatic oil	10 "
(8) Black mass, very compact; has a strong resemblance to pitch or coal; fracture conchoidal in all directions; lustre shining	4 "
Total depth of bog	$38\frac{1}{4}$ feet

Underlaid by 3 feet of marl containing 64% carbonate of lime, 4 feet of blue clay, and this in its turn by clay mixed with limestone gravel of an unknown thickness.

---

[1] Proc. Geologists' Association, Vol. VI, No. 5, January, 1880.

[2] T. Rupert Jones, Proc. of the Geologists' Association, London, Vol. VI, No. 5, January, 1880. This authority classifies the peat bogs, swamps, and marshes, as follows: —

I. Peat bogs and turf moors on such plateaux as flat mountain tops and wide hill moors.

Deposits of the cumulose type pass by all gradations into the paludal, swamp, or marsh type and these in turn into ordinary alluvium. Or it would perhaps be better to reverse this order, since, as in the gradual silting up of an enclosed lake, we may have, in the first stages, stratified alluvium, then when the waters become sufficiently shallowed, swamp and muck deposits, and lastly the deposits of pure organic, or cumulose material.

2. TRANSPORTED MATERIALS

Because of the constant action of gravity, the well-known transporting power of water, the wind or moving ice, few residual products retain for any length of time their virgin purity, but become more or less contaminated with materials from near or distant sources. The avalanches of mountain regions afford an illustration of the bodily transfer of, it may be, millions of tons of matter from the mountain slopes to be debouched into the valley below; the slow-creeping glacier brings down its load and deposits its moraine when, succumbing to the blandishments of warmer climes, it is no longer able to bear it further: spasmodic winds catch up the smaller particles as clouds of dust to be transported, assorted, and redeposited as their

II. Peat bogs of valleys: (1) At the heads of valleys; (2) at the salient angles within river curves; (3) in deserted beds of rivers; (4) in plains and lakes of expanded valleys; (5) special peat bogs of Denmark and the black earth of Russia; (6) river deltas; (7) maritime peat marshes, where certain valleys and plains open to the sea.

Regarding the black earth of Russia, it should be stated that this is now regarded by at least one authority (Hume, Geol. Mag., Vol. I, No. 2, 1894) as being but a local phase of the loess, the color being due to the prevalence of organic matter.

Shaler (Ann. Rep. U. S. Geol. Survey, 1888-80), on a basis of physical characters, classifies the inundated lands of the United States as below: —

Marine marshes	Above mean tide . . .	Grass marshes. / Mangrove marshes.
	Below mean tide . . .	Mud banks. / Eel-grass areas.
Fresh-water swamps	River swamps	Terrace. / Estuarine.
	Lake swamps	Lake margins. / Quaking bogs.
	Upland swamps . . .	Wet woods. / Climbing bogs.
	Ablation swamps.	

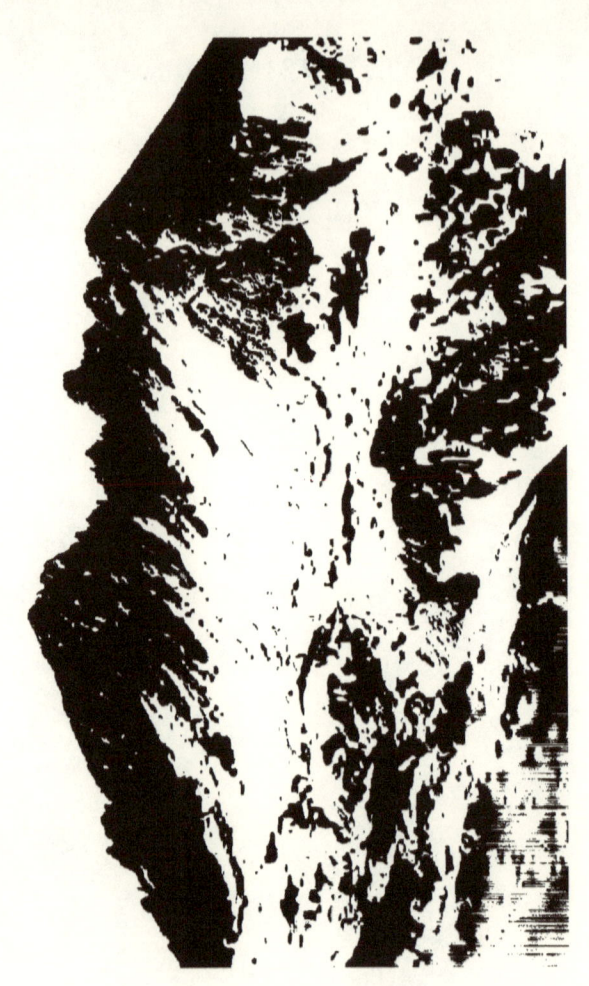

COLLUVIAL DEPOSITS 319

force is spent. It is, however, through the continual transportation of running streams, both in the past and present, and through the action of moving ice in ages gone, that have been brought about the great amount of transportation and admixture characteristic of that part of the regolith comprised under the general name of drift. According to which of the agencies enumerated prevailed, we may subdivide our subject as follows : (1) Colluvial deposits, (2) alluvial deposits, (3) æolian deposits, and (4) glacial deposits, though as we proceed we shall find that the lines of separation are not in all cases sharply drawn, and in many an area the regolith bears impress of compounded agencies.

(1) **Colluvial Deposits.**[1] — Under this head it is proposed to include those heterogeneous aggregates of rock detritus commonly designated as talus and cliff débris. The material of avalanches may also be classed here. Such result wholly from the transporting action of gravity. The deposits in themselves are comparatively limited in extent, ever varying in composition, and are composed of an indiscriminate admixture of particles of all sizes, from those as fine as dust to blocks it may be of hundreds of tons' weight. Such are necessarily limited to the immediate vicinity of the cliffs or mountains from which they are derived. As loosened by heat or frost from the

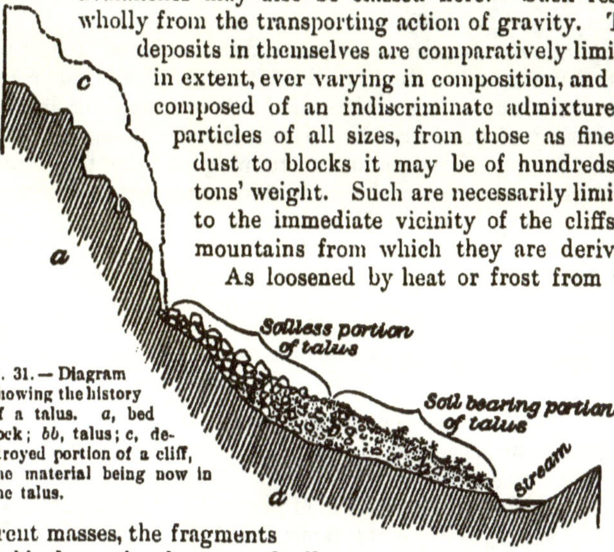

FIG. 31. — Diagram showing the history of a talus. *a*, bed rock; *bb*, talus; *c*, destroyed portion of a cliff, the material being now in the talus.

parent masses, the fragments tumble down the slopes, gradually accumulating in beds the slope of which is limited only by the laws of gravity and the character of the débris. (See Pl. 23.) Inclinations of 30° are common ; less commonly of 40°. From

[1] From the Latin "colluvies," a mixture. The term as here used is more restricted in its meaning than as used by Professor Hilgard.

their mode of origin it is natural that the individual particles should be mainly angular and comparatively fresh. In fact, they represent rock-weathering through disintegration, and not decomposition, which will come later. Below, *i.e.* further down the slopes and in the edges of the valleys, these coarse, illy assorted deposits pass gradually into soils; above, they consist simply of masses of loose rock wholly unfitted for the support of vegetable life. (Fig. 31.) Through becoming saturated with water, ice, or snow, such at times become loosened from the steep slopes on which they lie and slide down in the form of avalanches into the valleys. Although comparatively limited in their extent, these latter, owing to the resistless energy and suddenness of their advance, are sometimes appallingly destructive, as has been repeatedly illustrated in the Swiss Alps, and other mountain regions. The geographic distribution of talus deposits as controlled by climatic conditions has been already noted (p. 283).

(2) **Alluvial Deposits.** — The deposits included under this head differ structurally from those thus far described in that they are always more or less distinctly stratified, or bedded. In writing of the formation of sedimentary rocks, and again when treating of the action of running water, a few figures were given relative to the amount of transported débris deposited yearly in the Gulf of Mexico. In a similar way the amount of débris carried annually to the ocean by some of the chief rivers of the world has been estimated as below : —

	Cubic feet		Cubic feet
Mississippi	7,468,694,400	Rhone	600,000,800
Upper Ganges	6,368,077,440	Danube	1,253,738,000
Hoang-Ho	17,520,000,000	Po	1,510,147,000

The muddy condition of the water, caused by this suspended matter, is so conspicuous a feature of certain rivers that they have received special names on this account. Hoang-Ho means simply yellow river; Missouri is the Indian name for Big Muddy; while the famous Red River of the North is so called merely because of the red mud it carries. Such silt-bearing streams, flowing into lakes and tideless seas, begin depositing their loads so soon as their currents are checked, building up thus the so-called delta deposits for which the Mississippi, the Po, Ganges, and the Nile are noted.

The character of the material in the delta deposits is variable only within certain limits, consisting always of siliceous sand and mud intermingled with organic matter.

Professor Judd, who examined samples from borings in the alluvial deposits of the Nile delta, found the materials to vary abruptly in texture from the surface downward, the variations following no recognizable law. The percentage amounts of constituents classed as sand and mud, as obtained from (I) borings at Kasr-el-Nil, Cairo, (II) Kafr-ez-Zayat, and (III) Tantah, are given in the table below.

Depth	I		II		III	
	Sand	Mud	Sand	Mud	Sand	Mud
	%	%	%	%	%	%
3' 0"	....	....	2.35	97.65	....	....
4' 0"	....	....	30.42	69.58	1.71	98.29
6' 0"	5.77	94.33	....	....	....	....
8' 6"	....	....	....	....	7.27	92.73
11' 0"	....	....	50.99	49.01	....	....
16' 0"	86.27	13.73	....	....	....	....
17' 6"	79.65	20.35	....	....	....	....
18' 0"	....	....	....	....	8.78	91.22
19' 0"	....	....	87.41	12.59	....	....
22' 6"	....	....	....	....	31.16	68.44
26' 0"	....	....	90.19	9.81	....	....
31' 0"	....	....	....	....	39.43	60.57
35' 0"	....	....	86.42	13.58	....	....
38' 6"	65.05	34.95	....	....	....	....
40' 0"	....	....	81.94	18.06	80.70	19.30
40' 6"	80.83	19.17	....	....	....	....
45' 0"	68.72	31.28	....	....	....	....
46' 0"	....	....	....	....	95.90	4.10
48' 0"	....	....	87.23	12.77	....	....
55' 0"	....	....	0.25	99.75	97.71	....
56' 0"	....	....	....	....	99.53	2.29
58' 0"	....	....	..*..	....	99.09	0.47
60' 0"	....	....	12.60	87.40	....	40.01
66' 0"	....	....	62.07	37.93	....	....
68' 0"	....	....	....	....	7.76	....
73' 0"	....	....	....	....	59.95	92.24
75' 0"	....	....	66.38	36.02	....	40.05

The material described as sand consists of rounded, angular, and sub-angular grains. The well-rounded granules are mainly of quartz and feldspar; the angular and sub-angular of quartz, feldspars, hornblende, and augite, with smaller quantities of mica, tourmaline, sphene, iolite, zircon, fluor-spar, and magnetite

y

all in a nearly unaltered condition. The feldspars are mainly orthoclase and microcline — rarely a soda-lime variety — and in a state of surprising freshness. The quartz is in part the quartz of granitic rocks and the larger grains well rounded, best described as microscopic pebbles. He says: "It is evident that these sand grains have been formed by the breaking up of granitic and metamorphic rocks, or of older sandstones derived directly from such rocks. The larger grains exhibit the perfect rounding and polishing now recognized as characteristic of æolian action; the smaller ones from their larger surfaces in proportion to their weight, have undergone far less attrition in their passage through the air; but it is fair to conclude that they are really desert sand, derived from the vast tracts which lie on either side of the Nile valley, and swept into it by the action of the wind." The material described as mud is composed of essentially the same materials as the sands, but in a more finely divided state. There is an entire absence of anything like kaolin, though there are present particles of organic matter and frustules of diatoms. The surprising freshness of the materials and lack of kaolin is regarded as indicative of an origin through the action of heat and frost; *i.e.* through mechanical agencies rather than through the processes of rock decomposition.[1]

But, as has been already noted, only a part of the sediment carried by any stream reaches its mouth. A comparatively small, but, from our present standpoint, very important portion is carried during seasons of high water beyond the usual channels and spread out over the flood plains, as described on p. 287. Such deposits are, as a rule, plainly stratified, and consist of mineral matter in a finely comminuted condition derived, it may be, from the breaking down of a great variety of rocks. Their physical and chemical properties, as well as the periodic character of their deposition, are favorable to the formation of soils possessing great strength and fertility. Both fertility and rate of deposition in such cases are augmented through plant growth, which takes place with great rapidity wherever climatic conditions are favorable. So soon as the water leaves the flood plain, a host of moisture-loving plants, as reeds and rushes, spring up in countless numbers to die down again in the fall, and yield the carbon and nitrogenous constituents to serve as fertilizers,

---

[1] Proc. Royal Soc. of London, Vol. XXXIX, 1885, p. 213.

and augment the crop of the following year. Moreover, the remaining stems and fallen leaves of the plants serve to retard the running waters of each succeeding flood, catching in their meshes the floating sediments which might otherwise be carried seaward. The Anacostia, which empties into the Potomac River east of Washington, serves as a good illustration of the working of these agencies. A century ago the stream was navigable by coasting crafts as far as Bladensburg. Now, owing to shallow waters, nothing but rowboats can navigate beyond the Navy Yard at Washington. Each season the stream, murky with suspended silt from cultivated fields along its shores, comes down, till, ponded back by tides, it begins to deposit its load. As year by year its bed was thus raised, water plants, encroaching more and more from shallow shores, still further dammed

FIG. 32.

its sluggish current till now, during summer months, it is little more than a stagnant pond full of rank vegetation, and a source of odors foul and atmospheres enervating. The so-called "Potomac Flats" south of the city of Washington owed their origin and unhealthy conditions to similar processes.

The method of alluvial deposition in the flood plain, or delta, of the lower Mississippi has been worked out by McGee,[1] from whom we cannot do better than quote in considerable detail.

In length this flood plain reaches from the mouth of the Ohio 1100 miles measured along the river, or half as far measured in an air line, to the Gulf, and is bounded on the east by the bluff rampart separating it from the contiguous district; it is bounded on the west by a less continuous and less conspicuous rampart crossing the Arkansas River at Little Rock and gradually failing southward until this district and its more westerly

[1] The Lafayette Formation, Ann. Rep. U. S. Geol. Survey, 1890-91.

neighbor nearly blend. The surface of this otherwise monotonous district is relieved by a few small tracts of higher land. Most conspicuous of these is Crowley Ridge in eastern Arkansas, a long belt of upland stretching from the southeastern Missouri southward between the White and St. Francis rivers to the Mississippi at Helena. This belt of upland rises 100 or 200 feet above the insulating flood plain, and in its steepness of slope and rugosity of outline fairly simulates the eastern rampart overlooking the "delta" in corresponding latitudes.

The vast lowland tract comprised in and constituting most of this district is at once the most extensive and most complete example of a land surface lying at base-level or a trifle below that the continent affords.

It is trenched longitudinally by the Mississippi, and transversely by the White, Arkansas, Red, and other large rivers; between these greater waterways it is cut into a labyrinth of peninsulas and islands by a network of lesser tributaries and distributaries, the former gathering the waters from its own surface and from adjacent country, and the latter aiding the main river to discharge its vast volume of water and its immense load of detritus into the Gulf. The whole surface lies so low that it is flooded by periodic overflows of the Mississippi and its larger tributaries, and with each flood receives a fresh coating of river sediment; and much of the flood plain, fertilized by freshet deposits, is clothed with luxuriant forests and dense tangles of undergrowth, or with brakes of cane, or with subtropical shrubbery, only a few of the broader inter-stream tracts being grassed. Partly by reason of this mantle of vegetation, the current of each overflow is checked as the river rises above its banks, and most of the sediment is dropped near by; and so the Mississippi, the White, the Arkansas, and the Red, as well as each lesser tributary and each distributary from the great Atchafalaya down, are flanked by natural levees of height and breadth proportionate to the depth and breadth of the stream. The network of waterways is thus a network of double ridges with channels between; and each inter-stream area is virtually a shallow, dish-like pond in which the waters of the floods lie long, to be drained finally, perhaps, through fresh-made breaks in the natural dikes, weeks after the stream flood subsides. In the southern part of the district the inter-stream basins approach tide level and drain still more slowly; in the sub-coastal zone many of the basins are permanent tidal marshes. In the western part of the district is an area in which the inter-stream basins lie so high that they are invaded only by the highest floods and veneered with only the finest sediments; in some cases these sediments are so fine and so compactly aggregated and the surface is so ill drained and watered that trees may hardly take root, and these are either drowned by the floods or withered by the sun in the drought. Such portions of the surface are but scantily covered with coarse grass and form the "black prairies" of southern Arkansas and northwestern Louisiana.

It is to just such processes as those described that the Nile valley owes its remarkable fertility. The sediments deposited over these plains during the season of freshets consist of fine sand brought down by the Blue Nile and the Atbara from the decomposing siliceous rocks of mountainous Abyssinia. The gneisses and granites yield their detritus to the lixiviating influence of the mountain torrents and majestic Nile, the clayey particles being borne seaward, while the fresh quartzose, feldspathic and other siliceous particles, and smaller traces of apatite and alkaline carbonates remain in just the right stage of subdivision to yield a soil, which has brought forth for a period of over 4000 years crop after crop without artificial fertilization.

The following table will serve to show the physical characteristics of alluvial deposits, a portion of which are but reassorted materials from the glacial drift.

Approximate Number of Grains of Sand, Silt, and Clay in One Gramme of Alluvial Sub-soil from Illinois

Diameter mm.	Conventional Names	(a) Chillicothe	(b) Rockford	(c) American Bottoms
2-1	Fine gravel	0	1	0
1-.5	Coarse sand	83	48	0
.5-.25	Medium	6,755	3,428	5
.25-.1	Fine sand	18,000	29,300	104
.1-.05	Very fine sand	53,470	212,400	151,400
.05-.01	Silt	4,070,000	5,888,000	12,230,000
.01-.005	Fine silt	86,860,000	115,100,000	195,600,000
.005-.0001	Clay	2,537,000,000	3,842,000,000	14,080,000,000
	Total	2,628,008,908	3,963,232,177	14,887,981,509

(a) Terrace of Glacial age. (b) Flood deposits. (c) Post-glacial terrace (bottom land of Mississippi).

The same processes active in delta formation are manifested on a smaller scale in the gradual silting up of many inland lakes, particularly those of glacial origin, the rapidity of the filling being augmented by aquatic plants.

These lakes lie not infrequently between high hills, being fed by one or more streams flowing through narrow valleys, and having outlets at the opposite extremity. Soon after the close of the Glacial epoch, we may imagine one of these to have existed as a lake of clear blue water of varying depths, filled with abundant fish and wild fowl. But the little streams which fed it brought down continually sand and silt to be deposited at varying distances so soon as the currents fall to sleep within the bosom of the lake. Hence each year it shallows, and the pure white water-lily, reeds, and the rotting trunks of trees and shrubs encroach upon its shores until in course of time there remains but a flat plain, for a time subject to annual inundations, but ultimately permanently above the level of but the most severe floods, and through which flow in a meandering course the sluggish streams that first gave it birth and then wrought its extinction. This is the story of thousands of the so-called meadows, swales, swamps, and intervals throughout the northern portion of the United States, and the process in some easily recognizable stage may be found in almost any lake or pond now remaining.

It is a striking thought that all our lakes are but transient enlargements of pre-existing streams, and will in time, perhaps even before our own species is extinct, become converted into broad expanses of meadow lands; and that our children's children may yet sow and reap from rich and fertile areas which now echo only to the cry of water-fowls, and whose blue expanse is broken but by wind-born waves and leaping fish.

The lithological character of the deposits thus formed vary within certain limits almost indefinitely, since everything depends on the character and quantity of the silt brought down by the streams. Rarely, if ever, are they clayey, since the finer particles are carried beyond. In nearly all instances they are found to consist of very fine sand, largely siliceous, permeated, often quite blackened, through the presence of organic matter. Such are the mucks or mucky soils of New England.

So abundant is this organic matter that, when dried, such are not infrequently used locally for mulching purposes, though in their fresh condition they are sour and almost worthless except for growing sedges and the ranker kinds of forage grass. During the later stages of the process of filling up, deposition of sediments may almost entirely cease, since the water no longer rises above the level of past accumulations. In such cases the final stages consist simply in the accumulation of organic matter and the deposits come to closely resemble, or are even superficially identical with, the cumulose deposits already described. This same statement holds good also for the closely related salt-water marsh or paludal deposits, to be noted later.

Loess and Adobe. — Under the head of transported deposits, we must also consider the so-called loess of the Mississippi valley in our own country; of the Rhine valley, and other parts of Europe; of northern China and the Russian steppes, though, as we shall see, the name includes deposits which, while having many physical properties in common, may vary widely in composition as well as in method of deposition. It is more than doubtful, indeed, if the name, through misapprehension, has not been so loosely applied as to rob it of its proper geological significance.

The loess of China, made famous through the researches of Richtofen, is now regarded by some authorities[1] as of the same nature as our adobe. Richtofen himself, it will be remembered, regarded the Chinese loess as largely an æolian deposit, as due to the action of wind in transporting for long distances the fine detritus swept by rain and wind from mountain slopes into enclosed basins, to ultimately become entangled and deposited among the growing vegetation. This foreign material, intermingled with the collective residue of herbaceous plants, with the inorganic residuum from the decay of prairie vegetation for countless generations, makes up its mass over many hundreds of square miles of territory, and to depths in places of thousands of feet. The characteristics of the loess, as found in China, are those of a fine calcareous silt or clay, of a yellowish or buff color, so slightly coherent that it may be readily reduced to powder between the thumb and fingers, and yet possessing such tenacity as to resist the ordinary weathering action of the atmosphere, and, wherever cut by stream erosion or other means, to stand with vertical walls, even

[1] See I. C. Russell, Subaerial Deposits of North America, Geol. Mag., August, 1889.

though they may be hundreds of feet in height. The loess country is described as thus cut up by an almost impassable system of gorges, so that to cross it in any fixed direction is almost an impossibility. "Wide chasms are surrounded by castles, towers, peaks, and needles, all made up of yellow earth, between which gorges and chasms radiate labyrinthically upwards into the walls of solid ground around. High upon a rock of earth — steeper than any rock of stone — stands the temple of the village, or a small fortress which affords the villagers a safe retreat in times of danger.. The only access to such a place is by a spiral stairway dug out within the mass of the bluff itself. In this yellow defile there are innumerable nooks and recesses, often enlivened by thousands of people, who dwell in caves dug in the loess."[1]

One of the striking features of the loess, both in China and elsewhere, is the abundance of minute tubes or canals — lined with carbonate of lime — which traverse it from above downward, and which are assumed by some to be due to root fibres. It is the presence of these presumably that causes the vertical cleavage, and at the same time the remarkable absorptive qualities for which the loess is noted. Such is the material which for more than three thousand years has brought forth crops continuously, and without exhaustion, over many square miles of the Chinese Empire. Its distribution in Europe is given as extending from the French coast at Sangatte, eastward across the north of France and Belgium, filling up the depressions of the Ardennes, passing far up the valleys of the Rhine and its tributaries, the Neckar, Main, and Lahr; likewise those of the Elke above Meissen, the Weser, Mulde, and Saale, the upper Oder and Vistula. Spreading across upper Silesia, it sweeps eastward over the plains of Poland and southern Russia, where it forms the substratum of the tschernoseun, or black earth. It extends into Bohemia, Moravia, Hungaria, Galicia, Transylvania, and Roumania far up into the Carpathians, where it reaches heights of from 2000 to 5000 feet above sea-level. In northern China it spreads over a large portion of the region drained by the Hoang-Ho. For nearly a thousand miles from the borders of the great alluvial plain of Pechele, through the provinces of Shansi, Shensi, and Kansu, everywhere to the

[1] The Chinese Loess Puzzle, by J. D. Whitney, American Naturalist, December, 1877.

northern base of the range of the Tsing-ling-shan, the loess may be followed to the very divide which separates the basin of the Hoang-Ho from the region destitute of drainage, into the sea. Toward the north it reaches almost to the edge of the Mongolian plateau. The entire area covered continuously is stated to be as large as the whole of Germany, while it is found in more or less detached portions over an area in addition, nearly half as large. In the United States the loess covers thousands of square miles throughout the drainage basin of the Mississippi River. It is found in Ohio, Indiana, Michigan, Iowa, Kansas, Nebraska, Illinois, Tennessee, Alabama, Mississippi, Louisiana, Arkansas, Missouri, Kentucky, and the Indian Territory. According to Professor Aughey it prevails over at least three-fourths of Nebraska, to a depth ranging from 5 to 150 feet, and furnishes a soil of extraordinary strength and fertility. As here found, however, the æolian hypothesis fails to satisfactorily explain all the existing conditions, and there is little doubt but that it represents in large part the fine silt, the glacial flour brought down by the ice of the Glacial epoch, borne southward by streams, and deposited in water just sufficiently in motion to carry the fine clay farther away. The loess, in fact, illustrates in a remarkable manner the wonderful assorting power of water.

Microscopic and chemical examinations of loess sustain this hypothesis. The particles are as a rule quite fresh and sharply angular. Out of 150,000 particles examined under the microscope only about 3 % measures above .0025 of a millimetre and 1 % over .005 of a millimetre. Quartz is the preponderating material, with lesser amounts of orthoclase and plagioclase feldspars, white and dark micas, hornblende, augite, magnetite, dolomite, and cal-

Fig. 33. — Showing outlines of particles in Chinese loess.

ANALYSES OF LOESS FROM VARIOUS SOURCES

Constituents	I Dubuque, Iowa	II Galena, Illinois	III Kansas City, Missouri	IV Vicksburg, Mississippi	V Valley of the Rhine	VI Valley of the Rhine	VII Neubad Switzerland
Silica (SiO_2)	72.68 %	64.01 %	74.46 %	60.69 %	68.97 %	62.43 %	71.09 %
Alumina (Al_2O_3)	12.03	10.04	12.26	7.95	0.97	7.51	16.78
Iron sesquioxide (Fe_2O_3)	3.53	2.01	3.25	2.01	4.25	5.14	
Iron protoxide (FeO)	0.96	0.51	0.12	0.67	....	....	....
Titanium oxide (TiO_2)	0.72	0.40	0.14	0.52	....	....	0.11
Phosphoric anhydride (P_2O_5)	0.23	0.06	0.09	0.13	....	....	....
Manganese oxide (MnO)	0.06	0.05	0.02	0.12	....	....	1.81
Lime (CaO)	1.59	5.41	1.60	8.96	11.31	9.88	None
Magnesia (MgO)	1.11	3.69	1.12	4.56	2.04	1.65	1.23
Soda (Na_2O)	1.63	1.35	1.43	1.17	0.84	1.75	1.30
Potash (K_2O)	2.13	2.06	1.83	1.08	1.11		1.06
Water (H_2O)	(a) 2.50	(a) 2.05	(a) 2.70	(a) 1.14	(c) 1.37	(c) 2.31	0.80
Carbon dioxide (CO_2)	0.39	6.31	0.40	9.63	11.08	9.33	....
Sulphurous anhydride (SO_3)	0.51	0.11	0.06	0.12	....	....	....
Carbon (C)	0.09	0.13	0.12	0.19	....	....	(b) 2.87
	100.21 %	99.99 %	99.78 %	99.54 %	100.84 %	100.00 %	....

(a) contains H of organic matter. (b) organic matter dried at 100° C. (c) ignition. Analyses I, II, III, and IV from Russell's Subaerial Deposits of North America, Geol. Mag., August, 1889; V and VI from Bischof's Chemical Geology.

cite. The loess of the Rhine valley and of China offers no differences that can be readily described, though, as will be noticed by reference to the analyses, there may be a wide difference in chemical composition. Indeed, the essential characteristic of the loess is a physical rather than a chemical one, and it is doubtless to this that is due its uniform fertility. On p. 330 are given analyses of loess from the United States, the Rhine valley, and from Switzerland.

The following table will serve to show the fine state of subdivision in which the particles exist in loess as well as in the dust brought down by snow, which will be described on p. 344.

Constituents	I Upland Loess: Virginia City, Illinois	II River Loess: Virginia City, Illinois	III Loess: Nebraska	IV Dust from Snow: Rockville, Indiana
Moisture	....	....	5.40 %	3.17 %
Organic matter	....	....	4.90	11.08
Gravel	0.00 %	0.00 %	0.00	0.00
Coarse sand	0.00	0.00	0.00	0.00
Medium sand	0.00	0.01	0.00	0.00
Fine sand	0.01	0.10	0.00	0.00
Very fine sand	7.68	24.84	23.14	0.00
Silt	61.85	60.98	54.81	69.37
Fine silt	9.60	2.80	2.46	5.80
Clay	15.15	6.15	9.45	0.08

Aughey[1] gives the following section of the loess and soil in Nebraska.

(1) Loess . . . . 4 feet
(2) Black soil . . . . 2 "
(3) Loess . . . . 4 "
(4) Black soil . . . . 1½ "
(5) Loess . . . . 5 "
(6) Black soil . . . . 1½ "
(7) Stratified loess . . . . . . 15 "

This alternation is accounted for on the assumption of frequent changes of level during the loess-forming period. It would seem that the loess was deposited in shallow water and that as the lake became filled plant life came in as in modern

[1] Physical Geology and Geography of Nebraska, p. 270.

bogs and marshes, and throve until sufficient organic matter was formed to make the black soil layer. A period of subsidence followed, more loess was deposited and the previous condition repeated, this process going on till all the layers were formed. The material of the loess, in this case, would seem most likely to have been of æolian origin.

The name *adobe* is given to a calcareous clay of a gray-brown or yellowish color, very fine-grained and porous, which is sufficiently friable to crumble readily in the fingers, and yet, like loess, has sufficient coherency to stand for many years in the form of vertical escarpments, without appreciable talus slopes. The material of the adobe is derived from the waste of the surrounding mountain slopes, the disintegration being largely mechanical. According to Prof. I. C. Russell,[1] from whose descriptions is drawn a portion of what is given here, it is assorted and spread out over the valley bottom by the action of ephemeral streams, where it becomes mixed with dust blown by the winds from the neighboring mountains, and rendered more or less coherent by the cementing action of interstitial carbonate of lime.

Hilgard[2] limits the name *adobe* to the distinctly clayey soils of the arid regions, and divides them into two classes,—the upland and the valley adobes, the first being derived mainly from the disintegration, in place, of clay shales, while the second are mostly paludal or swamp formations, and represent either the finest materials that remain suspended in slack water, from any source, or sometimes the direct washings of the clayey soils of the hills. Whichever authority we follow, it is evident the name includes materials alike not in mode of origin or composition, but only in physical characteristics.

Adobe forms the soil of a large portion of the rainless region of the United States. It is found therefore in Colorado, Utah, Nevada, southern California, Arizona, New Mexico, and western Texas, as well as in the southern portion of Idaho, Wyoming, and Oregon. It has also a wide distribution in Mexico. In the United States it occurs from near the sea-level in Arizona, and even below it in southern California up to an elevation of at least 6000 or 8000 feet along the eastern border of the Rocky Mountains, and in the elevated valleys of New Mexico, Colorado, and Wyoming.

[1] Subaerial Deposits of North America, Geol. Mag., August, 1889.
[2] Bull. 3, U. S. Weather Bureau, Dept. of Agriculture, 1892.

The maximum thickness of the various deposits grouped under this name is not in all cases readily determined, for the reason that it is still accumulating and has not been sufficiently dissected by erosion to expose sections to any considerable depth. Many of the valleys of the arid region have been filled by it to a depth of 2000 or 3000 feet. In the larger valleys there are rocky crests, called "lost mountains," which project above the broad level desert surface, and which are in reality the summits of precipitous mountains that have been almost completely buried beneath these recent accumulations. The prevailing color of adobe is light buff to gray, excepting when contaminated with organic matter. In its typical form it is so fine as to be quite without grit when rubbed between the fingers. When examined under the microscope, it is seen to be composed of irregular unassorted flakes and grains, principally quartz, but fragments of other minerals are also present. The adobe of Salt Lake shows flocculent masses of amorphous matter, which, when thoroughly disintegrated, are found to consist of minute sharply angular fragments of quartz and feldspar with much calcareous matter, and only rarely a shred of micaceous or hornblendic material. In size the particles vary from those too small for measurement up to .08 millimetre in diameter.

The valuable characteristics of the adobe are its extreme fineness, great depth, and wonderful fertility.

Although comprising the soil of almost the entire region that was but recently known as the Great American Desert, it needs but water to make it laugh with harvests. While its physical properties undoubtedly have much to do with its fertility, this quality must also be in part due to the fresh and undecomposed condition of its constituent parts. Originating doubtless by purely mechanical agencies, it has been swept by winds and spasmodic rains into closely adjacent basins occupied by but temporary lakes, where, spread out over a floor sometimes almost absolutely level, it has been subjected to a minimum amount of leaching and retains until to-day its youthful strength and powers of recuperation.[1] The analyses given on p. 334 will serve to show the varying character of the deposits included under this name. Especial attention need only be called to the relatively high percentages of lime and the alkalies.

Under the head of alluvial deposits we may also consider

[1] See further on p. 309.

those clay accumulations which result from the deposition of fine aluminous sediments sorted by running streams from glacial débris and like the loess laid down in quiet water, though usually estuarian rather than lacustrine. These are the well-known Leda clays [1] of glacial regions, and which on genetic grounds might well be classed as aqueo-glacial deposits.

Constituents	I	II
Silica (SiO_2)	66.69 %	44.04 %
Alumina (Al_2O_3)	14.16	13.10
Ferric oxide (Fe_2O_3)	4.38	5.12
Manganese oxide (MnO)	0.09	0.13
Lime (CaO)	2.49	13.91
Magnesia (MgO)	1.28	2.06
Potash (K_2O)	1.21	1.71
Soda (Na_2O)	0.57	0.50
Carbonic acid (CO_2)	0.77	8.55
Phosphoric acid (P_2O_5)	0.29	0.04
Sulphuric anhydride (SO_3)	0.41	0.64
Chlorine (Cl)	0.34	0.14
Water (H_2O)	4.94	3.84
Organic matter	2.00	3.43
	99.72 %	99.84 %

I. Adobe from Santa Fé, New Mexico. II. Adobe from Fort Wingate, New Mexico.

Such are very abundant along all the lower valleys of the principal rivers of New England, sometimes coming to the immediate surface or overlaid with a thin layer of sandy material which, together with a little organic matter, forms the true soil. They form, according to Dawson,[2] the sub-soils over a large part of the great plains of Lower Canada, varying in thickness up to 50 or even 100 feet, usually resting upon the boulder clay. They are, as a rule, of almost impalpable fineness, unctious, and extremely plastic. Excepting where superficially oxidized to buff or brown, they are of a blue-gray color and may show on analysis considerable quantities of lime carbonate and alkalies, features whereby they are readily distinguished from the residual clays, and which are regarded as indicative of an origin by mechanical rather than chemical means. When dried, they become greatly indurated, and when unmixed with other materials, bake so hard during seasons of drought, or are so plastic

[1] So called from their most characteristic fossil, Leda. [2] The Canadian Ice Age.

during seasons of rainfall, as to be quite unsuited for cultivation. Mixed with varing proportions of siliceous sand to counteract shrinkage, they form the common brick-making materials of the Northeastern states, burning red and brown.

The materials of the Leda clays naturally vary in different localities, being dependent on the characteristics of the rocks from which they were derived. Those of Canada, according to Dawson, were derived from the waste of the Utica and Quebec groups. This authority believes that when the clay was in suspension, it was probably of a reddish or brown color from the iron peroxide it contained, but that, like the bottom mud now forming in the deeper parts of the St. Lawrence, the coloring matter became deoxidized by organic matter so soon as deposited, the iron being converted into a sulphide or protoxide carbonate. Inasmuch, however, as the materials were so largely derived by the grinding action of the glaciers on fresh rocks, it is not impossible that they may have been again deposited as clay without having ever undergone the oxidizing process.

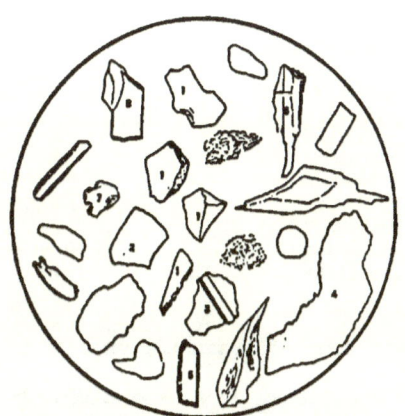

FIG. 34.—Showing particles from Leda clays. 1, quartz; 2, orthoclase; 3, plagioclase; 4, mica; 5, tourmaline; 6, pyroxene; 7, chlorite; 8, hornblende.

Unlike the till or boulder clays, these Leda clays are distinctly stratified, as shown in the accompanying illustration. (Pl. 24.) An analysis of a sample from this locality yielded the author results as given in column I on p. 336. In column II is given that of the portion (33.26 %) soluble in hydrochloric acid and sodium carbonate solutions, while in column III is given the composition of a "semi-assorted glacio-lacustrine" clay bordering on Lake Michigan near Milwaukee, Wisconsin, and in IV a glacial pebbly clay underlying II at the same locality.[1]

[1] Analyses II and III from Chamberlain and Salisbury's paper, 6th Ann. Rep. U. S. Geol. Survey, 1884-85.

Analyses of Stratified Clays

Constituents	I	II	III	IV
Silica (SiO_2)	56.17 %	10.98 %	40.22 %	48.81 %
Alumina (Al_2O_3)	24.25	8.06	8.47	7.54
Phosphoric acid (P_2O_5)	Not det.	Not det.	0.05	0.13
Titanic oxide (TiO_2)	Not det.	Not det.	0.35	0.45
Ferric iron (Fe_2O_3)	Not det.	Not det.	2.83	2.53
Ferrous iron (FeO)	3.54	5.19 [1]	0.48	0.05
Manganese oxide (MnO)	Not det.	Not det.	Trace	0.03
Lime (CaO)	2.00	1.02	15.65	11.83
Magnesia (MgO)	2.57	2.19	7.80	7.05
Potash (K_2O)	4.06	1.12	2.36	2.60
Soda (Na_2O)	2.25	0.75	0.84	0.92
Water (H_2O)	4.69	3.05	1.95 [2]	2.02 [2]
Carbonic acid (CO_2)	None	None	18.76	15.47
Organic carbon (C)	None	None	0.32	0.38
Sulphuric anhydride (SO_3)	None	None	0.13	0.05
	99.56 %	33.26 %	100.21 %	100.46 %

Related to the delta deposits already described, but differing in that their inorganic materials are in large part derived immediately from the sea, are due to the transporting and assorting power of tide and wave action, are the salt-water marsh, or *paludal* deposits so common along the Atlantic border of North America. In discussing the formation of these and their gradual transitions into arable lands, we cannot do better than follow Professor N. S. Shaler.[3]

The formation of a sea-coast swamp is due mainly to wave action and plant growth. It is dependent upon the configuration of the coast. Wave action upon an irregular coast such as that of New England nearly always results in a breaking or wearing away of the exposed headlands and the transportation of the débris from these into intervening inlets, and thrown upon, or at least in a direction toward, the beaches. On these beaches, as one may any day observe, the rock fragments are ever being ground smaller and smaller, and must in time be reduced to the condition of the finest sand and mud. Each incoming

---

[1] All iron determined as FeO.
[2] Contains H of organic matter dried at 100° C.
[3] Ann. Rep. Director of the U. S. Geol. Survey, 1884-85.

wave hurls more or less of this fragmental material upon the beach, whence a considerable portion of it may be again carried seaward by the bottom current or undertow as the wave recedes. One who has stood upon a high rock on the sea-shore and watched the waves come tumbling at his feet and then go creeping silently oceanward once more cannot have failed to notice the continual seething sound due to the constant drag of the rock fragments one over the other as they are impelled inward and outward by the alternating currents. A considerable part of this mud is taken out to sea by the undertow, or bottom current, which always sets from a storm-beaten beach along the bottom, but another part is urged by the movement of the water caused by the waves and of the tidal flow into the fjords, where it falls to the bottom. In this process of carriage the mud is generally conveyed along the shores and is most commonly deposited in the parts of the inlets near the shore line. Wherever there is a bay within which the tidal current is deadened and where the waves have little play, this sediment is most rapidly laid down. If the process of deposition begins on a pebbly bottom, it is at first aided by the irregularities between the stones and the friction of the water among the seaweeds, which frequently attach themselves to the stones. As soon as a sheet of mud is established, it commonly becomes occupied by a dense growth of cel-grass. This plant, by its habit of growth, greatly favors the deposition of sediment. The separate stems are set very closely together, the interspaces not generally exceeding 1 or 2 inches. A tidal current of 2 miles an hour, swift enough to carry much sediment, is almost entirely deadened in this tangle of plants.

At half tide on the New England coast these eel-grass fields are generally covered with water to the depth of several feet; at this stage the tidal currents are commonly strongest. The water above the level of the grass has its usual freedom of motion and brings much sedimentary matter above the level of the foliage. As the tide falls, a part of this waste is entangled and held until it gradually sinks to the bottom, so that each run of the tide gives a certain contribution of sedimentary matter, which goes to shallow the water. This process is easily observed from a boat floating over a field of these plants. The deadening of the current when the lowered tide brings the tops of the plants near the surface is very noticeable. The mass of

z

floating matter — mud, fronds of sea-weed (often with shells or small pebbles attached to their bases), dead fish, and a mass of other refuse, is seen to collect in the mesh of foliage and sink to the bottom. The dead stems of the eel-grass and the bodies of many small crustaceans and mollusca which live on its stalks or on the bottom contribute to the deposit, so that it thickens with considerable rapidity.

When the bed formed on the sea-bottom by the action of the eel-grass and its associated plants has risen to the point where it is dry at low water of the ordinary run of tides, the eel-grass can no longer maintain itself, but gives place to other groups of sea-weeds and grasses.

These species of plants find their place first near the shore line, where the eel-grass platform is naturally the highest. At first their vegetation is quite sparse, owing to the difficulty with which they endure the depth of water at high tide. There is often, indeed, a considerable difficulty in establishing the growth of the second group of plants, and for a while the deposit takes the shape of bare mud-flats, dependent in the main for their accumulation of detrital matter on the growth of certain mollusca, especially of the genera *Mytilus* and *Modiola*.

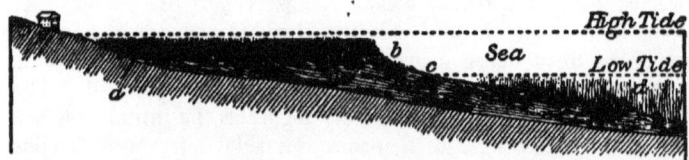

FIG. 35. — Cross-section of marine marsh. *a*, original surface of shore line; *b*, grassy marsh; *c*, mud-flats; *d*, eel-grass; *e*, mud accumulated in eel-grass.

When, as is usually the case, the more highly organized plants have difficulty in establishing themselves over the broad surface of the mud-flat, they win their way to it in the following manner.

From the vantage ground of the shore line, where these plants easily find the conditions of submergence which suit their needs, the plants slowly extend the front of their bench out over the mud-flats. (See Fig. 35.)

This process of growth can be more easily studied than that of the earlier or eel-grass stage of the marshes, for it is visible along miles of our sea-shore. The higher grasses have even more thick-set stems than those of the eel-grass flats; they entangle sediment even more effectively. At first their stems are covered for a few hours at each ordinary tide; they gather waste rapidly, and soon lift the plain which they are constructing up to the point where only at the highest tides are the tops covered by water. At this stage the growth of the deposit is practically arrested, there being no means of increase save from the decay of the grasses themselves.

"On the central parts of the New England shore, as about Boston, the mud-flat occupies at most 2 or 3 feet in the altitude above mean low tide and the annual addition to its mass in a year is very small," perhaps not so much as the tenth of an inch in a year. "On the other hand, in the Basin of Minas, one of the principal inlets leading from the Bay of Fundy, the contribution of sediment is so great that vast areas have been easily reclaimed from the sea by building a rude enclosure around an area of the higher parts of the mud-flat, so that the speed of the sediment-laden waters is checked and they are made to lay down their burdens. In a few years, often in a few months, this enclosed area is raised to near the level of high tides. It is then only necessary to erect a barrier sufficient to exclude the tide, with gates for the rain water, in order to have the land completely reclaimed from the sea. In this simple way there has been an area of many thousand acres of excellent arable land created along these shores."[1]

The lithological and chemical character of deposits of this nature have been but little studied, and we are here able to give only two analyses, as below, in which, however, it is probable that the matter tabulated as insoluble silica includes as well all *silicates* insoluble in acid.

Column I of the table is mud from the marshes of Newport River, a few miles above Beaufort, in Carteret County, North

---

[1] As the total reclaimable area between New York and Portland (Maine) probably exceeds 200,000 acres, their money value in their best state will amount to at least $40,000,000,000. The cost of reclaiming these areas and reducing them to cultivation should not exceed the fifth part of that sum. It may be noted that from the chemical composition of these soils, they are practically inexhaustible, and that from their position they are often well placed for irrigation. South of the New England shore the marsh area is much more extensive than in that region. It is probable that the improvable marshes of the Atlantic coast amount to at least 3,000,000 acres and they may exceed double this amount. (Shaler, p. 380.)

Carolina. This marsh, formed by the filling up of the old river channel, several miles wide, is continually enlarging at the expense of the water surface; and similar formations, to the extent of hundreds of square miles, are accumulating in very many shallow bays and sounds and rivers near the sea.

Constituents	I	II
Silica, insoluble	54.42 %	72.70 %
Silica, soluble	....	1.02
Oxide of iron and alumina	16.45	5.60
Lime	1.18	1.39
Magnesia	0.07	0.05
Potash	1.18	1.82
Soda	0.79	0.85
Phosphoric acid	0.25	0.13
Sulphuric acid	1.46	0.33
Organic matter	....	10.35
Water	20.92 }	3.05
Oxide of manganese	0.54 }	
Sulphide of iron	1.09	0.11
Common salt	1.03	1.71
	99.98 %	100.10 %

Column II is the sea mud or slime which is deposited in the shoal waters of Beaufort Harbor and along the sounds and estuaries of the North Carolina coast. It is a fine, dark-colored salt mud, formed of the silt brought down by the rivers, mixed with decaying vegetable matter (mostly sea-weed and marsh grass), and animal remains, — of fish, molluscs, and all sorts of marine organisms.[1]

What is described by Whitney[2] as a typical swamp bog or peat soil, from a rice field near Georgetown, South Carolina, yielded the results as below, in columns I, II, III, and IV, the last two being simply recalculated from columns I and II on an organic and water-free basis. These are the so-called sob-field soils, in themselves poor, but responding readily to fertilizers. When exhausted by cultivation, they recuperate quickly through the aid of silt deposits from the rivers, brought about by the continual ebb and flow of the tides.

[1] Geology of North Carolina, Vol. I, 1875, p. 214.
[2] Rice, Its Cultivation, Production, and Distribution, Rep. No. 6, Misc. Series, U. S. Dept. of Agriculture, 1893.

Diameter of Particles mm.	Conventional Names	I Soil 0–6 inches	II Sub-soil 6–9 inches	III Soil 0–6 inches	IV Sub-soil 6–9 inches
2–1	Fine gravel	0.00 %	0.00 %	0.00 %	0.00 %
1–.5	Coarse sand	0.71	0.08	1.30	0.14
.5–.25	Medium sand	2.70	0.25	5.18	0.43
.25–.1	Fine sand	0.83	0.13	1.59	0.23
.1–.05	Very fine sand	0.37	0.15	0.71	0.26
.05–.01	Silt	10.32	13.97	19.79	24.30
.01–.005	Fine silt	5.32	7.10	10.20	14.09
.005–.0001	Clay	31.90	34.85	61.17	60.65
	Total mineral matter	52.15 %	57.53 %	100.00 %	100.00 %
	Organic matter, water loss	47.85	42.47		
		100.00 %	100.00 %		
	Loss by direct ignition	47.36	39.65		

Beach Sands. — Although differing radically in composition from the sea-coast swamp deposits already described, we must, on account of their intimate geological relationship, include here a brief description of those fragmental deposits formed by wave action along beaches and in many instances almost absolutely free from organic matter of any kind. Such are the clean white beach sands, the delight of the summer visitor at the seasides. These are found here and there in isolated stretches along the Atlantic slopes, particularly where, as at Old Orchard, Maine, they receive the full sweep of wave and tide from the open sea. In many instances the material forming these beaches is siliceous sand from glacial deposits which the ocean has reassorted according to its own liking. In other cases it is sand brought down by rivers, and which has undergone fractional separation through the varying strength of transporting agencies. In still others it is material derived immediately from the shore rocks through the weathering action of atmospheres and the hammering of the waves. In other cases yet, as along the coasts of Florida, the source is problematical. We can only say, knowing the character of rocks forming the mainland, that they could not have here originated, but must have been transported and probably down the coast, from the areas of crystalline rocks to the northward. It is sometimes, though not always, possible to gain an idea of the probable source of these sands through a study of their mineralogical nature and the physical condition of the individual particles.

Sorby, who devoted careful attention to the microscopic appearance of granules of quartz sand belonging to various geological periods, divided them into five types, "which though characteristically distinct, gradually pass into one another."[1] These types are: —

1. Normal, angular, fresh-formed sand, as derived almost directly from granitic or schistose rocks.
2. Well-worn sand in rounded grains, the original angles being completely lost, and the surface looking like fine ground glass.
3. Sand mechanically broken into sharp angular chips, showing a glassy fracture.
4. Sand having the grains chemically corroded, so as to produce a peculiar texture of the surface, differing from that of worn grains or crystals.
5. Sand in which the grains have a perfect crystalline outline, in some cases undoubtedly due to the deposition of quartz over rounded or angular nuclei of ordinary non-crystalline sand.

The material of most beach sands is largely quartz, though this is not invariably so. Those of the Bermudas are, as a matter of necessity, calcareous. Those of isolated deep-sea islands like the Hawaiians, are derived in part from the volcanic rocks of the islands, and in some instances are composed almost wholly of minute shells of the size of a pin's head. These last from their faculty of emitting a crunching sound when disturbed, are known as "sounding" or "singing sands."

The beach sand at Diamond Head, Oahu, is mainly of olivine and magnetite granules, with smaller amounts of calcareous matter. As usual, the grains in samples from the same level are of fairly uniform dimension, varying from 0.5–1.0 millimetre, the larger forms being often fairly well rounded, while the smaller may still show crystal outlines. The granules, even in the same sample, however, vary greatly in the amount of rounding they have undergone. Like the quartz granules from the Florida beach, these show conchoidal chippings due to the shock of impact as one granule strikes against another.

[1] Proc. Geol. Soc. of London, Anniversary Address, Session, 1879–80, p. 58.

The beach of Santa Rosa island, south of Pensacola, Florida, is composed of clear white quartz sand of almost ideal purity. The grains, though water-worn and with the lesser angles rounded, are still in many cases angular, and of very uniform size (about .5–1.0 millimetre), as shown in Fig. 36. These granules offer a very beautiful illustration of Sorby's type No. 2, the surface of each one, through abrasion, being reduced to the condition of ground glass. Examination with a high power brings out minute fractures and conchoidal chippings, at once suggestive of the preliminary stages of manufacture of the quartz spheres for which the Japanese are so noted. It is as though each granule had been held in the hand of some pigmy aboriginal, and its surface reduced by hammering with another pebble, after the manner known among archæologists as "pecking."

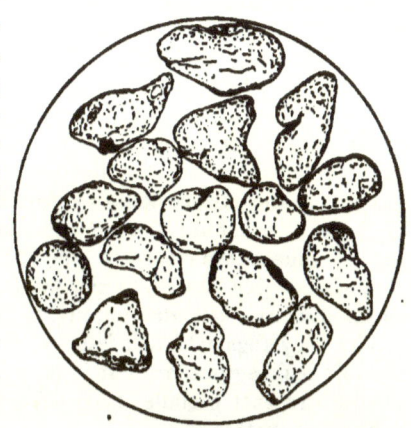

FIG. 36.—Quartz granules in sand from beach, Santa Rosa island.

The shape assumed by a rock or mineral fragment subjected to wave action varies somewhat with the nature of the material, schistose rocks and easily cleavable minerals naturally giving rise to pebbles or granules of quite unequal dimensions in three directions. The schist on the coast of Cape Elizabeth, Maine, for instance, gives rise to pebbles in the form of a greatly flattened oval, while the more homogeneous quartz, with which it is associated, yields nearly spherical forms. "But of whatever character the material, the normal shape of a beach-formed boulder or pebble is oval, and this for the reason that the wave action is a dragging rather than a carrying one; the stone is not lifted bodily and hurled toward the shore to roll back with the receding wave, but is rather shoved and dragged along. Gravity tends to hold the fragments in one position so that the wear is greatest on the side which is down, and this in itself would cause them to assume an oval or flattened form even were they spherical and of homogeneous material at the start."[1]

(3) **Æolian Deposits.** — That no sharp lines can in all cases be drawn between alluvial and æolian deposits has been made evident in our discussion of the loess and adobe. We will now consider those deposits which owe their origin and present structural features almost altogether, if not entirely, to wind action.

The efficacy of the wind as an agent of transportation was dwelt upon in considerable detail on pp. 184 and 292. The material thus carried into the air, often to great heights, is brought to the surface again by gravity, though the normal rate of descent is not infrequently greatly accelerated by rain or snow. Indeed, the clearness, limpidity, of the atmosphere after a rainfall is due simply to the fact that it has been washed, is cleansed of its suspended impurities.

The very fogs which infest our cities, particularly those of the soft coal regions, are but indices of the dust particles in the atmosphere, each globule of fog being condensed about a nucleus of floating matter.

The amount of this dust brought down even from moderately clear atmospheres is often sufficiently abundant to attract the attention of the most casual observer. Professor H. L. Bruner of Irvington, Indiana, has stated[2] that during a snowstorm in February, 1895, a layer of snow about one-fourth of an inch in thickness was colored distinctly brown by the dust it contained. One sample of snow collected yielded .37% of dust, by weight, and it was calculated that dust was thus deposited at the rate of 30.7 pounds avoirdupois for each acre. Another observer calculated the fall as taking place at the rate of 12.77 pounds per acre.

From a gallon of water melted from a snowfall of but 4 inches, which fell in London in January, 1895, there was obtained 10.65 grains of solid matter, 5.75 grains being inorganic and 4.90 grains carbonaceous. Water from a snow collected near the centre of the city, January 30 of this same year, gave 6.25 grains of mineral and 11.07 grains of carbonaceous matter. It was also found that 75% of these impurities were brought down with the first 2 inches of the snowfall.

Dr. Whitney, who examined samples of the black earth

[1] Merrill, Preliminary Handbook, Dept. of Geology, U. S. National Museum, 1889, p. 23.
[2] Monthly Weather Review, U. S. Dept. of Agriculture, January, 1895.

PLATE 24

FIG. 1. Section of beds of Leda clay, Lewiston, Maine.
FIG. 2. Beds of volcanic dust, Reese Creek, Gallatin County, Montana.

brought down near Rockville, Indiana, during a snowfall of the winter of 1895, reported [1] it as consisting of material almost identical with the prevailing loess of that region, from whence it was doubtless derived. The individual particles varied in size between .10 and .05 millimetre. The results of a mechanical analysis of the dust are tabulated with those of loess on p. 331. Samples of the same dust submitted to microscopic examination were found to consist of fully 96 % silt and 4 % organic matter, the latter consisting mainly of fresh-water algæ, diatoms, fungi, cells from decayed grasses, and shreds of woody tissue.

Hilgard, who has examined the so-called "dust soils" of Oregon, California, and Washington, and which during the dry seasons are so loose and fine as to rise in clouds at the merest puff of wind, gives the following tables to show their chemical and physical natures, and which he regards as fairly typical for soils of the arid regions of the United States.[2]

CHEMICAL ANALYSES OF DUST SOILS

CONSTITUENTS	I ATATHNAM PRAIRIE, YAKIMA COUNTY, WASHINGTON	II RATTLESNAKE CREEK, KITTITAS COUNTY, WASHINGTON	III PLATEAU ON WILLOW CREEK, MORROW COUNTY, OREGON
	%	%	%
Insoluble matter	71.67 } 76.78	78.33 } 80.53	79.21 } 81.51
Soluble silica	5.11	2.20	2.30
Potash (K_2O)	1.07	0.70	0.89
Soda (Na_2O)	0.35	0.24	0.05
Lime (CaO)	2.00	2.08	1.37
Magnesia (MgO)	1.34	1.47	1.08
Brown oxide of manganese (Mn_3O_4)	0.04	0.07	0.06
Peroxide of iron (Fe_2O_3)	6.88	6.13	5.63
Alumina (Al_2O_3)	7.91	6.12	6.02
Phosphoric acid (P_2O_5)	0.13	0.18	0.18
Sulphuric acid (SO_3)	0.02	0.02	0.03
Water and organic matter	2.82	2.35	2.55
Total	99.33 %	99.90 %	99.35 %
Humus	4.10	....	0.44
Hygroscopic moisture	4.98	3.20	4.92

[1] Monthly Weather Review, U. S. Dept. of Agriculture, January, 1895.
[2] Bull. No. 3, Weather Bureau, U. S. Dept. of Agriculture, 1892.

MECHANICAL ANALYSES OF DUST SOILS

Conventional Name	Diameter of Particles	I	II	III
Clay	.0023 mm.	0.03 %	3.59 %	1.27 %
Fine silt	.005–.011 "	30.93	13.06	32.29
Silt	.013–.027 "	3.20	5.82	12.75
Very fine sand	.027–.05 "	7.18	27.37	37.51
Fine sand	.05–.122 "	21.88	43.78	10.02
Medium sand	.122–.5 "	32.30	4.57	3.97
		96.57 %	98.18 %	98.72 %

Sand Dunes. — The influence of the wind in the formation of sand hills or *dunes*, as they are commonly called, has received attention on p. 184. A few words more regarding their physical qualities and lithological nature are here essential.

The effect of the single whirlwind or it may be that of the more constant air current for days, weeks, or even months, may be from a geological standpoint comparatively insignificant; but they are, nevertheless, interesting, and at times important. In certain regions of the West, and notably in parts of the Colorado desert, as described by W. P. Blake, in 1853, all the fine loose sand on the surface of the ground is blown away, leaving every pebble and boulder standing out in strong relief from the hard sun-baked soil, or ledge of bed-rock.

Under favorable conditions the material thus blown along may gather in the form of dunes, which themselves travel slowly across the country, ever changing their outlines like drifts of snow. A few miles' north of Winnemucca Lake, in western Nevada, is a belt of these dunes described by geologist Russell[1] as fully 75 feet in thickness and about 40 miles in length by 8 miles in breadth. These, under the restless goading of the winds, are constantly varying in shape, and though moving in mass probably but a few feet a year have already, in more than one instance, made necessary the splicing of telegraph poles to prevent the burial of the wires. Another range of sand dunes, at least 20 miles in length, and forming

[1] Geological History of Lake Lahontan, Monograph XI, U. S. Geol. Survey, 1885.

hills 200 to 300 feet high, occurs on the eastern end of Alkali Lake in the same state. On the eastern shore of Lake Michigan are also dunes of sand sometimes 200 feet in height, and which at Grand Haven and Sleeping Bear have drifted over the adjacent woodlands, leaving only the dead tops of trees exposed. Similar dunes occur frequently on the Atlantic coast, as at Hatteras, Long Island, and Cape Cod. The island of Bermuda is made up almost altogether of coral and shell fragments. These are washed by the waves upon the beach, dried by the winds, and blown gradually inland, thus forming hills in some cases, as stated by Professor Rice,[1] not less than 250 feet in height. In other instances, as at Elbow Bay, on the south shore of the main island, the sand, like a huge glacier, has quite filled a valley, and still progressing in a mass some 25 feet in thickness, is covering houses, gardens, and even woodlands, leaving, as at Lake Michigan, only the trunks of dead trees standing partially exposed in the midst of sandy plains.

One of the most interesting and remarkable of the many regions for the observation of sand dunes, lies between Bordeaux and Bayonne in Gascony, and which has been admirably described by Reclus.[2] The sea here throws every year upon the beach along a line 100 miles in length some 5,000,000 cubic yards of sand. The prevailing westerly winds, continually picking up the surface particles from the seaward side, whirl them over to the inland or leeward slope, where they are again deposited, and the entire ridge by this means alone moves gradually inland. In the course of years there have thus been formed a complex series of dunes all approximately parallel with the coast and with one another, and of all altitudes up to 250 feet. These are still marching steadily inward, though at the rate of but 3 to 6 feet annually, and whole villages have more than once been torn down to prevent burial, and rebuilt at a distance, to be again removed within 200 years.[3]

The lithological nature of the dunes is widely variable,

[1] Geology of Bermuda, Bull. 25, U. S. National Museum.
[2] The Earth, Atmosphere, and Life.
[3] The church of Lege, owing to the encroachment of the sand dunes, was torn down in 1090, and rebuilt at a distance of 2½ miles from its first site. By 1850 the dunes had traversed the intervening space, and again necessitated its removal.

though naturally siliceous sand is the prevailing constituent in the majority of cases. J. W. Retgers describes[1] the dune sands of Holland as consisting principally of quartz granules, together with those of garnets, augite, hornblende, tourmaline, epidote, staurolite, rutile, zircon, magnetite, ilmenite, orthoclase, calcite, and apatite; and, more rarely, microcline, cordierite, titanite, sillimanite, olivine, kyanite, corundum, and spinel. The majority of these minerals occur in the form of well-rounded granules, though many of the garnets, zircons, and magnetites show quite well-preserved crystal outlines. It is noticeable that these sands contain no mica, although the mineral occurs in the sea-sand, from whence the dunes are derived. Retgers accounts for this on the supposition that during the transportation of the material the mica folia become so finely shredded as to be sifted out from the heavier particles of sand, and quite dissipated. It is well to note that the abrasive power of wind-blown particles is greater than among those carried by water, since, as noted by Daubree, a thin intervening film of water may serve to buoy up the granules, and keep them apart. To this fact is ascribed the angular nature of many of the wind-blown grains, they having become shattered through the shock of impact. This same authority seems to think that with wind-blown sand, as with water-worn material, there is a minimum limit, beyond which reduction in size of particles rarely goes. This minimum he places at about .25 millimetre in diameter. It seems, however, more probable that attrition may go on to an almost indefinite limit, but that the finer and lighter materials are driven farther away — perhaps not collecting in the form of dunes at all — leaving, as one would naturally expect, the sands of any one series of dunes of nearly uniform size.

It was noted by Blake during the surveys of the railway routes to the Pacific that the wind-blown sands of the Colorado desert were sometimes in the form of almost perfect spheres, all their sharp edges and asperities having been worn away by mutual attrition. The grains were composed mainly of quartz, agate, garnet, and dark granules derived from the débris of volcanic rocks. In places there is a black iron sand, and usually a considerable proportion of lime carbonate, as indicated by its brisk effervescence when treated with acid. The sand dunes of

[1] Neues Jahrbuch für Mineralogie u. Geologie, etc., 1895, 1 B. 1st Heft, p. 22.

the Bermudas, as elsewhere noted, are composed wholly of calcareous material from finely comminuted shells and corals, while those of the Sevier desert region of Utah, as described by Gilbert,[1] are of fine gypseous sand formed by the evaporation of the water in the neighboring playa lakes.

Volcanic Dust. — The finely comminuted materials ejected from volcanoes and caught up by atmospheric currents, as described on p. 153, are sometimes carried long distances to be again deposited either on land or in the water, forming loose, often flour-like deposits of varying thickness. At various points in Colorado, Kansas, Nebraska, Montana, and other of the Western states, are remnant beds of fine volcanic dust such as must originally have covered many square miles of territory, and the materials of which were derived from sources now wholly obscured.[2] The illustration given on Pl. 24 is from a photograph, taken by the writer, of one of these beds in the lower Gallatin valley, Montana. From the height of the man's shoulder to his feet the bed is of pure glassy dust, very light gray in color, and so fine and light that when thrown into the air it floats away at the slightest breath. The figure given shows the appearance of this glass as seen under the microscope. Beds of this nature upwards of 4 feet in thickness occur underlying the loess or surface soil along the Republican River in Nebraska and Kansas and even as far east as Omaha in the first-named state. The source of their materials is problematical.

FIG. 37. — Showing outlines of shreds of volcanic dust, as seen under the microscope.

Deposits of this nature thus far described are of very recent origin, and the beds loosely coherent. There are, however, good

[1] Monograph I, U. S. Geol. Survey, 1890.
[2] See On Deposits of Volcanic Dust and Sand in Southwestern Nebraska, Proc. U. S. National Museum, Vol. VIII, 1885, p. 99.

reasons for supposing that similar processes were carried on in the earlier stages of the earth's history, but that the peculiarly susceptible deposits have since undergone such extensive alteration as to be no longer recognizable as wind-drifted materials. Where the material still exists as a surface deposit, it undergoes ready decomposition on account of its porosity and easy permeability. The character of the resultant soil is dependent somewhat upon the character of the material, which varies indefinitely. The volcanic dusts are as a rule siliceous, more nearly allied to the acid potash rocks than to the basalts.

The analyses given below show the chemical nature of (I) a fine, white, almost flour-like pumice dust from Harlan County, Nebraska, and (II) of dune sands from the Pamlico Peninsula, North Carolina. This last is described[1] as a tolerably fine, nearly white sand consisting of smooth, well-rounded grains, mainly quartz, but containing also occasional shell fragments and black granules of iron ore.

Constituents	I	II
Silica (SiO_2)	69.12 %	92.12 %
Alumina (Al_2O_3) } Iron oxide (Fe_2O_3) }	17.04 }	5.20
Lime (CaO)	0.86	1.13
Magnesia (MgO)	0.24	0.03
Potash (K_2O)	0.04	0.04
Soda (Na_2O)	1.69	0.35
Sulphuric acid (SO_3)	....	0.33
Ignition	4.05	0.60
	100.23 %	100.49 %

(4) **Glacial Deposits.** — Under this name are included those drift deposits which are the product mainly of glacial action, though their immediate deposition may have been brought about in part through the instrumentality of water. The strictly aqueo-glacial materials have been noted under the head of alluvial deposits.

Allusion has been already made to the manner in which glaciers erode and transport. During a comparatively recent period in geologic history, there appears to have come over a

[1] Geology of North Carolina, Vol. I, 1875, pp. 182–183.

portion of North America a gradual lowering of the normal temperature or increase in the annual precipitation, or perhaps both, until the condition of affairs now existing in northern Greenland prevailed as far south as the 39th parallel of north latitude. Now whether the ice sheet extended at any one time over the area outlined below or whether there were periods of advancement and retreat; whether the glaciation was produced by floating ice and local glaciers as argued by certain Canadian geologists, or by a truly continental ice sheet thousands of feet in thickness, are for our present purposes matters of slight concern. We have more to do with results than methods. Suffice it for the moment, that over the entire northeastern part of the United States and eastern Canada, all the existing loose materials from rock decay that had been gathering for untold ages was carried bodily northward, westward, or southward, as the case might be. From over a considerable part of southern New England the original residual soils were stripped and dumped into the Atlantic, portions of the transported material still protruding above sea-level in the forms known now by the names of Nantucket, No Man's Land, and Block Island. In process of this transfer the rocks were planed down to hard fresh surfaces, over and upon which were deposited new materials from the north. It follows that over this entire glaciated area, estimated by Upham [1] as some 4,000,000 square miles, with the exception of a few comparatively insignificant patches here and there, scarcely a foot of clastic matter is to be found that is truly native. Wherever road cuts or stream erosion favors, the regolith in various conditions of compactness may be found lying directly upon the hard, smooth, and striated rock with which it has perhaps no affinity in composition or structure. The rotten and mechanically triturated detritus of many rocks from many sources more or less admixed by the moving glacier or commingled by resultant streams, is spread out to form the soils on lands to which it is as truly foreign as are the emigrants who land to-day upon our shores. The stone wall, built of boulders found loose in the field, may consist of granites, diabases, schists, or shales even though the underlying rock may be a limestone; or the wall may be of limestone though the country rock be a gneiss, or slate. A similar distinction exists in the soil itself, which, while it may in part consist of the material

[1] Ice Age in North America, p. 579.

of these boulders in a finely divided state, is more likely to consist of detritus of softer rocks which yielded more readily to the abrasive force. Sand and gravel or clay, dust or mud, black with organic matter or red-brown from iron oxides, the admixture is ever varying, dependent only on the nature of the materials to the north. But the material of the glacial drift is spread out over the land in a manner far from uniform and under conditions widely variable. Following Professor Salisbury[1] and others, we may, according to its physical characters and method of deposition, separate the deposits into two general groups: (1) the stratified or assorted drift,[2] and (2) the unstratified or unassorted, the first having been laid down under the influence of water and hence showing a more or less stratified condition, while the second, deposited directly from the ice, consists of a heterogeneous aggregate of coarse and fine materials without evident marks of stratification. The two forms are not always readily separable nor is their relative position always the same, either one not infrequently occurring uppermost, and "not rarely they alternate with each other several times between the surface and the bottom of the drift."

A large part of the drift is composed of this unstratified and unassorted material, consisting of clay, sand, gravel, and boulders in ever-varying proportions, and to which the name *till* or *boulder* clay is commonly applied, or from its mode of deposition, that of *ground moraine*. As already noted, it is the material carried along bodily beneath the ice sheet and left in the position it now occupies on its final retreat. This, entirely unmodified except upon the immediate surface where it has become converted into soil through the agencies elsewhere described, forms the regolith over large areas of the northeastern portion of America and of northern Europe as well. Where as yet unaffected by oxidation, it is of a gray or blue-gray color, and often so intensely tough and hard as to necessitate, in process of excavation, recourse to blasting. The upper portion, through percolation of meteoric waters, is as a rule of a buff or brownish color, owing to oxidation of the ferruginous constituents. Through the combined agencies of this oxidation, of plant and animal life and of cultivation, considerable contrasts in both physical and chemical properties

---

[1] Ann. Rep. State Geologists of New Jersey, 1801.
[2] Here included in large part with the aqueo-glacial deposits.

are brought about between the superficial and deeper-lying portion, which are commonly recognized by the terms *soil* and *sub-soil* respectively applied to them, though originally they may have been one and the same thing. The composition of this till naturally varies with the character of the rocks from whence it was derived. It may have, and indeed probably *has*, in most cases travelled but a short distance, and its constituent particles may be the same as that of the rocks which it overlies, though in a finely divided condition, only the harder and tougher rocks retaining their lithological identity, while the more friable, like the shales and sandstones, have been ground to the condition of clay and sand. To attempt to give then the composition of the till would necessitate its study and analysis in innumerable localities, — an endless and profitless task. It will be sufficient to here describe a few representative occurrences. In nearly all till the boulders, consisting of the harder and more resistant of the materials, are in a more or less rounded or rhomboidal form, with their surfaces scarred and with other marks of the rough treatment to which they have been subjected. They are in fact the tools with which the glacier has done its work, and these scars are but the signs of wear. Intermingled with these is an ever-variable amount of finer detritus largely a result of mechanical abrasion. Professor W. O. Crosby, who has studied in great detail the physical properties of the till about Boston, states[1] that, excluding the larger stones, it consists of 25 % of coarse material which may be classed as gravel; 20 % of sand; 40 to 45 % of extremely fine sand, or rock flour, and less than 12 % of clay. The gravel in these cases consists mainly of pebbles of the harder and more massive rocks of the region, such as granite, diorite, diabase, quartzite, and sandstone. In passing from sand to gravel, there is noted an increase in the proportional amount of quartz, in clear and angular or subangular forms, due mainly to the disintegration of the granite, quartzite, and sandstone pebbles. The "rock flour" also consists essentially of quartz. The most striking feature here brought out is the very small proportion of actual clay material, which varies from one-tenth to one-eighth of the total bulk.

The following table, as given by F. Leverett, shows the approximate physical condition of the till as represented by the sub-soil in various parts of Illinois.

[1] Proc. Boston Soc. of Natural History, 1890, p. 123.

APPROXIMATE NUMBER OF GRAINS OF SAND, SILT, AND CLAY IN ONE GRAMME OF GLACIAL TILL
(BOULDER CLAY) FROM ILLINOIS

Diameter	Conventional Names	Charleston	Sheldon	Marshall County	San José	Champaign
2-1 mm.	Fine gravel	0	1	3	0	2
1-.5 "	Coarse sand	7	13	31	0	28
.5-.25 "	Medium sand	284	472	950	25	955
.25-.1 "	Fine sand	2,507	5,750	13,060	822	8,753
.1-.05 "	Very fine sand	73,480	132,900	239,000	156,600	104,100
.05-.01 "	Silt	13,530,000	11,860,000	6,816,000	12,800,000	7,907,000
.01-.005 "	Fine silt	238,100,000	242,800,000	200,000,000	231,700,000	276,100,000
.005-.0001 "	Clay	10,800,000,000	10,830,000,000	11,520,000,000	12,410,000,000	14,090,000,000
	Total	11,051,706,338	11,084,799,136	11,733,070,253	12,054,657,457	14,374,320,838

The till is not, however, always spread out evenly over the land, but though partaking in a general way of the topography of the slopes which it covered, lies much deeper in certain places than others. Indeed, it thickens and thins out very irregularly and in many places fails entirely either through having never been deposited, as over many a rocky hillside in New England, or through having been removed by running water. Moreover, there are found in certain parts of the drift-covered areas rounded hills of very symmetrical form, composed of material identical with the till, but which must have been deposited under slightly different conditions. These range in height up to 200 or 300 feet, though rarely more than half that amount. Such forms are known as *drumlins*.

The moraines, as already noted, represent those portions of the ice drift which gathered near the edge of the ice sheet in the form of submarginal accumulations, to be left as broad belts or ridges of sand and gravel on its retreat. Such with reference to their position to the margin of the ice are known as *terminal, marginal,* and *frontal* moraines. The materials of which they are composed represent (1) that which accumulated beneath the edge of the ice while it was practically stationary for a considerable length of time; (2) that dumped from the surface at its margin; and (3) that pushed up by the ice sheet, in front of itself during its forward movement. Such ridges are not sharp as a rule, but broad and low, it may be from a fraction of one to several miles in width. Unlike the subglacial drift,—the till,—the materials are but loosely consolidated, and but a small part, if any, of the boulders show the scarred and abraded surfaces so characteristic of those of the till proper.

This frontal moraine, occupying the southern and western margin of the glaciated area, forms one of the most striking and unique of geological bodies. Composed of materials of a most heterogeneous nature, ever varying, and limited in range of variation only by the lithological character of the rocks to the northward and eastward; in all degrees of coarseness and fineness, from boulders of many tons' weight to particles too small to be visible to the unaided eye, only obscurely and sometimes scarcely at all stratified excepting where subsequently modified by running water; in the form of broad low hillocks, domes, and ridges,—the moraine sweeps in an interrupted, sinuous belt from eastern Massachusetts to North Dakota and over

400 miles into British America, having a length, in all its windings and turnings, of not less than 3000 miles.

The water arising from the melting ice sheet flowed off, in part, over the surface, forming superglacial streams, or in part upon the surface of the ground beneath as subglacial streams, of which last the river Rhone of to-day is a good example. Presumably also a portion of the water became concentrated and flowed for short distances in the mass of the ice itself, forming thus englacial streams. In all cases the running water would collect, reassort, and variously modify the rock débris found either in immediate connection with the ice itself or at its extremity, in the terminal moraines. There were thus formed hillocks and ridges or low fan-shaped masses of "modified drift." The sand, gravel, and boulders which collected in the troughs of superglacial streams would, on the final melting of the ice, be deposited as ridges running essentially parallel with that of the movement of the ice on which they formed. Such are known as *eskers*, or *osars*. Other deposits closely resembling these and sometimes confounded with them, but formed, it is believed, only by swift and changeable currents near the frontal margin of the ice, present often a rude and disturbed and distorted stratification, and are known as *kames*. They differ from the eskers in their outlines as well as positions with reference to the glacier from whence their materials were derived, being as a rule in the form of hills, rather than ridges, and with their longer axes at right angles with that of the ice motion.

Beyond the margin of the ice and its terminal moraines are found still other loosely aggregated deposits of a similar heterogeneous nature which are likewise due to swiftly running water caused by the melting ice. Such, according to their position and form, are known as *valley drift, morainic* or *frontal aprons*, and *overwash plains*.

The thickness of these glacial deposits varies greatly, as has been already indicated. Variations of upwards of a hundred feet may occur within the limits of even less than one square mile. Professor Newberry estimated that the area south and west of the Canadian highlands covered with glacial drift was not less than 1,000,000 square miles, and that its average depth would not be less than 30 feet. Other estimates on deposits in Ohio, Indiana, and Illinois give an average thickness

PLATE 25

1

2

Fig. 1. Section of glacial till.   Fig. 2. Glaciated landscape.

in these states of 62 feet. In extreme cases the deposit has been found to extend to a depth of 300 to 500 feet. Bell has stated [1] that glaciation of the surface of British America has been almost universal in the regions east of the Rocky Mountains, and all over the Palæozoic districts west and south of Hudson and James Bay the average depth of the till is 100 feet, and perhaps 200 feet in Manitoba and the northwest territories.

The following section is given by James Geikie [2] as showing the varying character of the glacial drift and its interstratified interglacial lacustrine deposits: —

	Feet	Inches
Sandy clay	5	0
Brown clay and stones (till)	17	0
Mud	15	0
Sandy mud	31	0
Sand and gravel	28	0
Sandy clay and gravel	17	0
Sand	5	0
Mud	6	0
Sand	14	0
Gravel	30	0
Brown sandy clay and stones (till)	30	0
Hard red gravel	4	6
Light mud and sand	1	8
Light clay and stones	6	6
Light clay and whin block	26	0
Fine sandy mud	36	0
Brown clay, gravel, and stones	14	4
Dark clay and stones (till)	68	0
	355	0

3. THE SOIL

There remains now to be summarized a few of the characteristics of those superficial portions of the regolith to which the name *soil* is commonly applied, and these, too, only in direct relation to their properties as soils, since as integral portions of the regolith they have already been sufficiently touched upon.

(1) **The Chemical Nature of Soils.** — The prevailing constituent of any soil, whatever its source, is nearly always silica, with varying amounts of alumina, oxides of iron, lime, magnesia, and the alkalies.[3] A small amount of organic matter, from

---
[1] Bull. Geol. Soc. of America, Vol. I, 1890, p. 289.
[2] The Great Ice Age, 3d ed., 1894, p. 120.
[3] The peat deposits furnish almost the only exception to this rule.

extraneous source, is usually present. This prevalence of silica, as may be readily understood, is an essential consequence of soil formation through the breaking down of rocks by the processes of weathering, whereby all but the most indestructible portions are lost.

The predominantly inorganic nature of any soil may easily be shown by fractional separations, made either by washing, or by sieves of varying degrees of fineness, whereby it is brought into portions of like size and weight such as can conveniently be submitted to microscopical and chemical analyses. All portions, from the finest dust to particles of such size as to be classed as pebbles, will thus be found to be but mineral matter, particles of quartz, feldspar, shreds of mica, and other silicates in ever-varying proportions and stages of alteration or decomposition.

Owing to the destructive nature of their formation, it is but natural that a soil, particularly one of considerable antiquity, should but slightly resemble the parent rock. This fact was more than suggested in the chapter on rock-weathering. In order that its significance may be fully comprehended, the analyses of fresh rock and corresponding residual material from various sources are given in the table on p. 359.

The most striking of the dissimilarities shown by this table are, as is to be expected, those of the limestone soils, as in columns I and II, where the proportional amounts of silica, iron, and alumina are increased, roughly speaking, nearly one hundred fold, while the amount of lime carbonate is correspondingly diminished. This condition of affairs is still further exaggerated in the case of the red soil of Bermuda (columns III and IV) and which offers particularly favorable opportunities for study, owing to the isolated condition of the islands and the consequent freedom from danger of contamination by other than local drift.

The shells and corals which in a more or less consolidated condition form the entire mass of these islands, although essentially of carbonate of lime, are nevertheless not entirely so, carrying, aside from the magnesia, about 1% of inorganic impurities, chiefly oxides of iron and alumina and earthy phosphates, which are practically insoluble in the water of rainfalls, with which alone we have to do here. As time goes on, the lime is slowly leached out and carried away into the ocean, the

CHEMICAL NATURE OF SOILS

ANALYSES OF ROCKS AND RESIDUAL SOILS

Constituents	I	II	III	IV	V	VI	VII	VIII
Silica (SiO_2)	0.44 %	43.07 %	0.052 %	45.156 %	60.09 %	45.31 %	46.75 %	42.44 %
Alumina (Al_2O_3)	} 0.42	} 25.07	0.64	{ 15.473	16.89	26.55	17.61	25.51
Ferric oxide (Fe_2O_3)		15.16		{ 13.898	0.10	12.18	16.79	19.20
Lime (CaO)	34.77	0.03	54.406	3.948	4.44	Trace	9.46	0.37
Magnesia (MgO)	Trace	0.03	1.751	0.539	1.06	0.40	5.12	0.21
Potash (K_2O)	Not det.	2.50	0.006	0.133	4.25	1.10	0.55	0.49
Soda (Na_2O)	Not det.	1.20	0.252	0.007	2.42	0.22	2.56	0.50
Carbonic acid (CO_2)	42.72	Trace	44.251	2.533	....	0.00	0.00	0.00
Phosphoric acid (P_2O_5)	....	....	....	....	....	0.47	0.25	0.29
Water (H_2O) and Ignition	1.08	12.98	0.328	18.265	0.62	13.75	0.02	10.92

I. Trenton limestone, unaltered. II. Residual soil derived from the same. III. Coral and shell limestone from Bermuda. IV. Residual soil derived from the same. V. Gneiss: Albemarle County, Virginia. VI. Soil derived from the same. VII. Diorite: Albemarle County, Virginia. VIII. Soil derived from the same. All iron calculated as sesquioxide.

insoluble parts remaining. Throughout the centuries of decay, this 1% of insoluble impurities, representing but one ton of earth to every 99 tons removed, slowly accumulates until it forms the common red earth of the islands. Though usually fertile, in numerous instances where the leaching has been excessive the resultant soil is so rich in iron and other deleterious constituents as to be quite barren.

There are few more impressive facts in agricultural geology, than that each foot in depth of such soil, as it now lies at our feet, may indicate the removal of at least 100 feet in actual thickness of limestone. In other words, assuming that nothing has been lost by mechanical erosion, the surface of the ground has been lowered this much in bringing about the present conditions.

From what has gone before, it is obvious that soils derived by purely mechanical agencies will, if unmixed with other materials, show a composition closely resembling the mother rock, as in the case of that derived from granite as described on p. 207 or those derived from argillites and siliceous sandstones; others in which chemical agencies prevailed may by solution and other changes have so far lost important constituents as to be scarce recognizable as rock derivatives at all. Obviously a rock mass containing in itself none of the elements of plant food cannot, merely through its decay, furnish soil of appreciable fertility. This fact is well illustrated in the region known as the Bare Hills north of Baltimore, Maryland, or the Chester County Barrens in southern Pennsylvania. Both regions are underlaid by peridotites — rocks rich in iron-magnesian silicates, but almost wholly lacking in lime, potash, or other desirable constituents. Such rocks not merely decompose very slowly, but the stingy product of such decomposition consists only of hyaline forms of silica, magnesian carbonates, or silicates and ferruginous products quite devoid of nutrient matter, affording food and foothold to scanty growths of grass and stunted shrubs. That, however, a rock contains all the desired materials, is no certain indication as to character of its decomposition product, since in this process of decomposition much desirable matter may have become lost. Nevertheless most soils retain what we may call inherited characteristics, and a direct comparison whenever possible is by no means uninteresting, as will be noted later.

It need scarcely be remarked that the value of any soil depends wholly upon its capacity for plant growth. Hence a satisfactory treatise on the subject should be written with a view to showing to what this capacity is due, and what are the laws governing its fertility and its rejuvenation when that fertility becomes exhausted. Such a method of treatment is, however, far beyond the limits of the present work, and we must content ourselves with merely touching upon a few of the most salient points, leaving the at present little understood subject of fertility for other and abler writers. It may be well to remark, however, that a soil left to itself and nature's processes rarely becomes barren or exhausted except it may be under changed geological conditions. A growing organism takes temporarily from the soil that which is essential, but restores it again with accrued interest in the form of carbonaceous and nitrogeneous matter derived from the atmosphere, when it dies. Thus, under normal conditions, the soil grows yearly richer and richer and capable of supporting larger and more luxuriant crops. It is only when the husbandman comes in, and by his improvident harvesting robs the soil not merely of its interest due, but of a part of the principal as well, that bankruptcy results.

For a long period the fertility of a soil was felt to be dependent very largely upon its chemical composition, and older treatises and reports of geological surveys are filled with tables of analyses which the acquired knowledge of years now shows us to be almost as worthless as can be, either for the purposes for which they were first intended, or as indicative of the mineral nature of the soil itself.[1] A soil which, under certain conditions of climate or moisture, is utterly barren may, under changed conditions, be fruitful in the extreme, as has been repeatedly demonstrated in the case of the so-called American deserts, dreary stretches of aridity given over to sage brush and a few degraded forms of animal life, but which need only moisture to cause them to laugh with harvests.

[1] The common practice of making soil analyses, whereby the results are tabulated as soluble and insoluble (meaning by soluble the portion extracted by boiling hydrochloric acid) and putting down the latter as silica (or sand) and insoluble silicates, cannot be too strongly condemned. It means nothing. A growing plant is capable of extracting only a small, and as yet unknown, portion of that taken out by the acid, and as to what silica and insoluble silicates may be, we are left in ignorance. Such analyses can be of use to neither the student of soils or of geology.

Naturally, a soil containing in itself nothing in the way of available plant food can be made to produce crops only when the needed constituents are supplied. Investigations have, however, shown that, though varying in different species, the proportional amount of food demanded by plants which can be supplied by the atmosphere and meteoric waters is very large.

It seems to be now pretty well conceded that of all the constituents found in soil aside from moisture, only potash, lime, magnesia, phosphoric and sulphuric acids, can be considered absolutely essential as plant food. The ash of all plants, to be sure, contains silica, soda, — and it may be iron and other mineral ingredients, — but such are to be regarded as accidental rather than otherwise. Of the constituents enumerated as essential, magnesia and sulphuric acid are almost invariably present in sufficient quantities, while potash, lime, and phosphoric acid, even though sufficiently abundant in a virgin soil, are liable to exhaustion under the ordinary methods of cultivation. The source of these materials has been shown in the previous pages and need here be only touched upon. The potash and the lime must have come originally from the decomposition of potash-lime-bearing silicates, as the feldspars and micas, amphiboles and pyroxenes. The original source of the phosphoric acid was undoubtedly the apatite of the eruptive rocks, though now to be found in bones and skeletons of animals, whose remains become entombed in sedimentary rocks of all ages. How small and proportionally insignificant are the percentages of these constituents in any soil, fertile or barren, is shown in the table below,[1] in which are given the general average composition of a large number of soils, sedentary and transported. The sulphuric acid, which is not given in this table, rarely amounts to more than from 0.05 % to 0.5 % when calculated as sulphuric anhydride (SO_3).

So small, comparatively, are these percentages, that it is rare, indeed, to find a soil which on complete analysis will not be shown to contain them in sufficient proportion. The varying degrees of fertility in such cases are due then, not to differences in ultimate composition, but to difference in combination of these elements whereby they are or are not available for plant food, and to physical and climatic differences as well.

[1] From Part A, Vol. II, Part II, Chemical Analyses, Geological Survey of Kentucky, p. 113.

CHEMICAL NATURE OF SOILS

ESSENTIAL CONSTITUENTS OF SOILS

Constituents	I 8 Ohio Valley Alluvial Soils	II 2 Mississippi Valley Alluvial Soils	III 21 Quaternary (Loess) Soils	IV 60 Coal-Measures Soils	V 42 Upper Sub-Carboniferous Soils	VI 10 Waverly Soils	VII 9 Black Slate Soils	VIII 15 Corniferous Limestone Soils	IX 16 Upper Silurian Soils	X 32 Trenton ("Blue") Lime-stone Soils
	%	%	%	%	%	%	%	%	%	%
Sand and insoluble silicates	84.310	74.840	88.008	87.608	87.058	89.013	80.131	83.617	82.905	73.380
Alumina, iron and manganese oxides	9.835	10.437	0.941	0.160	8.308	4.200	10.587	9.060	10.493	11.200
Carbonate of lime ($CaCO_3$)	0.102	1.385	0.370	0.208	0.245	0.145	0.476	0.409	0.322	0.749
Magnesia (MgO)	0.189	0.401	0.292	0.103	0.235	0.204	0.524	0.620	0.422	0.044
Phosphoric acid (P_2O_5)	0.118	0.198	0.118	0.134	0.125	0.101	0.234	0.270	0.100	0.328
Potash in acid extract	0.450	0.142	0.257	0.244	0.232	0.148	0.178	0.343	0.242	0.404
Potash in the insoluble silicates	1.405	1.880	1.706	1.208	1.005	N.E.	N.E.	N.E.	N.E.	N.E.
Organic and volatile matters	3.472	0.305	2.937	4.160	3.031	4.102	5.929	5.071	0.224	0.211

363

Naturally a growing plant can take up only that which is soluble by the means at its command. A high percentage of any of the above constituents counts for little when they are combined in the form of difficultly soluble silicates. A granitic rock, as has already been noted, contains locked up in its mass all the mineral elements necessary for a fertile soil, but remains barren simply because these are in a condition of slight solubility and its physical structure is such that even the soluble portions are unavailable. Pulverize this rock sufficiently, and it will become immediately available for soil, though naturally its fertility is slight, and rendered enduring only by gradual decomposition. It is of course possible, that by nature's methods, decomposition and incident leaching may have gone so far that a soil on the immediate surface, though derived from rocks rich in essential constituents, has become quite impoverished and barren. This is especially true with limestone residuals, as has been already noted. It is doubtless to this fact that is due the enduring qualities of the glacial till as a soil, though its immediate fertility may not be as great as one of sedentary origin. The undecomposed feldspathic and other mineral particles contained by the till, due to its mechanical origin, yield up slowly but continually their supply of plant food, and such a soil may long outlast the residual clays of non-glaciated regions.

The soils derived from deposits of modified glacial drift are almost invariably sandy or gravelly in their nature. Such, on account of their easy working qualities, great porosity, and ready permeability, are commonly known as light soils, even though their actual specific gravities may be greater than the so-called heavy soils of the ground moraine.[1]

[1] Mechanical analysis of a glacial soil from an old pasture, Cape Elizabeth, Maine, yielded results as below. The portion selected was of just the thickness turned up by the plough, — about 7 inches. In color it was dark gray, at the immediate surface almost black from organic matter, and penetrated throughout by grass roots. Fine angular grains of white quartz were the most conspicuous feature on macroscopic examination. Eight hundred and thirty grammes of this soil on sifting yielded: (1) 2.5 grammes gravel, which failed to pass a sieve containing 8 meshes to the lineal inch. This consisted mainly of angular quartz and cleavage bits of feldspar with occasional rounded lumps of impure limonite, and not completely disintegrated particles of granitic rock. (2) 40 grammes coarse sand retained by 20-mesh sieve and consisting of clear glassy and white opaque quartz in angular and sub-angular fragments, the largest forms being some 3 millimetres in greatest diameter; cleavage bits of white and pink feldspar, rarely

There is many an humble homestead throughout the glaciated areas of North America whose lack of worldly prosperity is due to the dry and barren soil supplied by these deposits of modified drift. On the other hand, there are numerous regions, like those of northern Ohio, where a light, barren, residual soil derived from sandstone has become enriched by an admixture of glacial clays from the north, and thus brought prosperity to thousands of happy homes. Nature works out her own compensations, impoverishing, it may be, here but correspondingly enriching there.

R. H. Loughbridge has shown [1] that the percentage of soluble

folia of white mica, a few bits of mica schist, and lastly hard, rounded pellets of indurated silt and organic matter. (3) 170 grammes retained by 40-mesh sieve and consisting of a clean sand composed of some two-thirds its bulk white quartz particles and one-third opaque, partially kaolinized feldspathic particles; rarely any mica or free iron oxides. (4) 180 grammes retained by 60-mesh sieve and consisting, like the last, of clean quartz and feldspar sand, the quartz particles in excess of the feldspar, and rarely a little mica. (5) 82 grammes retained by the 80-mesh sieve. This, very clean sand of quartz and feldspar, in the proportion of about ⅓ quartz and ⅔ feldspar. (6) 150 grammes retained by a sieve of silk bolting cloth of 120 meshes to the lineal inch. Like the last, composed almost wholly of bright quartzes and somewhat kaolinized feldspars with scarcely a trace of other silicates. (7) 185 grammes which passed the silk bolting cloth. This was submitted to washing, the lighter finer material being poured off as silt. By this means were obtained 118 grammes very fine sand and 67 grammes silt. The fine sand, as before, showed under the microscope only quartz and feldspars, the quartzes still in excess. The silt to the naked eye consisted of a light brown, almost impalpable material, which the microscope revolved into quartz and feldspar particles with shreds of ferruginous products evidently derived from the decomposition of iron-magnesian silicates, such as micas or amphiboles. (8) Organic matter, 19.5 grammes.

A bulk analysis of the air dry-soil, excluding all grass and roots, yielded results as below: —

Ignition (water and organic matter)	2.72 %
Silica	76.80
Alumina and iron oxides	14.04
Lime	0.78
Magnesia	Traces
Potash	2.87
Soda	1.18
	98.39 %

Such a soil is plainly little more than a highly quartzose granite or gneiss in a pulverulent condition and in which the agencies of decomposition have scarcely begun their work. Its composition could have been almost foretold by the microscopic examination.

[1] On the Distribution of Soil Ingredients among the Sediments obtained in Silt Analysis, Am. Jour. of Science, Vol. VII, 1874, p. 17.

material in a soil rapidly increases with the degree of comminution; *i.e.* the finer the material the larger the proportional amount of soluble matter, and hence of matter available as plant food. This is well brought out in the following table abridged from the one given in Mr. Loughbridge's original paper, the figures in the upper space of each column indicating the size of the particles, and the percentage amount of each as determined by fractional separations.

Conventional Name:	Clay	Finest Silt	Fine Silt	Medium Silt	Coarse Silt
Diameter of Particles:	21.64 % ?	23.56 % min. .005–.011	12.54 % min. .013–.016	18.67 % min. .022–.027	18.11 % min. .033–.038
Constituents	%	%	%	%	%
Insoluble residue	15.96	73.17	87.96	94.13	96.52
Soluble silica	33.10	9.95	4.27	2.35	....
Potash (K$_2$O)	1.47	0.53	0.20	0.12	....
Soda (Na$_2$O)	(1.70)[1]	0.24	0.28	0.21	....
Lime (CaO)	0.09	0.13	0.18	0.09	....
Magnesia (MgO)	1.33	0.46	0.26	0.10	....
Manganese (MnO$_2$)	0.30	0.00	0.00	0.00	....
Iron sesquioxide (Fe$_2$O$_3$)	18.76	4.76	2.34	1.03	....
Alumina (Al$_2$O$_3$)	18.10	4.32	2.64	1.21	....
Phosphoric acid (P$_2$O$_5$)	0.18	0.11	0.03	0.02	....
Sulphuric acid (SO$_3$)	0.06	0.02	0.03	0.03	....
Volatile matter	0.00	5.61	1.72	0.02	....
Totals	100.14	99.30	100.00	100.21	....
Total soluble constituents	75.18	20.52	10.32	5.16	....

According to Hilgard,[2] the substance which assumes commanding importance as controlling the fertility of a soil, aside from physical conditions, is lime, in the presence of which, in adequate proportions, smaller percentages of the other plant foods will suffice for high and lasting productiveness, than would otherwise be the case. Since lime is the essential constituent of the rock limestone, it follows that, other things being equal, a "limestone country is a rich country." As elsewhere noted, however, a limestone soil may have become so

[1] An excess of original amount, due to the addition of sodium chloride to produce flocculation of clay in suspension.
[2] The Relation of Soil to Climate, Bull. No. 3, U. S. Weather Bureau, 1892.

leached of its lime, through prolonged decay, as to be benefited by artificial applications of this same constituent. Lime is, moreover, so generally distributed throughout the great majority of rocks that few soils would be lacking in this constituent, were even a small proportion of the original amount left in the residue from rock decay, instead of being so largely removed in solution.

It would follow from this that the composition and fertility of a soil is dependent not more upon the character of the rock mass from which it is derived, than upon the prevalent climatic conditions under which it originated, the general average temperature and the amount and distribution of the rainfall being particularly important factors. This branch of the subject has also been considered in some detail by Hilgard, to whom we are indebted for the only satisfactory résumé. Concerning conditions of temperature, this author says: —

"Within the ordinary limits of atmospheric temperatures all the chemical processes active in soil formation are intensified by high and retarded by low temperatures, all other conditions being equal. This being true, we would expect that the soils of tropical regions should, broadly speaking, be more highly decomposed than those of the temperate and frigid zones. While this fact has not been actually verified by the direct comparative chemical examination of corresponding soils from the several regions, yet the incomparable luxuriance of the natural as well as the artificial vegetation in the tropics, and the long duration of productiveness, offer at least presumptive evidence of the practical correctness of this deduction. In other words, the fallowing action, which in temperate regions takes place with comparative slowness, necessitating the early use of fertilizers on an extensive scale, has been much more rapid and effective in the hot climates of the equatorial belts, thus rendering available so large a proportion of the soil's intrinsic stores of plant food that the need of artificial fertilization is there restricted to those soils of which the parent rocks were exceptionally deficient in the mineral ingredients of special importance to plants that ordinarily form the essential material of fertilizers."[1]

[1] While the action of frost in bringing rock masses into the condition of soil is, in temperate climates, of very great importance, there seems to be a limit beyond which it accomplishes little in the way of directly promoting decomposi-

Concerning the concentration and leaching out of certain constituents by the action of meteoric waters, the same authority says : —

" When, however, the rainfall is either in total quantity or in its distribution insufficient to effect this leaching, the substances which otherwise would have passed into the sea are wholly or partially retained in the soil stratum, and when in sufficient amount may become apparent on the surface in the form of efflorescences of 'alkali' salts. One of the most important modifications produced by scantiness of rainfall on soil formation is the great retardation of formation of clay from feldspathic rocks (kaolinization) and the sediments derived therefrom. As a result, it is observed that the soils of the Atlantic slope are prevalently loams, containing considerable clay, and even in the case of alluvial lands, oftentimes very heavy, while the character of the soils of arid regions is predominantly sandy or silty with but a small proportion of clay, unless derived directly or indirectly from clay or clay shales. In the former case, the clay, becoming partially diffused in the rain water when a somewhat heavy fall occurs, percolates through the soil in that condition and tends to accumulate in the sub-soil, the result being that almost without exception the sub-soils of the humid regions are very decidedly more clayey than the corresponding surface soils. Not only does this clay water tend to make the sub-soil more compact and heavy, making it less pervious to water and air, but it is assisted materially in this by the action which tends to leach the lime carbonate out of the surface soil into the sub-soil. The accumulated clay is thus frequently more or less cemented into a 'hardpan' by lime partly in the form of carbonate and partly in that of zeolitic (hydrous silicate) compounds, adding to the compactness of the sub-soil, and therefore to the usual specific difference between the soil and sub-soil; viz. the deficiency or absence of humus and the difficulty of penetration by an aeration of the roots of plants."

For these reasons the soils of arid regions, even though containing the same materials, are often of uniform physical and chemical character to great depths. The soluble salts, as car-

tion, and presumably disintegration as well. Collier's (8th Ann. Rep. New York Exp. Station, 1889) experiments showed that 47 successive freezings and thawings of a soil did not perceptibly increase the percentage of *soluble* potash.

bonate of lime and salts of potash and soda, which are leached away in regions of great average humidity, remain in those where the annual precipitation is less, or where, on account of its uneven distribution throughout the warmer months of the year, its permeability and consequent leaching action is less. Hilgard brings out this fact prominently in tables from which that below is condensed, the original being compiled from several hundred analyses of soils from the humid regions of North and South Carolina, Georgia, Florida, Alabama, Mississippi, Arkansas, Kentucky, and the arid regions of California, Washington, Montana, Utah, Colorado, Wyoming, and New Mexico.

SHOWING THE PROPORTIONAL AMOUNTS OF SOLUBLE SALTS IN SOILS OF ARID AND HUMID REGIONS

CONSTITUENTS	ARID REGION	HUMID REGION
Insoluble residue	69.081 %	84.472 %
Soluble silica	0.280	3.873
Potash	0.825	0.187
Soda	0.251	0.071
Lime	1.645	0.112
Magnesia	1.384	0.200
Brown manganese oxide	0.056	0.120
Iron peroxide	5.431	3.455
Alumina	7.309	4.008
Phosphoric acid	0.144	0.114
Sulphuric acid	0.035	0.005
Water and organic matter	5.585	3.557
Total	99.978 %	100.093 %

Discussing these figures, Professor Hilgard says : "Concerning this table with reference to the lime, a glance at the columns for the two regions shows a surprising and evidently intrinsic and material difference approximating to the proportion of 1 to 14½. This difference is so great that no accidental errors in the selection of analysis of the soils can to any material degree weaken the overwhelming proof of the correctness of the inference drawn upon theoretical grounds ; viz. that the soils of the arid regions must be richer in lime than those of the humid countries." These remarks hold good also for the percentages of magnesia and the alkalies. From the fact that in humid regions the more soluble constituents are leached out, we may safely infer a corresponding proportional increase in the insoluble constituents. This is also made manifest by the tables, there being a difference of nearly 15% in favor of the humid regions. The table shows, further, a probably greater proportion of zeolitic material in the soil of arid regions, the assumption being based upon the percentages of soluble silica. Concerning this difference, the author says: —

"Nor should this be a matter of surprise when we consider the agencies which are brought to bear upon the soils of the arid regions with so much greater intensity than can be the case where the solutions resulting from the weathering process are continually removed as fast as formed by the continuous leaching effect of atmospheric waters. In the soils of regions where summer rains are insignificant or wanting, these solutions not only remain, but are concentrated by evaporation to a point that in the nature of the case can never be reached in humid climates. Prominent among these soluble ingredients are the silicates and carbonates of the two alkalies, potash and soda. The former, when filtered through a soil containing the carbonates of lime and magnesia, will soon be transformed into complex silicates in which potash takes the precedence of soda, and which, existing in a very finely divided (at the outset in a gelatinous) condition, serve as an ever-ready reservoir to catch and store the lingering alkalies as they are set free from the rocks, whether in the form of soluble silicates or carbonates.[1] The latter have still another important effect. In the concentrated form, at least, they themselves are effective in decomposing silicate minerals refractory to milder agencies, such as calcic carbonate solutions, and thus the more decomposed state in which we find the soil minerals of the arid regions is intelligible on that ground alone. But it must not be forgotten that lime carbonate, though less effective than the corresponding alkali solutions, nevertheless is known to produce, by long-continued action, chemical effects similar to those that are more quickly and energetically brought about by the action of caustic lime. In the analysis of silicates we employ caustic lime for the setting free of the alkalies and the formation of easily decomposable silicates by igniting the mixture; but the carbonate will slowly produce a similar change, both in the

[1] See author's remarks on page 374.

laboratory and in the soils, in which it is constantly present. This is strikingly seen when we contrast the analyses of calcareous clay soils of the humid region with the corresponding non-calcareous ones of the same. In the former the proportions of dissolved silica and alumina are almost invariably much greater than in the latter, so far as such comparisons are practicable without assured absolute identity of materials."

It is evident from the above that, provided the amount of decomposition be the same, the soil of an arid region may contain a larger proportion of desirable constituents than one in a region of considerable annual precipitation. It may, also, and for the same reasons, contain a larger proportion of constituents that are positively deleterious. This is particularly true of arid and semi-arid regions of poor drainage, like the Great Basin regions of the United States, where salts of sodium not infrequently accumulate to such an extent as to render the land sterile and barren in the extreme.

The primary origin of the sodium in these salts lies in the soda-bearing silicate minerals forming the rocks of the region and from which they have been set free through their decomposition.

It should be stated, however, that the so-called "alkali" is not composed wholly of sodium compounds, but contains also salts of magnesia, lime, iron, and potash. Nor is the form under which the salts exist at all constant. As a rule, the larger portion of the alkali is in the form of sulphate of soda, though a considerable portion may exist as carbonate or chloride, and smaller proportions in the form of nitrates. Concerning the formation of these carbonates, Hilgard says:[1]—

"There seems to be a consensus of opinion that the carbonation of the soda is connected in some way with the presence of limestone or carbonate of lime, and that an exchange has occurred in which either common salt or Glauber salt have transferred their acidic components to lime and have become carbonates instead. . . . Yet the simple explanation of the contrary reaction was given and published as early as 1826 by Schweigger. In 1859 it was again observed by Alex Muller, in a different form, but neither of these chemists, nor any of their readers, appear to have perceived the important bearing of this reaction, not only upon the formation of the natural depos-

[1] Bull. No. 3, Weather Bureau, U. S. Dept. of Agriculture, 1892.

its of carbonate of soda, but also upon a multitude of processes in chemical geology. Without going into details . . . it may be broadly stated that the formation of carbonated alkalies occurs whenever the neutral alkaline salts (chlorides or sulphates) are placed in presence of lime or magnesia carbonates and carbonic acid, or of alkali 'supercarbonates' (hydrocarbonates) containing even a slight excess of carbonic acid above the normal carbonates, the latter being the actual condition of all natural sodas."

We have thus far considered, only those elements of the soil that are derived directly from the rocks from which they are formed.

To this list we should add the element nitrogen, not so much on account of its quantity, as its value as plant food and of the great economic value of some of its compounds. The common forms under which this element exists, are (1) atmospheric nitrogen, a colorless, tasteless, and innocuous gas and which forms some three-fourth by bulk of the air we breathe, and (2) the nitrogen of the soil, where it exists in at least three distinct forms, (1) organic nitrogen, (2) as ammonia or ammonia salts, and (3) as nitric acid.

The average amount of nitrogen present in agriculture soils is given by authorities as varying from 0.1% to 0.3%, though occasionally, as in certain soils rich in organic matter, 4 or 5%. Of these forms only the ammonia salts and nitric acid are of direct value for plant food. Nitrogen, in the form of nitrate of soda, forms an important mineral fertilizer, as noted on p. 71.

The extraordinary richness in nitrates of the soils in tropical countries, and particularly in South America, has often been remarked since the subject was first broached by Humboldt and Boussingault. According to Müntz and Maracano, nitrates occur in the soils of Venezuela, the valley of the Orinoco, and other localities sometimes to the amount of 30% of their mass. These nitrates they show to be due to the oxidation of organic nitrogen through the agency of bacteria. They state that in the caverns of the regions, a guano composed mainly of the excreta of birds and bats, but admixed also with the dead bodies of these and other animals, has accumulated to the amount of millions of cubic metres. Through the gradual nitrification of this guano, and a combination of the nitrogen with the lime of bones, or existing as a carbonate in the soil, a gradual transition is brought about wherever there is free access of air or the temperature is sufficiently high to stimulate the nitrifying organisms to their fullest activity. There is thus a gradual change in the character of the nitrogeneous combination from the interior to the exterior portions of the cave, as shown in the following analyses: —

Constituents	Guano from Interior of Cave	Earth from the Entrance	Earth from some Distance from Entrance
Organic nitrogen	11.74 %	2.41 %	0.80 %
Nitrate of lime	0.00	3.03	10.36

These authorities would account for the presence of extensive deposits of nitrates as in Chili, on the assumption that the soluble nitrate, originally derived from decomposing organic matter, as noted above, had been leached out from its place of origin by percolating water and redeposited elsewhere on evaporation. The invocation of atmospheric electricity to account for any part of the nitrates of the soils, they regard as quite unnecessary, the same being of indirect influence only, furnishing first nitrogen for growing plants which in their turn serve as food for animals. These same authorities give the following figures relative to the amounts of nitrates and nitrogen in South American soils: —

Constituents	San Juan	Los Morros de Parapara	El Encantado
Nitrate of lime	2.85 %	3.50 %	0.62 %
Organic nitrogen	0.15	0.27	0.21

(2) **The Mineral Composition of Soils.** — This is essentially the same as that of the regolith of which the soil forms a part. Fragmental quartzes and feldspars form the larger proportion of most soils. These are intermingled with shreds of mica, amphibole, pyroxene, calcite or aragonite, iron and manganese oxides, and in variable proportions, kaolin and other silicates, carbonates and oxides. The presence of these constituents is

usually somewhat obscured by iron oxides and carbonaceous matter; but when these are removed by acids or by ignition, and the residue submitted to microscopic analyses, the true mineral nature can be, as a rule, made out with approximate accuracy.[1]

From what has gone before, it must be evident that the constituents of any soil are almost universally in a finely fragmental condition, a few of the smaller more resisting minerals, like the rutiles, tourmalines, zircons, etc., having perhaps escaped the comminuting processes. Of the silicate minerals we may be sure that many are in an advanced stage of hydration and the ferruginous constituents in a state of peroxidation. It is possible that under favorable conditions new minerals of fairly perfect crystalline development may be temporarily formed. Since the work of Lemberg was made public,[2] it has been very commonly assumed that various minerals of the zeolite group were present and exercised an important function in the conservation of soil fertility. Notwithstanding the somewhat enthusiastic endorsement by Hilgard, of this idea, as set forth in the previous pages, the writer can but feel that too much has been assumed, both regarding their actual presence and their possible utility.

We must not lose sight of the fact that the actual occurrence of zeolites in soils, *where they have been formed*, is as yet not proven. Their presence is inferred from the fact that weak acids, such as are known to be capable of decomposing zeolitic minerals, will extract from the soil, among other constituents, certain ones which are characteristic of minerals of the zeolitic group; and it is assumed, purely for lack of a better reason, that these elements are those thus combined. Even if this be true, their efficacy as potash holders may well be questioned, since potash is not as a rule an element of great importance in zeolitic minerals. Out of the 23 known species of zeolites (including apophyllite), in but five is potash considered an essential constituent. These five, as already noted on p. 32, are apophyllite, ptilolite, mordenite, phillipsite, and harmotome, of which phillipsite alone carries upwards of 6 % (theoretically),

[1] See Anleitung zur Mineralogischen Bodenanalyse, etc., by Franz Steinriede, Inaug. Dis. Friedrichs-Universität Halle-Wittenberg. Halle, 1889.
[2] Zur Kenntniss der Bildung und Umbildung von Silicaten, Zeitschrift der Deutschen Geolischen Gesellschaft, Vols. XXXVII and XXXVIII, 1885 and 1887.

the other smaller amounts, the average for the five being about 4 %. Now assuming that all the zeolites in the soils belonged to these five groups and none to the 18 potash-free varieties, and that 10 % of any soil consisted of zeolitic material, even then we have thus combined only 0.4 % of K_2O.

We must remember, further, that the zeolites are invariably secondary minerals, as already noted, and as such are commonly regarded as decomposition products. This does not necessarily mean, however, that they are products of superficial weathering. Indeed, in the majority of cases the evidence is all to the contrary, they being plainly a result of deep-seated processes going on in the rock masses long before atmospheric action began to manifest itself. (See under Hydrometamorphism, p. 161.) It is even questionable if the conditions prevalent in soil are not unfavorable rather than otherwise to the formation of zeolitic compounds, and if such traces as there exist are not rather residuals from the breaking down of rock masses in which they had been previously formed.

In this connection it is well to remember that zeolites as a whole are characteristic of basic eruptive rocks, such as have yielded but a proportionately small amount of our soils. Also that the mutual chemical reactions that may go on in a rock mass due to close juxtaposition of the various minerals may largely, if not entirely, cease in a soil where the amount of interspace is so enormously exaggerated.

The researches made during the *Challenger* Expedition[1] show, it is true, that even at so low temperatures as from 2° to 3° C. phillipsite is being formed in the deep-sea muds of the Central Pacific and Indian oceans. But in these cases the mud is the finely comminuted débris from basic eruptive rocks, itself peculiarly liable to decay, and containing all the materials necessary for zeolitic formation. It is, moreover, in a condition of continual moisture and under the weight of the thousands of fathoms of overlying water which is here in a state of extreme quiescence, being beyond the influence of superficial movements, as waves, tides, and currents. These conditions are so widely different from those which exist in the superficial parts of land areas, that they can be regarded as merely suggestive. The same may be said relative to the zeolite (phillipsite and apophyllite) formations at Plombières as described by Dau-

[1] Rep. on the Scientific Results, 1873-76, Deep-sea Deposits, 1891, pp. 400-411.

brée.[1] Another fact which mitigates against the theory of zeolitic formation in soils, is the almost universal absence of these minerals in such secondary, unmetamorphosed rocks as are the product of the reconsolidation of the same class of materials as in their unconsolidated condition form soils. If they once existed, it would seem strange they have not in some cases at least survived. If formed in soils, why should they not be formed in secondary rocks where the conditions are apparently so much more favorable?

It would, to the writer at least, seem more probable that the soluble potash of soils exists, not in zeolitic combination, but in some of the numerous decomposition products of feldspar, nepheline, scapolite, etc., to which the name *pinite* is commonly applied. Such at least is the case in the potash-rich soils of Maryland, examined by R. L. Packard.[2] It is possible also that it may exist in compounds allied to glauconite.

The writer has elsewhere[3] pointed out that, particularly among basic rocks, there may be actually a larger percentage of matter soluble in hydrochloric acid and sodium carbonate solution in rocks ordinarily designated as fresh, than in the débris resulting from their decomposition. This fact he has since emphasized in a paper read at the December (1896) meeting of the Geological Society of America, and from which the following statements are drawn. Rock-weathering, it must be remembered, is in the majority of instances accompanied by a leaching process, whereby original soluble compounds, or new soluble compounds formed during the process of decomposition, are gradually removed. The final result is therefore, as already many times noted, a residue consisting of the least soluble constituents, and which forms the ordinary surface soil. Even in cases where the actual amount of soluble matter is greatest in a soil, the apparent excess may be due to water of hydration and to the large amount of sesquioxide of iron, the latter being practically insoluble in meteoric waters so long as there is a free supply of oxygen, though readily soluble in hydrochloric acid. These conclusions are based upon the following table, in which the total percentage loss on ignition, minus the ignition in the insoluble residue, is tabulated with the soluble matter.

[1] Geologie Expérimental, pp. 180 et seq.
[2] Bull. 21, Maryland Agricultural Experiment Station, 1893.
[3] Bull. Geol. Soc. of America, Vol. VII, 1895, p. 355.

SOLUBLE CONSTITUENTS OF ROCKS 377

TABLE SHOWING PROPORTIONAL AMOUNTS OF FRESH AND DECOMPOSED ROCKS SOLUBLE IN BOILING HYDROCHLORIC ACID AND SODIUM CARBONATE SOLUTIONS

CONSTITUENTS	PHONOLITE: MARIENFELS, BOHEMIA		DIABASE: MEDFORD, MASSACHUSETTS		BASALT (ABSAROKITE): FORT ELLIS, MONTANA		GNEISS: ALBEMARLE COUNTY, VIRGINIA		GRANITE: DISTRICT OF COLUMBIA	
	Fresh	Decomposed	Fresh	Decomposed	Fresh	Decomposed	Fresh	Decomposed	Fresh	Decomposed
SiO_2	21.64%	17.08%	10.85%	9.50%	20.90%	19.90%	10.09%	17.69%	9.40%	25.37%
Al_2O_3	10.37	11.26	4.74	4.86	3.80	5.80	13.54	24.86	8.30	
Fe_2O_3	2.23	2.72	10.91	10.00	FeO 4.28 / 2.21	7.81		11.80		
CaO	1.07	1.01	3.09	1.50	1.01	1.59	1.64	0.06	0.00	
MgO	0.40	0.44	2.20	1.84	10.42	6.88	0.89	0.37	0.71	
K_2O	0.28	0.11	1.21	0.68	Not det.	0.08	2.40	0.75	1.08	
Na_2O	5.45	0.06	0.50	0.17	Not det.	0.22	1.10	0.22	0.08	
H_2O	4.10	8.78	2.73	3.07	5.42	12.34	0.62	13.40	1.23	
	45.54%	42.36%	36.23%	31.62%	54.13%	54.62%	30.28%	69.18%	22.75%	25.37%

(3) **Physical Condition of the Soil.** — Chemically, as previously noted, a soil differs from the parent rock in the amount of leaching it has undergone, and in the finely comminuted and more or less decomposed condition of its particles. There are other distinctions, not the least important of which are its state of loose coherency and porous condition due to interstitial air spaces. It has been estimated by Whitney[1] that the approximate number of grains in one gramme of soil varies between 2,000,000 and 15,000,000, the lowest estimate being that for a sandy soil containing only some 4.77 % of material in such an extremely fine state of comminution as properly to be classed as clay, while the highest number is that in a sub-soil containing some 32.45 %. Our interest in these remarkable figures is still further heightened when we are called upon to remember that these grains are not in actual contact, but each separated from the other by thin films of moisture, or, in a dry soil, by actual air spaces. That such spaces exist is easily proven by the fact that any soil may be greatly diminished in bulk by pressure. The amount of this empty space is naturally quite variable, but it is estimated to constitute on an average some 50 %, by volume, of the soil. That is to say, a cubic foot of soil, in its natural condition, contains an amount of space between its grains, filled by air or water, equal to one-half the entire mass.

These extraordinary figures are given, not merely to illustrate the wonderful degree of comminution reached in rock-weathering, but also, and what is of more importance from the standpoint of an agriculturist, the amount of surface exposed to the solvent action of roots and percolating waters. Indeed, it has been estimated that the total surface areas of the grains in a cubic foot of soil amounts, on the average, to 50,000 square feet. The amount is of course greater in a fine than a coarse soil, but in any case sufficiently large to enable us to understand how, under the ordinary conditions of cultivation, all the materials essential to plant growth may in a brief time be removed, unless renewed by artificial fertilizers.

Further than this, the amount of space between the grains is of very great importance in determining the circulation of water in the soil, and its capacity for retaining the right proportion essential to plant growth as noted later.

The experimental work of late years goes to show that fertil-

[1] Bull. No. 4, U. S. Dept. of Agriculture, Weather Bureau.

ity is dependent upon these physical properties perhaps even more than upon chemical composition. If the structure, *i.e.* the manner of arrangement of the soil particles, is such as to be most favorable to root action and conservation of moisture, there are few soils but may be made fertile by proper treatment, even cannot the desired physical properties be imparted by artificial means. A soil which contains too large a proportion of fine clay matter may hold so large a proportion of moisture as to be quite unsuited for cultivation when saturated, and become equally unfitted by induration when dry. A light, porous, sandy soil on the other hand, though fertile during seasons of abundant precipitation, parts with its moisture so readily as to be quite barren in seasons of drought. Porosity and capillarity, two properties dependent wholly on the size and shape of the soil particles, are therefore very essential items in this consideration. Moisture precipitated in the form of rain soaks into the ground or flows off upon the surface in varying proportions, according to local conditions, an open porous soil naturally absorbing more rapidly than one that is close and compact.

When, after the rain ceases, evaporation sets in from the surface, the water which has soaked in is brought back again in part by capillarity, though a part escapes through leaching downward beyond the reach of capillarity, ultimately coming to the surface again, at lower levels, in the form of springs. The capacity of a soil to care for the water it receives from rains is, perhaps, the most important of any one property. It has been demonstrated that the soils of the semi-arid regions of the West will produce abundant crops of wheat and corn, though receiving but about half the amount of water from rainfall that would be requisite in the East. This is accounted for wholly on physical grounds, and is explained as follows:[1] Water falling upon a perfectly dry soil descends very slowly, and indeed, in extreme cases, may continue to fall for hours without wetting the mass for more than a few inches below the surface, while it will be absorbed very rapidly by a soil already wet, but not saturated. This is due to the fact, as explained by Whitney, that in a dry soil the tension or contracting power of the surface of the water is greater than the attraction of the soil grains. If, on the other hand, there is any appreciable

[1] Conditions in Soils of the Arid Region, by Milton Whitney, Yearbook U. S. Dept. of Agriculture, 1894.

amount of moisture in the soil, the tension of the water surface will cause it to contract and pull the water from above into the sub-soil. It follows, then, that the water of rains falling in semi-arid regions would not penetrate into the dry sub-soil, until the overlying portions had become successively so far saturated that they could no longer hold the water back, and it would pass downward, therefore, very gradually into the lower depths, saturating, or nearly saturating, each successive depth as it progressed. Unless, then, as rarely happens in this region, the rainfall was so great and so continuous as to saturate the soil to a considerable depth, the whole supply of moisture absorbed would remain within a short distance of the surface, either immediately within reach of plant roots, or where it can be brought upwards once more by capillarity when evaporation from the surface begins. With a continuously wet sub-soil, however, as in the East, a very considerable portion of the water passes at once to depths beyond the reach of roots or capillary attraction, and is, so far as our present considerations are concerned, completely lost until, in the course of nature's endless cycle, it shall once more be returned as rain. Within certain limits, a small rainfall, equitably distributed, is more advantageous to agriculture than are the heavier precipitations which characterize the Atlantic slopes of the American continent.

The capacity of soils for moisture has been the subject of experiment, and is found to vary widely, being naturally largely dependent upon the size of the individual particles and the consequent amount of interspace. Whitney states [1] that sub-soils of Maryland truckland having 45 % of interspace will hold but 22.41 % by weight of water, when this space is completely filled. The sub-soil of the Helderberg limestone, having 65 % of space, will hold 41.22 %. King [2] gives the following table to show the actual amount of water held by field soils when their surfaces are only 11 inches above standing water, this water having been lifted into them by capillarity: —

[1] Some Physical Properties of Soils, Bull. No. 4, U. S. Dept. of Agriculture, Weather Bureau, 1892.
[2] The Soil, p. 159.

Soil	Per Cent of Water	Pounds of Water	Inches of Water
Surface foot of clay loam contained	32.2	23.9	4.59
Second foot of reddish clay contained	23.8	22.2	4.26
Third foot of reddish clay contained	24.5	22.7	4.37
Fourth foot of clay and sand contained	22.6	22.1	4.25
Fifth foot of fine sand contained	17.5	19.6	3.77
Total	....	110.5	21.24

According to Meister, different soils show water-holding capacities as follows :[1]—

Kind of Soil	Per Cent of Water Imbibed	Kind of Soil	Per Cent of Water Imbibed
Clay	50.0	Chalk	49.5
Loam	60.1	Gyseous	52.4
Humus	70.3	Sandy (82% sand)	45.4
Peat	63.7	Sandy (64% sand)	65.2
Garden	69.0	Pure quartz sand	46.4
Lime	59.9		

(4) The Weight of Soils. — This is dependent upon (1) the character of the particles composing the soil and (2) their degrees of compactness. The figures given below are those of Schubler.[2]

Weight per Cubic Foot in Pounds, of Various Soils

Dry siliceous or calcareous sand	110
Half sand and half clay	90
Common arable soil	80–90
Heavy clay	75
Garden mould, rich in vegetable matter	70
Peat soil	30–50

(5) Kinds and Classification of Soils. — Being derived from rocks of all kinds and under greatly varying conditions; in

[1] Handbook of Experiment Station Work, U. S. Dept. of Agriculture, 1893, p. 317.
[2] Handbook of Experiment Station Work, U. S. Dept. of Agriculture, 1893, p. 315.

almost infinitely variable conditions of comminution, decay, and proportional amounts of their various constituents, no hard and fast lines for soil classification can be laid down. All things considered, they are best classed with the regolith of which they form a part, the general divisions of which are given in tabular form on p. 299. We thus have the primary divisions of sedentary and transported soils, accordingly as they have been formed in place, or transported. Each of these is again subdivided according to the agencies involved in its transportation or original formation.

Many varietal names have been applied to soils, but as a rule in so loose and ill-defined a manner as to give them only a very general significance. A common practice is to name one of sedentary origin according to the rock from which it was derived, as *granite soil, limestone soil*, etc. Transported soils, on the other hand, are often designated either by the agencies involved in transportation, as *glacial*, or *æolian soils*, their position, as *terrace soils*, or their physical or chemical characteristics, as *sandy* or *clayey soils*. A *loam* is usually defined as an admixture of sand and clay with more or less organic matter, a clayey loam being one in which clay predominates and a sandy loam one in which sand prevails. The terms *peat, muck, loess, marl*, etc., have been already sufficiently defined. Local names indicative of suitability for particular crops, or sometimes of doubtful or obscure meaning, are frequently met with. The *bluegrass soils* of central Kentucky are limestone residuals celebrated for the luxuriant growths of *Poa pratensis* which they bear. The red "buckshot" soils of the Yazoo bottoms, Louisiana, are stiff clayey alluvial soils mottled with ferruginous spots.

Many names indicative of mode of formation have already received attention, but a few others may be here noted. The names *Endogeneous* and *Exogeneous* have been proposed for soils formed in place (sedentary) or derived from other sources (transported). It is presumably scarcely necessary to remark that such terms are quite inapplicable and inappropriate.

The name *regur* is locally applied to a fine dark argillaceous soil particularly suited for cotton growing and which has a wide areal distribution throughout southern India. Its origin appears to be mainly subaerial, though a part of the material so called is undoubtedly alluvial. The material is highly plastic when wet, and expands and contracts to a remarkable degree under varying conditions of moisture and dryness. This soil is very retentive of moisture and rarely requires to be irrigated artificially. It is, as a rule, of great fertility and of wonderful lasting powers, it being stated that in some localities it has borne crops for 2000 consecutive years, without the aid of manures. In depth this soil is rarely over 6 to 8 feet. The following analyses show the chemical character of the regur (from near Seoni) on the surface and at depths of (A II) 5 feet and (B II) 3 feet below. The analyses A, like those given on p. 306, are instructive as showing the large increase in the amount of lime from the surface downward. Although not so stated, the slight differences in B I and B II are probably due to the lesser depth below the surface from which B II was taken.

Constituents	A		B	
	I	II	I	II
Insoluble matter	62.7 %	47.61 %	62.8%	63.7 %
Organic matter	9.2	8.4	9.0	8.7
Water	8.4	7.6	8.2	6.5
Oxide of iron	11.0	15.9	10.9	11.8
Alumina	7.5	8.6	7.6	8.4
Carbonate of lime	1.2	11.80	1.5	1.3
	100.00 %	100.00 %	100.00 %	100.00 %

In many cases this regur is derived directly from basaltic rocks, by surface decomposition *in situ*, whilst other varieties were derived from other aluminous rocks, or are alluvial deposits in river valleys, lakes, lagoons, and marshes. The dark color, as is usual, is due to the presence of organic matter.[1]

The term *sub-soil* is applied to that portion of the regolith which immediately underlies the soil proper, and from which it differs mainly in compactness, and the lesser amount of oxidation and decomposition it has undergone. In a soil which has never been cultivated, the sub-soil may pass gradually upward into the soil without distinct lines of demarcation. Prolonged cultivation may, however, have so thoroughly oxidized and physically altered the superficial portions down to the limit of plough and root action, as to bring about a very marked differ-

[1] Manual of the Geology of India, 2d ed., by R. D. Oldham, 1893, p. 411.

ence, both in color and texture, as well as in actual composition. At times the sub-soil becomes so thoroughly compacted as to be almost impervious, forming a so-called *hardpan*.

(6) **The Color of Soils.** — The color of soils is due mainly to carbonaceous matter and iron oxides. To the first are due the dark gray to black colors characteristic of prairie and swamp soils. To iron oxides are due the buff, yellow, ochreous-brown, and red hues, the source of the oxides being mainly the silicate minerals from whence the soils were derived. It not infrequently happens, as abundantly demonstrated in the southern Appalachian states, that it is possible in passing over any section of the country to designate with a fair degree of accuracy the lithological nature of the underlying rocks from the color of the residual soils, even though the rocks themselves may be wholly obscured by decomposition products. In such cases rocks rich in iron silicates, like hornblende and augite, give rise to bright red soils, while those poor in these constituents yield soils of a gray or slightly yellowish hue. Much, however, apparently depends on extent of decomposition and on climatic conditions, as noted below.

One of the most striking features of the landscape observed in travelling southward along the Appalachian belt is the abrupt transition in color of the soil, as the limit of glacial action is past. Within the glaciated area, except where derived directly from rocks themselves highly colored, like the Triassic sandstones, the soils are everywhere dull in color, some shade of gray, drab, or brown. South of this limit ochreous-red and yellowish prevail. Along the line of the Virginia Midland railway, south of Washington, these colors prevail in hues of astonishing brilliancy. Although the soils throughout the region are residual, their colors seem in many cases quite independent of the kind of rock to which they owe their origin. Granite, gneiss, schist, or basic trappean rocks alike give rise to red and yellow highly tenacious soils of such depth and brilliancy of color that every gully, ravine, and roadway stands out against the green background of the landscape, as though painted by some Titanic hand with brushes dipped only in yellow, red, and vermilion ochres. These contrasts are particularly striking in the early summer and directly after a rain. But he who wishes to admire had best do so from his window, and without too much attention to detail.

The soil is plastic and adherent to an intolerable degree. The grass forms no compact sod, as in the North, and as a result the walls of houses, fences, feet, legs and clothes of pedestrians become uniformly stained a dirty ochreous color equally trying to the housewife and to ploughman.

The cause of this color variation has been the subject of speculation by Professors Crosby,[1] Dana,[2] Russell,[3] and others. So far as our knowledge now extends, it is apparent, as first stated by Crosby, that the difference is due to a spontaneous dehydration which takes place in the warmer regions, whereby the hydrous sesquioxides of the type of limonite and göthite are converted into the less hydrated or anhydrous forms turgite and hematite with corresponding changes in color from yellow or brown to red.

This view is rendered the more plausible from the fact that the most brilliant hues are entirely superficial, and below the surface gradually fade out into brown and yellow or even gray hues. Such a transition may be observed in any fresh road cut, but quickly becomes obscured by the washing down of the deeply colored material from the higher levels. Sometimes the brilliant red will be found a mere wash, but a fraction of an inch in thickness, or again it penetrates to the depth of a foot or more before giving way to more modest hues. In such cases the brilliant colors will be found to have penetrated deepest along joint lines, or the more porous portions, leaving the intervening compact masses of more sombre hue.

In discussing this matter, there is, however, one point that we should not overlook, although its importance seems not to have been fully realized by the authorities quoted, and that is, a change in color not due alone to a change in the conditions of the iron, but to the relative greater abundance of this constituent in the upper portions. The iron oxides, as already noted, owing to their less soluble nature accumulate in the residues, and as a rule, the more thorough the decomposition the greater the proportional amount of iron. A small percentage of free oxide disseminated throughout a relatively large amount of detritus imparts but little color; the more iron, the more color.

---

[1] Proc. Boston Society of Natural History, 1885, p. 219, and Technological Quarterly, Vol. IV, 1891, p. 36.
[2] Am. Jour. of Science, Vol. XXXIX, 1890, pp. 317–319.
[3] Bull. No. 52, U. S. Geol. Survey, 1889.

The residue from the Medford diabase described on p. 220 is of a deep brown color, as a whole, but the finest silt washed from it is several shades brighter, of a dull ochreous red. Had the entire mass decomposed to the condition of this silt, we might expect it to have the same color. This change in color, due to increased proportional amounts of iron oxides, is particularly marked in limestone residuals where the original rock may contain merely traces of free oxides, or ferruginous silicates. Neumayer has shown[1] that the snow-white Karst limestones contain only some 0.044 % of ferruginous silicates which themselves carry 20 % of iron oxides. Yet the residual soil left by the decomposition of this limestone is of so pronounced a color, as to have long ago received the name *terra rossa* or red earth.

Other things being equal, brilliancy of color may then be regarded as (1) indicative of advanced decomposition, and (2) of geological antiquity.

(7) **The Age of Soils.** — No sooner were the first rocks pushed above sea-level than the various agencies described under the head of weathering began the work of disintegration, decomposition, and transportation. Of this we have ample proof in the entire series of sedimentary rocks extending from the Archæan down to the most recent and which are but the reconsolidated residues of preëxisting masses. That such a breaking down resulted in the production of soils is a fair inference, though we have no absolute evidence of land plants and hence, *a priori*, of soils, before the beginning of the Upper Silurian period, when plants of the lycopod type appeared. Such soils, *as soils*, have, however, long since disappeared in the never-ending cycle of change and it is not until we reach the Carboniferous period that we meet with soils which have been preserved in place and in recognizable form even to the present day. Even here induration and partial metamorphism has rendered them no longer fitted for the support of plant life,

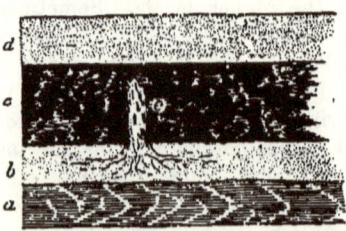

FIG. 38. — Trunk of tree still standing in soil of Carboniferous age. *a*, bed-rock; *b*, under clay or ancient soil; *c*, coal; *d*, bedded rock; *e*, fossil tree.

[1] Verhandl. k. k. Geol. Reichsandstalt, 1875-76, p. 50.

THE AGE OF SOILS

but that they once did so serve is amply proven by the occasional finding of still erect, though fossil, trunks with roots buried in their native soil, as they grew in the marshes and woodlands of the coal period. But as to the time of the beginnings of the formation of such soils as still retain their soil characteristics, we have not in all cases reliable data. Those which are but the unconsolidated sediments of recent geological time, like those of the eastern shore of Maryland, the loess and alluvium of the Mississippi valley, or the swamp and glacial soils of the north and east may, of course, be located with a reasonable amount of accuracy. But as for the residual soils, those which result from the breaking down in place of rock masses, we can only say that they must be younger than the rocks from which they were derived. The writer has shown that the granite soils of the District of Columbia are post-Cretaceous; in other parts of the Piedmont plateau of Maryland, they may be post-Tertiary. In but few instances, as at

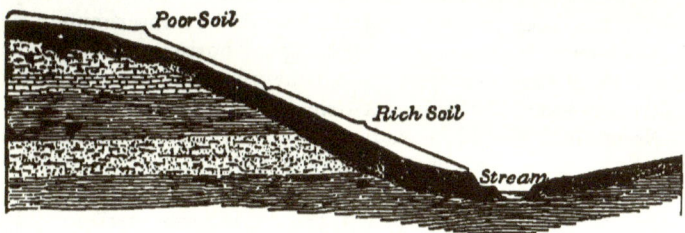

FIG. 39.

Medford in Massachusetts, have we evidence of any considerable amount of soil formation by decomposition and disintegration since the close of the glacial period. Obviously the older a residual soil, the greater the amount of decomposition and leaching it will have undergone and the less will it resemble the parent rock. Where horizontally lying strata of varying character have successively undergone decomposition and a loss of their soluble constituents, the resultant soil must periodically vary according to the nature of the rock undergoing decomposition and the inherited characteristics handed down from the strata earlier decomposed. In such a case as that here figured, we have a residual soil containing the least soluble constituents of the hundreds of feet of dissolved and disintegrated rock which once extended across the entire country becoming commingled with that now undergoing, in its turn, the soil-making process. Such a soil may, therefore, in extreme cases, contain materials of all ages from the first product of disintegration of the uppermost strata, which may have been Carboniferous, to that which formed to-day, and may be Cambrian.

It is, of course, true that through the erosive action of water these soils are continually losing their finer silt and clay-like particles, it may be almost as fast as formed, especially in hilly regions, and that as the soil drops lower and lower in the geological horizons indicated, it becomes more and more impoverished in those constituents derived from the upper beds. But as to what proportion of the material of one horizon is handed down to become admixed with that from the rocks below, we have no means of judging, and in fact it must be ever-varying.

The matter of the geological age of any soil, or the age of the rocks from which it was derived, is therefore of only very general interest, and may well be dismissed here. The attempt which has been made by another writer[1] to discriminate or classify soils according to the geological horizons of the rocks from which they were derived, is believed by the present writer to be futile and wholly inexpedient.

No attempt should be made, as has been done by at least one writer, to state the character of soil that may arise from the weathering of any particular class of rocks, since much depends upon the extent to which weathering has been carried. The ultimate product of weathering of rocks of any but the purely siliceous type is a more or less ferruginous clay, which may be contaminated or admixed with coarser, foreign particles. It is the extent of decomposition, more than its lithological derivation, that determines both the chemical composition and physical characteristics of any soil.

Rocks of essentially the same type so far as composition is concerned, regardless of structural modifications induced by metamorphism, have been formed and re-formed throughout every period of the earth's history, and the attempt made to classify those of igneous origin from the standpoint of geological age has invariably resulted in failure. As has been already indicated, the greater portion of the granitic, gneissic, or highly metamorphosed crystalline schists and calcareous rocks belong

[1] See Stockbridge's Rocks and Soils, p. 12.

either to the Archæan or older Palæozoic formations, but this merely because they, being older, have been longer subjected to metamorphosing agencies, and not because in themselves they possess essential differences. It is true that some authorities lay stress on the supposed abundance of animal remains in certain Palæozoic formations, but no one but the veriest amateur would now dream of attempting to discriminate between either igneous or aqueous rocks of the same nature, but of different geological ages, on purely chemical grounds.

It is a fact, however, that within certain climatic limits, the rocks of any one horizon may impart such characteristics to a residual soil as shall render it adapted to plant growth of a particular kind. Thus,[1] throughout the central portion of Kentucky, where, within the distance of a few miles, rocks occur of several distinct geological horizons, each bearing its mantle of residual soil, each horizon may be traced for long distances, though the rocks themselves are wholly obscured, merely by the character of its forest growth.

(8) **Soils as affected by Plant and Animal Life.** — There are various forms of animal and plant life the action of which is worthy of note in connection with the subject of decomposition; but since it is probable that they are of greater efficiency in promoting changes in soils once formed than in bringing about the preliminary rock disintegration, their consideration has been left to form a portion of the present chapter.

Ants, by means of their numerous borings, penetrating at times to depths of many feet, bring about not merely a rearrangement of soil particles through a transfer of materials from lower to higher levels, but also a condition of porosity whereby air and water gain access to the deeper lying portions, there to promote further chemical and physical changes.

Naturally these insects limit their work to dry and light soils, where their operations may be compared with that of earthworms whose operations are confined to moist ones. Shaler. has calculated[2] that over a certain field in Cambridge (Massachusetts) the ants have made an average transfer of soil matter from the depths to the surface sufficient to form a layer each year of at least a fifth of an inch over the entire four acres under observation. He further mentions a curious effect aris-

[1] As the writer is informed by Mr. J. R. Proctor.
[2] 12th Ann. Rep. U. S. Geol. Survey, 1890-01, p. 278.

ing from the interference of the ants with the original conditions, in the separation of the finer from the coarser particles. In certain parts of New England where sandy soils had laid for a long time uncultivated, fields were covered to a depth of some inches with a layer of fine sand without pebbles larger than the head of a pin, while below the level of perhaps a foot the soil was mainly pebbles, with very little finer material. This condition, it is argued, was brought about by the tens of thousands of ants which each year, over every acre, in the process of building their dwelling brought up the finer material and deposited it in the form of a mound about the surface opening, leaving behind the coarser particles, too heavy for them to move. The common black and brown ants of the United States (*Formica exsectoides*) build upon the surface mounds not infrequently from 1 to 2 feet in height and 3 to 5 feet in diameter and which are composed of materials brought up from below intermingled with twigs and shreds of bark and leaves from the surface. Shaler calculates the mass of some of these mounds as equal to 2 cubic yards. Being of unconsolidated, loosely coherent material, such are constantly being degraded by wind and rain and their particles distributed over the surrounding surface. "Where these structures are numerous, as they are in certain districts in the United States, by their constant deposits of matter on the surface of the ground, they bury a good deal of vegetable waste in the soil; at the same time the animals are constantly conveying into the earth large quantities of organic matter which serves them as food, and the waste of this, including the excreta of the animals themselves, is of considerable importance in the refreshment of the soil." The geological efficacy of insects of this and other types is undoubtedly greater in warmer climes, where not only are they found in greater abundance, but their period of activity extends over a larger portion of the year. Messrs. Mills and Branner, as already noted, are inclined to lay considerable stress on the work of ants and termites in bringing about soil

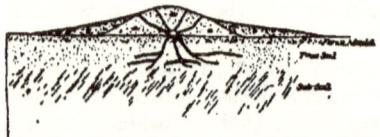

FIG. 40. — Effects of ant-hills on soils. *aa*, sand accumulated in hill; *bb*, material washed down the slopes, mingled with vegetable mould.

changes and rocks decomposition in Brazil. Branner states that in some parts of the Amazon valley, of Minas Goyaz and Matto Grosso, the soil "looks as if it had been literally turned inside out by the burrowing of ants and termites." The species popularly known as saubas excavate chambers and build galleries which are frequently from 50 to 100 feet long, from 10 to 20 feet across, and from 1 to 4 feet high, and contain tons of earth. The white ants or termites, like the true ants, burrow extensive channels in the ground, and build up huge nests upon the surface from the size of which one may gain some idea of the extent of the underground galleries. In the region extending from the state of Parana to north of the Amazon and along the upper Paraguay in Matto Grosso may be seen places where the nests are so close together that one can almost walk upon them for several hundred yards at a time, while no one of the nests is more than 10 feet from another over many acres of ground. Such nests vary in size from 1 to 12 feet in height and 1 to 10 feet in diameter, and do not seem to be confined to any particular kind of country, though especially noticeable in the interior and timberless regions. The constant transference of such quantities of soil from below to the surface, and of organic matter from the surface downward, cannot fail to bring about marked changes in its physical as well as chemical condition, while at the same time affording passageways for air and meteoric waters, as already noted.

Certain animals, like the crayfish, have likewise a habit of burrowing in the ground, though as they are wholly subterranean or aquatic in their nature, the results are less conspicuous to the casual observer. In searching for their food, these animals bore numerous horizontal channels or galleries sometimes an inch or so in diameter and extending for many feet, usually ending in an upward shaft extending to the surface, or at the margin of a pond or stream. These form natural drainage channels and allow a more ready access of air, converting what might under other conditions be a heavy, clayey or even marshy soil, unfit for cultivation, into one light and fertile.

By burrowing through dams and embankments, they have, however, in some instances so weakened these structures as to cause them to give way, whereby large districts have become inundated and for a time rendered unfit for cultivation.

Probably none of the forms of animal life thus far mentioned produce such wide-spread and beneficial results as have been ascribed by Darwin[1] to the common earthworm, the angleworm of the New England disciples of Izaak Walton. These insignificant creatures, as is well known, burrow in the moist rich soil, and derive their nourishment from the organic matter it may contain. In order, however, to obtain this comparatively small amount of nutritive matter, they devour the earth without any selective power, and pass it through their alimentary canals, rejecting the remainder, which nearly equals in bulk that first taken in. The numerous holes made, while in part perhaps to afford passage to the surface, are mainly excavated in this process of soil eating and actually represent the amount of material which the worms have passed through their digestive systems.

Darwin states that in certain parts of England these worms bring to the surface every year, in the form of excreta, more than 10 tons per acre of fine dry mould, "so that the whole superficial bed of vegetable mould passes through their bodies in the course of every few years." By actually collecting and weighing the excretions deposited on a small area during a given time, he found that the rate of accumulation was at the rate of two-tenths of an inch a year, or an inch in every five years. The importance of these worms, then, both as mellowers of the soil and as levellers of inequalities — by burying stones and filling hollows — is therefore very great, and we cannot afford to overlook it here.

While the main influence of the worm is manifested in a mellowing by burrowing and a transfer of material from a lower to a higher level, they bring about a slight admixture of organic matter through a habit of coming to the surface at night time, and dragging down into their burrows small shreds of leaves and grass, which, taken into account in connection with the excrementitious matter of the worms themselves, must tend, though it may be ever so slightly, to enrich the soil. The subject should not be dropped without referring to the abundance of these worms, which in England has been estimated as at the rate of 53,767 to each acre of garden land, and about one half that number for pasture land. It is scarcely necessary to remark that their distribution is very unequal throughout the

[1] The Formation of Vegetable Mould.

world, and that in dry sandy regions they are almost, if not wholly, unknown.

In northern temperate climates, such as that of New England, and particularly where the soil is of a clayey nature like the ground moraine, the burying action of the earthworm, as described above, may be wholly overcome through the heaving action of frost. Every farmer boy who has been condemned to pick the drift boulders from a field knows through bitter experience that, however well he may do his work in the fall, however clean the surface may be when winter sets in, the following spring, after the frost is out of the ground, will find a new crop in no way distinguishable from the old, and which, for all that he can see, may have rained down during the winter's storms. The fact is, however, that they have been actually thrown up, "heaved out," the farmers will say, from below the surface by the frost which here penetrates not infrequently to a depth of two or more feet. As the water-soaked clay underlying one of these buried boulders freezes, it expands upwards, since this is the direction of least resistance. The stone is carried up bodily for a distance dependent on the amount of expansion. When the frost leaves the ground, the soil sinks back nearly to its first position; but the boulder never quite regains its former place, being prevented by particles of soil, or clay or pebbles which fall into the cavity as the soil shrinks away from it. The amount of actual lifting for each season may be but slight, but as the process goes on unceasingly there is always an abundance of new material at the surface each succeeding spring. This heaving action of the frost is abundantly exemplified in these clay regions by the throwing out of fence posts and clover roots; sometimes, when the winter is one of frequent freezing and thawing, causing the destruction of a crop as completely as though it had been pulled up by the roots. In wet boggy lands this heaving action of frost, as exerted on partially buried boulders of small size, is sometimes exemplified in a peculiarly striking manner. The surface of the ground will be dotted here and there with small hummocks, each with a comparatively large crater-like opening at the top. Investigation reveals the fact that at a distance of but a few inches at most below the surface of this crater-like opening is a rounded boulder. The heaving action of the frost forces the boulder gradually upward, causing the turf to first rise with smooth rounded outline, till, through continual pressure from the boulder, it bursts at the top. When the frost leaves the ground, the boulder drops back a short distance, but enough to be quite out of sight, leaving the cavity at the top filled with mud, and looking — in outline — like a small mud volcano. So far as the writer's observations go, the heaving action rarely progresses, in these areas, to the point of actually throwing the boulder out upon the surface. Each summer the growing turf makes an attempt at healing the wound, but each winter's frost opens it once more, the alternating forces so nearly balancing that little is accomplished after this pseudo-volcanic stage is reached.

Insects like the boring bee, the burying beetle, or larger burrowing animals, like the "woodchuck" of the Eastern states, the prairie dogs, badgers, and spermophiles of the West, in the same way exert powerful though local influences in admixing the lower with the upper portions of the soil, and through allowing perhaps a more ready passage of water facilitating oxidation and decomposition at greater depths. (Fig. 2, Pl. 19.)

While the effect of these animals may be comparatively inconspicuous in the regions east of the Mississippi, in the drier regions of the West the surface is not infrequently so undermined by burrows as to make travelling on horseback at more than a very moderate pace a matter of grave difficulty. W. P. Blake, in the early reports of the Pacific Railroad Survey, states that the fine, silty soil of the Tulare valley in California is so undermined that it is almost impossible to travel over it. "Mules often break through the thin crust and sink to their shoulders in these holes."

The action of plant life in the accumulation of vegetable mould has been fully discussed under the head of cumulose and alluvial deposits. There is, however, one phase of action which may well be mentioned here. A growing tree, as already noted, sends its roots deep down into the earth in search of food and foothold. So long as the tree remains alive and standing, in firm soil the amount of change in the soil itself, except in the way of abstraction of certain constituents taken up by the growing plant, is presumably very small. When, however, the tree dies, the roots slowly decay, and besides yielding up their contents to form new soil, afford passageway for percolating water with all its attendant results. Moreover, cases are by no means infrequent in which trees are upturned by the winds, bringing entangled in their roots it may be tons of soil and boulders which in part gradually fall back into the hole and in part remain to form a mound which marks the spot long after the tree has decayed. Into the cavity thus formed, dead leaves and other organic débris accumulate, which in time form deep rich loam to be commingled with the stony matter of the soil. In sections of the country where heavy winds and hurricanes are of frequent occurrence, the efficacy of trees in thus burying organic matter, and producing a more complete intermingling of the soils, is by no means inconsiderable.[1] The influence of plants in adding carbon and incidentally carbonic and other organic acids to the soils has been described in previous pages.

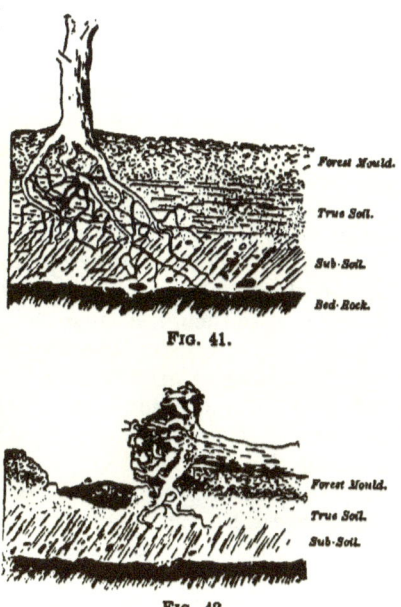

Fig. 41.

Fig. 42.

When plants die and decay upon the immediate surface, there is left only the inorganic matter or ash behind, the carbonic acid escaping into the air or being carried by rains into the soil. Hence it would seem to naturally follow that the soil where supporting an abundant vegetation should contain a larger percentage of carbonic acid than the atmosphere itself. That it does not contain, in all cases, a greater amount of free carbonic acid is apparently brought out in the table from the works of Boussingault and Lewy, as quoted on p. 178.

[1] Some of our archæologists go so far as to assert that the stone implements found buried several feet below the surface in glacial deposits, and brought forward as proving the existence of pre-glacial man, have been brought into that position by just such agencies. See Holmes, Early Man in Minnesota, American Geologist, April, 1893, p. 228.

Bacteria as agents of nitrification are undoubtedly efficacious in preparing nitrogeneous matter in the soils for assimilation by growing plants. Their influence as decomposers of rock masses was noted on p. 203. According to Wiley,[1] it is highly probable that organic nitrogen in the soil, in passing into the form of nitric acid, exists at some period of the process in the form of ammonia. The products of nitrification, he says, are ammonia, nitrous or nitric acid, carbon dioxide, and water. The ammonia and nitrous acid may not appear in the soils as the final products of nitrification, as the organism attacks the nitrous acid at once, converting it into the nitric form.

It may at first seem strange that man, who prides himself on being the highest type in the animal kingdom, as well as the only animal endowed with reasoning powers, should prove the most destructive; yet such is the case. Through prodigality, due in part to thoughtlessness and in part to a wilful disregard for any but immediate interests, man has, apparently from the very beginning of his existence, so conducted himself with relation to natural resources as to leave little less than ruin in his path. This is true not merely with reference to his treatment of the soil, but of the deeper lying rocks and their mineral contents. In the name of development he has squandered; through careless husbandry he has not merely impoverished the soil, but in many cases allowed it to run waste and be lost beyond recovery. So long ago as 1846, when Lyell made his second visit to America, he was struck by the rapid denudation of the land in our Southern states due to the reckless cutting away of the forests. He describes near Milledgeville, in Georgia, a washout in a lately deforested area. "Twenty years ago," he writes, "before the land was cleared, it [the washout] had no existence; but when the trees of the forest were cut down, cracks 3 feet deep were caused by the sun's heat in the clay; and during the rains, a sudden rush of water through the principal crack deepened it at its lower extremity, from whence the excavating power worked backwards, till in the course of 20 years, a chasm measuring no less than 55 feet in depth, 300 yards in length, and varying in width from 20 to 180 feet was the result. The high road has been several times turned to avoid this cavity, the enlargement of which is still proceeding, and the old line of road may be seen to have held its course,

[1] Principles and Practice of Agricultural Analysis, p. 464.

directly over what is now the widest part of the ravine. In the perpendicular walls of this great chasm appear beds of clay and sand, red, white, yellow, and green, produced by the decomposition *in situ* of hornblendic gneiss, with layers of veins of quartz, which remain entire, to prove that the whole mass was once solid and crystalline."[1]

The same lack of foresight or wanton disregard for coming generations is still manifested, and every muddy stream bears downward to the sea an increased load of silt from lands improperly cultivated and from which every rain removes a portion of the finest and richest of the soil, leaving behind but the barren gravel, channelled it may be beyond the possibility of cultivation. McGee[2] has more recently made observations of a similar nature in southern Mississippi, where the softer loam of the Columbia formation, which here forms the soil, has been allowed to become eroded down to the barren sandy loam of the Lafayette. "Old fields are denuded by the acre, leaving mazes of pinnacles divided by a complex network of runnels glaring red toward the sun and sky in strong contrast to the rich verdure of the hillsides never deforested; the plantations, mansions, and 'quarters' are undermined, and whole villages, once the home of wealth and luxury, are being swept away at the rate of acres for each year."

"The ravages committed by man," writes Marsh,[3] "subvert the relations and destroy the balance which nature had established between her organized and her inorganic creations, and she avenges herself upon the intruder by letting loose upon her defaced provinces destructive energies hitherto kept in check by organic forces destined to be his best auxiliaries, but which he has unwisely dispersed and driven from the field of action. When the forest is gone, the great reservoir of moisture stored up in its vegetable mould is evaporated, and returns only in deluges of rain to wash away the parched dust into which that mould has been converted. The well-wooded and humid hills are turned to ridges of dry rock, which encumbers the low grounds and chokes the watercourses with its débris, and — except in countries favored with an equable distribution of rain

---

[1] Lyell, Principles of Geology, 9th ed., 1846, p. 204.
[2] 12th Ann. Rep. U. S. Geol. Survey, 1890-91.
[3] The Earth as modified by Human Action, by Geo. P. Marsh, a new edition of Man and Nature, pp. 43, 44.

through the seasons, and a moderate and regular inclination of surface—the whole earth, unless rescued by human art from the physical degradation to which it tends, becomes an assemblage of bald mountains, of barren, turfless hills, and of swampy and malarious plains. There are parts of Asia Minor, of northern Africa, of Greece, and even of Alpine Europe, where the operation of causes set in action by man has brought the face of the earth to a desolation almost as complete as that of the moon; and though, within that brief space of time which we call 'the historical period,' they are known to have been covered with luxuriant woods, verdant pastures, and fertile meadows, they are now too far deteriorated to be reclaimable by man, nor can they become again fitted for human use, except through great geological changes, or other mysterious influences or agencies of which we have no present knowledge, and over which we have no prospective control. The earth is fast becoming an unfit home for its noblest inhabitant, and another era of equal human crime and human improvidence, and of like duration with that through which traces of that crime and that improvidence extend, would reduce it to such a condition of impoverished productiveness, of shattered surface, of climatic excess, as to threaten the depravation, barbarism, and perhaps even extinction of the species."

LIST OF AUTHORS CITED OR REFERRED TO

Agassiz, L., 175.
Alexander, H. F., 242.
Aughey, S., 331.
Bartlett, W. H., 180.
Bayley, W. S., 75, 78, 86.
Beaumont, Elie de, 160.
Becker, G. F., 235, 301.
Bell, Robert, 194, 243, 246, 275.
Belt, T., 175, 261, 277, 280, 298.
Berthier, P., 237.
Beyer, M., 178.
Bischof, G., 20, 27, 191, 192.
Blaas, J., 102.
Blake, W. P., 121, 126, 185, 247, 256, 346, 348, 394.
Bolton, H. Carrington, 202.
Bonney, T. G., 246.
Boussingault, J. B., 176.
Branner, J. C., 111, 175, 179, 188, 203, 278.
Brögger, W. C., 64.
Brongniart, A., 87, 175, 237.
Brown, A. P., 29.
Bruner, H. L., 344.
Buchanan, J. V., 204.
Caldcleugh, Alexander, 193.
Chamberlain, T. C., 278, 301, 303.
Choffat, P., 255.
Clark, W. B., 134.
Clark, W. C., 118.
Cloez, 176.
Collier, P., 309.
Comte de la Hure, 188.
Crosby, W. O., 138, 189, 255, 353, 385.
Cross, C. W., 35, 62, 71, 81.
Culver, G. E., 279.
Cushing, H. P., 279.
Dana, E. S., 31, 127.
Dana, Professor J. D., 49, 57, 117, 198, 235, 251, 253, 262.
Darton, N. L., 312.
Darwin, E., 175, 233, 292, 392.
Daubrée, A., 16, 197, 376.
Davis, W. M., 186.
Davidson, C., 287.
Dawson, J. W., 291, 334.

De Luca, 176.
Derby, O. A., 188, 277.
Diller, J. S., 87, 92.
Dutton, C. E., 196.
Dwight, 207.
Dyer, B., 202.
Ebelmen, M., 237.
Egleston, Thomas, 184.
Ewing, A. L., 194.
Failyer, G. H., 176.
Fernow, B. E., 282.
Fesca, Dr. Max, 243.
Forbes, 184.
Forschammer, J. G., 237.
Fournet, 175, 237.
Fulton, R. L., 280.
Furlonge, W. H., 277.
Geikie, A., 2, 146, 201, 288.
Geikie, James, 357.
Geldmacher, Max, 236.
Gesner, H. S., 317.
Gilbert, G. K., 50, 185, 256, 349.
Gordon, C. H., 168.
Griswold, L. S., 111.
Gumbel, C. W., 28, 88.
Hall, C. W., and Sardeson, F. W., 161, 250.
Harker, A., 39.
Hartt, C. F., 175, 280.
Hawes, G. W., 46, 75, 87, 170.
Haworth, E., 25.
Hayden, F. V., 252.
Hayes, C. W., 100, 194.
Heusser and Claraz, 175, 228, 251.
Hilgard, E. W., 333, 346, 366, 367, 369, 371, 374.
Hitchcock, C. H., 68.
Hitterman, 239.
Hobbs, W. H., 218.
Holmes, W. H., 395.
Hovey, E. O., 230.
Hunt, T. S., 86, 90, 124, 159, 258.
Iddings, J. P., 22, 39, 57, 60, 64, 71, 72, 81.
Irving, R. D., 278.
Johnson, S. W., 177, 178.
Johnson and Blake, 136.

LIST OF AUTHORS CITED OR REFERRED TO

Johnstone, Alexander, 189.
Joues, T. Rupert, 317.
Judd, J. W., 284, 321.
Julien, A. A., 190.
Kalkowski, E., 75.
Kemp, J. F., 81, 86, 87, 171.
Kerr, W. C., 286.
Keyes, C. R., 25.
Kidder, J. H., 179.
King, F. H., 381.
King, Clarence, 71.
Klement, M. C., 160.
Kletzinsky, W., 176.
Kuhn, M. Levy, 89.
Layard, A. H., 293.
Le Conte, J., 258.
Lemberg, J., 18, 217, 374.
Lindgren, W., 75, 274.
Livingstone, David, 183.
Loftus, 293.
Loughbridge, R. H., 365, 366.
Lyell, Sir Charles, 396.
Marsh, George P., 183, 297.
McGee, W. J., 301, 312, 323, 397.
Meister, 381.
Merrill, G. P., 37, 47, 54, 81, 87, 98, 113, 115, 154, 159, 206, 218, 349.
Mills, J. E., 175, 203, 273.
Muller, Alex, 371.
Müller, R., 192.
Munroe, C. E., 190.
Müntz, A., 203.
Müntz and Aubin, 179.
Müutz and Maracano, 373.
Murakozky, K. V., 238.
Neumayer, M., 302.
Newberry, J. S., 118, 356.
Nordenskiold, A. E., 242.
Oldham, R. D., 311.
Orton, Edward, 117, 118, 124.
Owen, D. D., 111.
Packard, R. L., 108, 376.
Pallarsen, 89.
Penrose, R. A. F., 231, 232.
Pirsson, L. V., 64.
Pliny, 73, 90.
Porter, J. B., 266.
Potter, W. B., 265, 275, 276.
Prestwich, Joseph, 65, 260.
Prichard, 30.
Pumpelly, R., 275, 277.
Purrington, C. W., 279.
Read, T. Mellard, 194.
Redwood, Boverton, 129.
Retgers, J. W., 349.
Reusch, H., 250.

Richthofen, F. von, 63, 85.
Rogers Brothers, 191.
Rohrbach, C. E. M., 89.
Roscoe and Schorlemmer, 4.
Rose, G., 79, 89.
Rosenbusch, H., 57, 62, 70, 72, 74, 82, 93, 97, 98.
Rosière, 73.
Roth, Justus, 72, 94, 101, 103, 208, 239, 256.
Russell, I. C., 112, 201, 266, 279, 280, 284, 296, 301, 333, 385.
Rutley, F., 111, 194.
Safford, J. M., 267.
Salisbury, R. D., 278, 287, 301, 303, 352.
Schlösing, 203.
Schutze, R., 228.
Shaler, N. S., 181, 197, 318, 336, 389, 498.
Smith, Angus, 179.
Sorby, H. C., 26, 38, 199, 243, 342.
Spurr, J. E., 107.
Stanley, H. M., 183.
Stejneger, L., 199.
Stone, G. H., 186.
Storer, F. H., 191, 202.
Strabo, 90.
Stroeruwitz, H. von, 182.
Streng, A., 86.
Teall, J. J. H., 24, 74, 90.
Thenard, P., 190.
Thompson, Wyville, 247.
Tornebohm, A. E., 87, 89.
Tschermak, G., 24.
Van Den Broeck, E., 178, 258.
Van Hise, C. R., 106.
Vom Rath, G., 255.
Von Buch, L., 83.
Wadsworth, Dr. M. E., 57, 68, 85, 97, 254.
Weed, W. H., 109.
Weruer, A. G., 73.
Whitaker, W., 267.
Whitney, J. D., 68, 127, 278, 328.
Whitney, Milton, 287, 307-309, 313, 340, 344, 353, 379.
Wichman, A., 170.
Widogradsky, 203.
Wiley, H. W., 178, 316.
Williams, G. H., 63, 72, 86, 96, 99, 100, 156, 216.
Williams, J. F., 64.
Willis, Bailey, 52.
Winchell, N. H., 297.
Wolff, Professor J. E., 93.
Woodward, J. B., 186.
Wurtz, H., 127.
Zirkel, F., 38, 57, 68, 80, 87, 89.

INDEX

Abrasive action of wind-blown sand, 185.
Acid rocks, meaning of term, 64.
Acmite, 22.
Adobe, 130, 332.
Ægerine, 22.
Æolian deposits, 344.
Æolian rocks, 153; defined, 58.
Agalmatolite, 116.
Age of soils, 386.
Air in motion, effects of, 189.
Alabaster, 117.
Alaska, rock-weathering in, 279, 284.
Albertite described, 127.
Albite as a rock constituent, 16.
Alkalies in soils, 371.
Alkaline carbonates, when formed, 372; in soils, 371; formed during weathering, 205.
Alkaline silicates in soils, 370.
Allanite, 25.
Allotriomorphic minerals defined, 41.
Alluvial cones defined, 54.
Alluvial deposits, 320.
Alteration defined, 174.
Alum shale, 138.
Aluminum as a constituent of the earth's crust, 5.
Amber, 128.
Amianthus, 115.
Ammonia in atmosphere, 177.
Ammonium sulphate, influence in decomposing feldspars, 178.
Amorphous, definition of, 40.
Amphiboles as rock constituents, 19.
Amygdaloidal structure, 34.
Anacostia, deposits of the, 323.
Analyses, calculations of, 210; discussion of, 212.
Anamesite, 92.
Andesites, 83.
Andesitic rocks, induration of surface, 235.
Anhydrite described, 118.
Animal life, effect on soils, 389.
Anorthite as a rock constituent, 17.

Anorthit-gesteine, 89.
Anthracite coal, 150.
Antique porphyry, 83.
Ants, effect on soils, 389; as promoters of rock decomposition, 204.
Apatite as a rock constituent, 27.
Apo-rhyolite, 72.
Appalachian Mountain system, material eroded from, 196.
Appomattox formation, 312.
Aqueo-glacial clays, 334.
Aqueous rocks, 105; defined, 58.
Aragonite as a rock constituent, 26.
Arenaceous group, the, 131.
Argillaceous rocks described, 135.
Argillites, 137; fissile, 170; Harford County, Maryland, weathering of, 229.
Arkansas River referred to, 280.
Asphaltum described, 125.
Atmosphere, action of, 176.
Augite, molecular alteration of, 39; relative durability of, 235.
Augite porphyrite, 90; Montana, disintegration of, 235.
Augite vitrophyrite, 90.
Augitite described, 101.
Auriferous sands, origin of, 266.

Bacteria, as agents of nitrification, 396; decomposing action of, 203.
Banding in gneisses, origin of, 165.
Barbadoes Island, volcanic dust on, 298.
Barite described, 118.
Barium as a constituent of the earth's crust, 7.
Basalt, described, 90; Bohemia, weathering of, 223; Haute Loire, France, weathering of, 223.
Basalts, geographical distribution in United States, 92; weathering of, 262.
Basanite described, 94.
Base, definition of, 40.
Basic rocks, meaning of term, 64.
Beach sands, 341.
Beauxite described, 108.

Bedded rocks defined, 53.
Bedded structure, 34.
Bermuda, weathering of limestones in, 247.
Binary granite, 68.
Biotite as a rock constituent, 23.
Bitumen, 125.
Bituminous coals, 149.
Bituminous dolomite of Chicago, 145.
Black earth, Russian, 318.
Bleaching of rocks on exposure, 257.
Bluegrass soil, 382.
Bog of Allen, 317.
Bogs, classification of, 317.
Boss, defined, 50.
Boss-like form accentuated by joints, 245.
Botryoidal structure, 37.
Boulder clay, 138.
Boulder clays, 353.
Boulders, of decomposition resembling those of the drift, 242; formed by weathering, 244.
Bowenite, 116.
Breccia, 133.
Brecciated limestones, 139.
Brecciated structure, 38.
Bronzitite, 100.
Brown hematite, 29, 107.
Brownstone, 133.

Cabook, formation of, 242.
Calc sinter, 112.
Calcareous group of rocks, 137.
Calcareous rocks, 143; rate of weathering, 272.
Calcite as a rock constituent, 25.
Calcium as a constituent of the earth's crust, 6.
Calcium carbonate, amount annually removed in solution, 194.
Camptonite, 8.
Cannel coal, 150.
Cape Cod, wind action on, 297.
Carbonates of alkalies, influence of, 238.
Carbonates of the alkalies formed during weathering, 205.
Carbonates, production of, during weathering, 205.
Carbonaceous rocks, 148.
Carbonic acid, influence of, in feldspathic decomposition, 237, 239; amount annually brought to the surface, 179; in air of soils, 178; in the atmosphere, 178.
Carboniferous soils, 386.
Catlinite, 139.
Cavernous structure, 38.
Cellular structure, 38.
Ceylon, rock disintegration in, 242.
Chalcedony, 110.

Chalk, 143; decomposition of, 267.
Chemical composition of rocks, 44.
Chemical elements constituting rocks, 4.
Chert, 110; of Arkansas, weathering of, 231; of Missouri, weathering of, 230.
Chilian nitrates, origin of, 373.
Chlorides, 119.
Chlorite as a rock constituent, 30.
Chrysotile, 115.
Citric acid, solvent property of, 202.
Classification of soils, 381.
Clastic rocks, 129; classification of, 130.
Clastic structure, 34.
Clay concretions, formation of, 37.
Clay, defined, 135; effect on soils, 368; protective influence of, 254.
Clay ironstone, 114.
Clay slates, 137.
Clays, aqueo-glacial, 334.
Climate, influence of, on weathering, 278.
Clinton iron ores, origin of, 266.
Coefficient of cubical expansion of minerals, 268.
Coking coals, 150.
Cold, effect on rocks, 180.
Colloidal structure, 33.
Colluvial deposits, 319.
Color; changes incidental to weathering, 257; of rocks, 45; of soils, 384; of soils, cause of, 385; variation, cause of, 47.
Colorado River, erosion by, 196.
Columbian formation, 312.
Columnar structure, 38.
Complexity of structure favoring disintegration, 250.
Concentric exfoliation, 244; not indicative of an original concretionary structure, 245.
Concentric structure inevitable to jointing, 245.
Concretionary structure, 35; in granite, 246; in crystalline rocks, 37.
Conductivity of rocks, unequal, 184.
Conglomerate, 133.
Conservative action of plants, 202.
Contact metamorphism, 157.
Contours incidental to weathering, 259.
Coprolite nodules, 152.
Coquina, 143.
Coral limestone, 143.
Corroded surfaces, irregularity of, 250.
Corsica, weathering of granite on, 250.
Crayfish, effects on soils, 391.
Creeping of soil cap, 287.
Crenic acid, 190.
Crystalline limestones and dolomites, 162.
Crystalline schists, the, 168.
Crystalline structure, 33.

INDEX

Crystallites defined, 41.
Cumulose deposits, 313.

Dacite, 84.
Daubrée's experiments in rock trituration, 17, 197.
Decay, time limit of, 272; of rocks, how characterized, 212.
Decomposition and disintegration, discrimination between, 283.
Decomposition, depth of, 278; following disintegration, 243; incident to erosion, 197; of fragmental rocks, 228; of greenstone dikes, effects of, 244; of rocks, chemical processes involved in, 238; of shells through the aid of salt, 233; natural acceleration of, 205.
Degeneration of rocks, 174.
Degradation of North America, rate of, 196.
Delta deposits, 320.
Delta of the Nile, section of, 321.
Deoxidation, 187; by marine animals, 204.
Desert varnish, 256.
Devil's Tower, origin of, 261.
Deweylite, 116.
Diabase, described, 87; mandelstein, 90; Medford, Massachusetts, weathering of, 218; porphyrite, 90; Venezuela, weathering of, 222.
Diallogite, 100.
Diamonds, origin of, 98.
Diatomaceous earth, 141.
Dichte diabase, 90.
Dike defined, 50.
Diluvium, rouge et gres, 258.
Diorite, Albemarle County, Virginia, weathering of, 224.
Diorite-andesite group, 81.
Diorites, 87.
Discoloration, above drainage level, 258; incidental to weathering, 257.
Discussion of analyses, 234.
Disintegration of rocks in Lower California, 183; prevented by surroundings, 252; without decomposition, 241.
District of Columbia, rock-weathering in, 283.
Ditroite, 79.
Dolerite, 92.
Dolomite, as a rock constituent, 26; described, 145; origin of name, 163; origin of, by metasomatosis, 159.
Dolomites, 102.
Dolomitic limestones, disintegration of, 250; weathering of, 239.
Drift, extent of, 291.
Drumlins, 355; defined, 55.
Dune defined, 55.
Dune sand, chemical composition, 350.

Dunite, 97.
Dust, in rain and snow falls, 344; volcanic, 298, 349.
Dust soils, 345.
Dust storms, 292: in Dakota, 293; in Montana and Nevada, 294.
Dynamic metamorphism, 156.

Earth's crust, thickness of, 2.
Earthworms, effects on soils, 392.
Eclogite, 170.
Effacement of characteristics by weathering, 262.
Effusive rocks, characteristics of, 61; defined, 60.
Elæolite as a rock constituent, 18.
Elæolite syenites, 78.
Elæolite syenite porphyry, 79.
Elaterite described, 126.
Elvanite, 70.
Eozoon Canadense, 159, 163; origin of, 116.
Epidiorite, 89.
Epidote, as a rock constituent, 25; alteration of, 25.
Erosion by rivers, 196.
Eruptive rocks, 59.
Eskers, 290, 356.
Eucrite, 89.
Eulysite, 97.
Eurite, 70.
Exfoliated rocks, shape and size of flakes, 182.
Exfoliation, attended by gun-like reports, 182; due to heat and cold, 181; of rocks on Cape Cod, 182.
Expansion through hydration, 188.
Extent of weathering, 276; in Brazil, Colorado, District of Columbia, Missouri, Nicaragua, South Africa, South America, 277.

Fault defined, 53.
Feldspars, as rock constituents, 13; decomposition of, 17.
Feldspathic decomposition, process of, 237; in Comstock Lode, 235; by fresh water, 238; influenced by ammonium sulphate and sodium chloride, 178.
Feldspar porphyry of Iron Mountain, weathering of, 265.
Feldspars, relative durability of, 235.
Felsitic structure, 33.
Felsite pitchstone, 70.
Felsophyr, 70
Felstone, 70.
Ferrous carbonate, solubility of, 239.
Fertility of soil dependent on physical condition, 379.

Fichtellite, 129.
Florite, 109.
Fissile argillites, roofing slates, 170.
Flagstone, 133.
Flexible sandstone, 134.
Flint, 110.
Flood plain of the Mississippi, 323.
Fluidal or fluxion structure, 34.
Fogs, indices of dust in atmosphere, 344.
Foliated or schistose rocks, 164.
Foliated structure, 34.
Forellenstein, 87.
Forests, buried by sand, 295; influence of, 280; protective action of, 282.
Fourchite, 79.
Foyaite, 79.
Foyaite-phonolite group, 77.
Fracture of rocks, 48.
Fragmental structure, 34.
Freestone, 133.
Freezing water, disintegrating action of, 198.
Frontal aprons, 356.
Frontal moraines, 355.
Frost, action in accelerating decomposition, 278; action on soil, 367; disintegrating action, 199; heaving effects on boulders, 393; supposed protective action of, 278.

Gabbro described, 85.
Gabbro-basalt group, 85.
Garnet rock, 170.
Garnerite, formation of, 226.
Garnetite, 170.
Geest, 301.
Gem sands of Ceylon, origin of, 266.
Genetic relationship of rocks, 64.
Geological age, of soils, 389; a basis for classification, 63.
Geyserite, 109.
Gilsonite, 127.
Glacial deposits, 351.
Glacial detritus, amount of, 201.
Glacial drift, extent of, 291.
Glacial lakes, extinction of, 289; filling of, 326.
Glacial landscape, 291.
Glacial moraine, 290.
Glacial soil of Cape Elizabeth, composition of, 364.
Glacier, the, as an erosive agent, 200.
Glaciers, as agents of transportation, 200, 289.
Glass abraded by wind-blown sand, 185.
Glauconite, 31, 134.
Glauconitic marl, 134.
Globulites defined, 41.

Gneiss, Albemarle County, Virginia, degeneration of, 213.
Gneisses, the, 164.
Grahamite described, 127.
Granite, described, 65; extent of weathering in District of Columbia, 276.
Granitell, 68.
Granite-liparite group, 65.
Granite porphyry, 68.
Granite soil defined, 383.
Granitite, 67.
Granofelsophyr, 70.
Granophyr, 70.
Granular structure, 34.
Granulite, 167.
Grauwacke, 133.
Graphic granite, 67.
Gravels superficially oxidized, 258.
Gravity, action of, in transporting débris, 286.
Greenland, rock-weathering in, 278.
Greensand marl, 134.
Greenstone, 81.
Greisen, 68.
Greywacke, 133.
Ground-mass defined, 40.
Ground moraine, 352.
Gruss, 301.
Guano, 151.
Gypsum described, 117.

Halleflinta, 167.
Hardpan, 368.
Harzburgite, 97.
Hatchettite, 129.
Hatteras and Henlopen, sand dunes of, 295.
Heat, action on pebbles in Arabia Petrea, 183; expansive action on rocks, 180.
Heat and cold, as agents of decomposition, 180; effects of, in Africa, 183; effects limited to surface, 183; most effective on slopes, 184.
Heavy spar, 118.
Hematite, 106; as a rock constituent, 28.
Holocrystalline, definition of, 40.
Hornblende, as a rock constituent, 19; decomposition of, 20; relative durability of, 235.
Hornblende picrite, 97.
Hornblendite, 100.
Humic acid, 189.
Humidity, weathering influenced by, 270.
Hyaline andesite, 85.
Hyalite formed during feldspathic decomposition, 238.
Hyalobasalt, 92.
Hyaloliparite, 72.
Hyalomelan, 92.

INDEX

Hyalotrachyte, 77.
Hydration, 187; importance of, 188, 234, 253, 278; of micas, 189.
Hydraulic limestone, 145.
Hydrocarbon compounds, description of, 121.
Hydro-metamorphism, 161.
Hyperite, 87.
Hypersthenite, 100.
Hypocrenic acid, 190.
Hypocrystalline, definition of, 40.

Ice, disintegrating action of, 198; influence in transporting rock débris, 287; mechanical action of, 195.
Idiomorphic minerals defined, 41.
Igueous rocks, 59; defined, 57.
Ilmenite as a rock constituent, 28.
Induration, cause of, 255; of rocks on exposure, 254; of sandstone by igneous contacts, 261.
Infusorial earth, 141.
Insects, effects on soils, 394.
Intrusive rocks defined, 60.
Inundated lands, classification of, 318.
Iron, as a constituent of the earth's crust, 5; removed in form of ferrous sulphate, 239; removed in form of protoxide carbonate, 239; variation in solubility, 239.
Iron Mountain, Missouri, pre-Silurian weathering of, 276.
Iron ores as rock constituents, 27.
Iron pyrites as a rock constituent, 29.
Itacolumite, 133.
Itacolumites, Brazilian, weathering of, 228.

Jasper, 110.
Joints, as aids to weathering, 244; cause of, 243; influence of, in producing boulders, 244; influence in producing boss-like forms, 245.

Kalk diabase, 90.
Kames, 290.
Kaolin, 116, 136, 265; composition of, 309; origin of, 308.
Kaolinite distinct from kaolin, 309.
Kaolinization defined, 18.
Keratophyr, 76.
Kersantite, 82.
Kimberlite, 98.
Kinds of rocks, 56.
Kinzigkite, 170.
Konlite, 129.
Krakatoa, dust from, 298.
Ktaadn Iron Works referred to, 107.
Kugel porphyry, 70.

Labradorite as a rock constituent, 17.
Laccolite defined, 50.
Lake Agassiz, deposits in, 290.
Lake Asphaltites, 126.
Lakes, filling of, 314; transient character of, 326.
Laminated or banded structure, 38.
Landscape, glacial, 291.
Lapilli, 140.
Laterite, 139, 310.
Laurvikite, 79.
Lava defined, 51.
Leda clays, 334.
Leopardite, 70.
Leptinite, 167.
Leucite as a rock constituent, 18.
Leucite basalt, 103.
Leucite-nepheline rocks, 102.
Leucite rocks described, 102.
Leucitite, 103.
Leucitophyr, 80.
Leucophyr, 88.
Leucoxeue, 28.
Lherzolite, 97.
Lichens, action of, 201.
Liebnerite, 79.
Lignite, 149.
Limburgite described, 98.
Lime carbonate, decomposing action of, 370.
Lime in soils, 366.
Limestone, unequal weathering of, 250; weathering of, 232.
Limestone residuals, character of, 303.
Limestone soils poor in lime, 259.
Limestones, 143; and dolomites, 162; corroded by acids, 194; corroded by meteoric waters, 259; unequal induration of, 247; variation in composition, 147.
Limit of diminution in size of particles by erosion, 197.
Limonite, 107; as a rock constituent, 29.
Liparite described, 70.
Litchfieldite, 79.
Lithophysæ, 72.
Loess, 139, 290, 327.
Logans, or tors, 252.
Lower California, rock-weathering in, 283.
Lumachelle, 143.
Lustre, 48.
Luxullianite, 70.
Lydian stone, 111.

Magma, definition of, 59.
Magnesian limestones, 145.
Magnesia removed in excess of lime, 239.

406　INDEX

Magnesium as a constituent of the earth's crust, 6.
Magnesite, 113.
Magnetite as a rock constituent, 27.
Man, has squandered in the name of development, 397; ravages committed by, 397.
Marbles, 163.
Marcasite as a rock constituent, 29.
Marginal moraines, 355.
Marine animals, influence of, on marine muds, 204.
Marl, 146.
Marmolite, 116.
Marsh gas, 121.
Marsh lands, reclaimable areas, 340.
Martite, 106.
Massive structure, 34.
Material lost through weathering, 208.
Materials lost during decomposition, proportional amounts, 234.
Mechanical action of water and ice, 195.
Mechanical disintegration most active in regions of extreme temperatures, 182.
Melaphyr described, 90.
Melaphyrs and angite porphyrites, 90.
Melilite basalt, 92.
Menaccanite as a rock constituent, 28.
Metamorphic rocks, 155; defined, 58.
Metamorphism defined, 155.
Metasomatosis defined, 158.
Miascite, 79.
Mica, relative durability of, 236.
Micaceous sandstone, cause of weathering, 189.
Micas, alteration and decomposition of, 23; as rock constituents, 22.
Microcline as a rock constituent, 16.
Microcrystalline structure, variation in, 41.
Micro-granite, 70.
Microlites defined, 40.
Microlitic structure, 33.
Micropegmatite, 70.
Microscope used in geology, 38.
Microscopic structure, 38; of rocks, 33.
Microscopic study of rocks, efficiency of, 39.
Mineral caouchouc, 126.
Mineral composition of soils, 373.
Mineral matter, dissolved by water, 191; in solution, removed annually from England and Wales, 194.
Mineral pitch, 125.
Minerals constituting rocks, 9; list of, 11.
Mineral variation of rocks, cause of, 9.
Mineral wax, 128.
Minette, 74.
Minnesota, wind action in, 297.

Mississippi, flood plain of, 323.
Mississippi River, amount of material transported by, 288.
Missouri River, muddy character of, 288.
Mode of occurrence of rocks, 49.
Monazite sands, origin of, 266.
Monoclinic feldspars, 14.
Monoclinic pyroxenes as rock constituents, 21.
Mouzonite, 74.
Moraine defined, 55.
Moraines, classified, 355; glacial, 290.
Mosses, action, 201.
Muck, 149.
Muscovite, as a rock constituent, 23; relative durability of, 236.

Natural gas, 121.
Nepheline as a rock constituent, 18.
Nepheline basalt, 107.
Nepheline dolerite, 104.
Nepheline rocks described, 103.
Nepheline syenites, 78; weathering of, 249.
Nephelinite, 104.
Nevadite, 72.
Niggerheads, how formed, 244.
Nile delta, section of, 321.
Nineveh, site obscured by sand dunes, 295.
Nitrates, influence of, in feldspathic decomposition, 239; in soils, 372; source of, 372.
Nitric acid, in atmosphere, 177; influence of, in feldspathic decomposition, 239.
Nitrogen, in atmosphere, 176; in soils, 372.
Non-coking coal, 150.
Norites, 86.
Noumæite, formation of, 226.
Novaculite, 111.
Nummulitic limestone, 143.

Obsidian, 72.
Oldest known rocks, 49.
Oligoclase, as a rock constituent, 16; disintegration of, 241; decomposition of, 237.
Olivine, as a rock constituent, 24; alteration into serpentine, 24; relative durability of, 235.
Onyx marbles, 113.
Oölites, English, coloration of, 258.
Oölitic limestone, 143; origin of, 53, 112.
Ophicalcite, 163.
Ophiolite, 89, 116, 163.
Organic acids, action of, 189; corrosive power on marble, 190; solvent power augmented by nitrogen, 190.

Oriental alabaster, 113.
Original constituents of rocks, 10.
Original structures preserved during decomposition, 264.
Orthoclase, relative durability of, 236.
Orthoclase porphyries, 75.
Orthoclase as a rock constituent, 14.
Orthophyr, 76.
Orthorhombic pyroxenes as rock constituents, 22.
Osars, 290, 356.
Ouachitite, 79.
Overwash plains, 356.
Oxidation, how manifested, 187; incidental to decomposition, 234.
Oxides, silica, 100.
Oxygen, as a constituent of the earth's crust, 5; influence in preventing loss of iron during rock decomposition, 239; of the atmosphere as an agent of decomposition, 180.
Ozokerite, 128.

Palagonite tuff, 140.
Paludal deposits, 336.
Pantellerite, 72.
Paraffin, native, 128.
Paramorphic minerals, 156.
Peat, 148.
Peat bogs, 317.
Pebble, normal shape of, 348.
Pegmatite, 67.
Pelites, 135.
Peperino, 140.
Peridotite, described, 95; weathering of, 225.
Peridotite-limburgite group, 95.
Perlite, 77.
Perlitic structure, 35.
Petroleum described, 122.
Petrosilex, 70.
Phenocrysts defined, 41.
Phlogopite, 23.
Phonolite, weathering of, 217.
Phosphates, 119.
Phosphates of Tennessee, origin of, 267.
Phosphatic sandstone, 152.
Phosphorite, 119.
Phosphorus, as a constituent of the earth's crust, 7; relative proportion of, in rocks, 8.
Phyllite, 169.
Physical and chemical properties of rocks, 33.
Physical condition of soils, 378.
Physical manifestations of weathering, 241.
Picrite, 97.
Picrite porphyrites described, 98.

Picrolite, 116.
Pic Pourri, decomposition of, by bacteria, 203.
Piedmontite, 25.
Pike's Peak, Colorado, weathering of granite, 243, 255.
Pisolitic limestone, 143.
Pitchstone, 77.
Placer deposits, origin of, 267.
Plagioclase feldspars, relative durability of, 236.
Plagioclases as rock constituents, 16.
Plant and animal life, effect on soils, 380.
Plant life, effect on soils, 394.
Plants and animals, agents of disintegration, 201.
Plutonic rocks, characteristics of, 60; defined, 60.
Porfido rosso antico, 83.
Porphyrites, 83.
Porphyritic structure, 35.
Porphyroid, 167.
Post-Cretaceous decay of granite, 272.
Post-Glacial decay of diabase, 273.
Post-Jurassic weathering of granodiorites, 274.
Post-Pliocene weathering of andesites, 274.
Potash, in soils, replacing power of, 370; soluble in soils, 376.
Potassium, as a constituent of the earth's crust, 6; relative proportion of, in rocks, 6.
Pot-holes, formation of, 196.
Potomac flats, 323.
Potomac formation, 313.
Potstone, 101.
Precious serpentine, 115.
Pre-Palæozoic decay of rocks, 275.
Primary rocks, 31.
Primary constituents of rocks, 10.
Principles involved in rock-weathering, 173.
Propyllite, 85.
Protective action, of plants, 202; of soil, 271.
Protogine, 67.
Psammites, the, 131.
Pseudotuffs, 140.
Psilomelane, 107.
Puddingstone, 133.
Pulaskite, 79.
Pyrite, as a rock constituent, 29; decomposition of, 29.
Pyroclastic rocks, 140.
Pyrolusite, 107.
Pyrophyllite, 116.
Pyrophyllite schist, 168.

Pyroxenes, alteration and decomposition of, 22; as rock constituents, 21.
Pyroxenite-augitite group, 99.
Pyroxenites, described, 99; weathering of, 225.

Quarrying by aid of fire, in India, 182.
Quarry water, 199, 254.
Quartz, 110; as a rock constituent, 12; the most refractory mineral, 234.
Quartz basalt, 92.
Quartz-free porphyries, 75.
Quartz porphyry described, 69.
Quartz veins, influence of contours, 260.
Quartzite, origin of, 158; feldspathic, disintegration of, 251; polished by wind-blown sand, 257.
Quartzites, weathering of, in the District of Columbia, 251.
Quaternary deposits, weathering of, 258.
Quitman Mountains, exfoliation of rocks, 182.

Rainfall, amount reaching the soil, 281.
Rain waters, temperatures of, 193.
Rapilli, 140.
Rate of weathering influenced by texture, 268; by composition, 269; by humidity, 270; by climate, 278; by position, 270.
Reaction rims, 240.
Regional metamorphism, 155.
Regolith, classification of, 300; origin of name, 299.
Regur defined, 382.
Relationship between plutonic and effusive rocks, 63.
Relative amount of material lost through weathering, 284.
Relative durability of minerals, 234.
Relative rapidity of weathering among eruptive and sedimentary rocks, 271.
Rensselaerite, 116.
Residual clays, 302; in caves, 233.
Residuary deposits, 301; analysis of, 306; names proposed for, 301.
Results, incidental to weathering, 266; of weathering due to position, 252.
Retinite, 70, 129.
Retinolite, 116.
Rhodochrosite, 114.
Rhombporphyry, 76.
Rhyolite, 72; weathering of, 255.
Ribbons in slates, 155.
River channels formed by rock-weathering, 243.
River erosion, 196.
Rivers, flood plains of, 289.
Rock, definition of, 1; disintegration of on Bering Island, 199.

Rock-forming minerals, classification, 10; list of, 11.
Rocking stones, 252.
Rock temperatures, in Africa, 183; at Edinburgh, Scotland, 184.
Rock-weathering, 206; a superficial phenomenon, 193; complexity of process, 240; early references to, 175; on Lone Mountain, Montana, 243.
Rocks, absorptive power of, 198; chemical composition of, 44; classification of, 57; color of, 45; composed mainly of inorganic material, 131; composed of débris from plants and animals, 141; expansion and contraction under natural temperatures, 181; formed through chemical agencies, 105; formed as sedimentary deposits, 129; kinds of, 56; mode of occurrence, 49; physical and chemical properties of, 33; specific gravity of, 43.
Roofing slate, microstructure of, 170.
Root action, how manifested, 202.
Roots, depth of penetration, in caves and soils, 202.
Rosso de Levante, 98.
Rottenstone, origin of, 267.

Salt, common, 119; disintegrating effects of, 198.
Sand, æolian, 346; Sorby's classification of, 342; of dunes, sources of, 296.
Sand blast carving, 186; natural, 185.
Sand dunes, 346; formation of, 295; rate of movement, 296.
Sand grains, lasting power of, 197.
Sandpipes, formation of, 260.
Sandstone, cause of disintegration, 247; cementing matter of, 132; induration of, 256; siliceous, weathering of, 228; spheroidal, weathering in, 247; unequal weathering of, 248.
Sandstone concretions, formation of, 37.
Sandstones, weathering of, 249.
Sanidin, kaolinization of, 238.
Sanidin-oligoclase trachyte, 77.
Saprolite, 301.
Satin spar, 117.
Saxonite, 97.
Scheerite, 129.
Schistose structure, 34.
Schists, the, 168; crystalline, weathering of, in Brazil, 251; of Cape Elizabeth, weathering of, 248; origin of, 156.
Seacoast swamps, 336.
Secondary constituents of rocks, 10.
Secondary minerals, influence of, 249.
Sedentary materials, classification of, 300.

INDEX

Sedimentary rocks, origin of, 52.
Selenite, 117.
Seneca oil, 123.
Septarian nodules, 36, 114.
Sericite, 23.
Serpentine, composition, 30; after peridotite, 97; origin of, 115, 159; origin of name, 31; Harford County, Maryland, weathering of, 226.
Shale, 137.
Sheet defined, 50.
Shell limestone, 143.
Shell marl, 146.
Shell sand, 143.
Shore ice, transportation by, 292.
Siderite, 114.
Silica, loss of, how accounted for, 237; lost during decomposition, 234; possibility of combination with iron during rock decomposition, 239; solubility of, 238.
Silicates, 114; most refractory, 235.
Siliceous sinter, 109.
Silicified wood, 110.
Silicon as a constituent of the earth's crust, 5.
Sill defined, 50.
Simplification of compounds incidental to weathering, 265.
Singing sands, 143.
Sink-holes, formation of, 259.
Slaggy structure, 34.
Slates, 137.
Slaty cleavage, origin of, 155.
Slickensides defined, 54.
Snow, effect in promoting decomposition, 280.
Snowfall, influence compared with rainfall, 280.
Soapstone, Amherst County, Virginia, weathering of, 226; Fairfax County, Virginia, weathering of, 227; origin of, 101.
Sodium as a constituent of the earth's crust, 7.
Sodium chloride, influence in decomposing feldspars, 178.
Sodium salts in soils, 371.
Soil, chemical nature of, 358; capacity for water, 379; definition, 3; mineral nature of, 373; nitrates in, 322; nitrogen in, 372; soluble matter of, 365; water content of, 281.
Soil cap, creeping of, 287.
Soil particles, movements of, 287.
Soil temperatures at Orono, Maine, 184.
Soils, age of, 386; affected by plant and animal life, 389; affected by winds, 296; as affected by man, 396; classification, 381; color of, 384; destructive process of formation, 360; essential constituents of, 362; fertility of, 361; fertility dependent on physical condition, 379; how affected by climates, 367; how affected by leaching, 368; inherited characteristics, 303, 360, 387; mineral composition of, 373; of arid regions, character of, 368; of arid regions, composition of, 369; of humid regions, composition of, 379; of Nile valley, cause of fertility of, 325; physical condition of, 378; resemblance to parent rock, 360; soluble salts in, 369; the, 357; weight of, 381.
Soluble matter in fresh and decomposed rocks, 377.
Soluble salts in soils, 369.
Solution, 189; rate increased by comminution, 192; relative amount of material removed in, 253.
Sounding sand, 143.
South Dakota, rock-weathering in, 279.
Specific gravity of rocks, 43.
Specular iron ore, 28.
Sphærosiderite, 114.
Sphagnous mosses, rate of growth, 317.
Spheroidal structure, 247.
Spheroidal weathering of sandstone, 247.
Spherulitic structure, 35.
Spilite, 90.
Stalactite, 113.
Stalagmite, 113.
Stamford dike, pre-Palæozoic decay of, 275.
Steatite, 116.
Stone implements, weathered, 273.
Stone Mountain, Georgia, weathering of, 245.
Stratification defined, 53.
Stratified rocks, weathering of, 248.
Stratified structure, 34.
Structure, as affecting weathering, 249; of rocks, 33.
Sub-soil defined, 383.
Succinite described, 128.
Sulphates, 117.
Sulphuric acid formed during rock-weathering, 205.
Swamp deposits, section of, 317.
Swamp soils, 315.
Swamps, cause of, 316; classification of, 317.
Syenite, Little Rock, Arkansas, weathering of, 214.
Syenite-trachyte group, 73.
Syenites described, 73.

Table Mountain structure, how produced, 252.

Tachylite, 92.
Talus, defined, 54; slopes, 319.
Temperatures, effect on soils, 367.
Tephrite and basanite described. 94.
Terminal moraines, 355.
Termites, effects on soils, 391.
Termites, or white ants, as promoters of decomposition, 204.
Terra rossa, 302.
Teschenite, 89.
Theralite-basanite group, 93.
Thin sections, preparation of, 42.
Till, 138.
Time considerations, 268.
Time limit of decay, 272.
Titanic iron as a rock constituent, 28.
Toadstone, 70.
Tonalite, 82.
Trachytes described, 76.
Transportation and deposition of débris, 286.
Transported materials, classification of, 318.
Trap rocks, 89.
Trass, 140.
Travertine, 113.
Trees, effect on soils, 394.
Triassic conglomerate, weathering of, 264.
Trichites defined, 41.
Triclinic feldspars, 15.
Tripolite, 142.
Trowlesworthite, 68.
Tufa, 112.
Tuffoids, 140.
Tuffs, 139.

Uintaite described, 127.
Ulmic acid, 189.
Unakite, 68.

Valley drift, 356.
Valleys, formed by decomposition of greenstone dikes, 244.
Valleys of solution, 253.
Variolite, 89, 90.
Vegetable matter, decomposing action of, 203.
Veins defined, 54.
Verd antique, 116.
Verde di Genora, 98, 205.
Verde di Pegli, 98.
Verde di Prato, 205.
Vesicular structure, 34.
Viridite, 30.
Vitreous or glassy structure, 33.
Vitrophyr, 70.
Vogesite, 74.
Volcanic ashes, 140.

Volcanic dust, 140, 298, 349.
Volcanic group of fragmental rocks, 139.
Volcanic mud, 140.
Volcanic neck defined, 51.
Volcanic necks, origin of, 261.

Wacke, 139, 311.
Wad, 107.
Water, action of, in dry soil, 379; amount absorbed by rocks, 198; apparent protective action of, 253; chemical action of, 186; contents of soil, 281; effects of freezing, 199; expansive force of freezing, 198; in cavities of quartz, 199; mechanical action of, 195; solvent power augmented, 186; solvent power tested, 191.
Water and ice, influence in transporting rock débris, 287.
Wave erosion, rapidity of, 198.
Waves, erosive action of, 198.
Weathering, character of, indicative of climate, 284; defined, 174; difference in kind in cold and warm climates, 283; effacement of characteristics of, 262; incidental results, 266; influenced by crystalline structure, 243; influenced by mineral composition, 248; influenced by position, 270; influenced by structure of rock masses, 244; irregular, due to lack of homogeneity, 251; of andesites, 274; of argillite, Harford County, Maryland, 229; of basalt, Bohemia, 223; of basalt, France, 223; of calcareous rocks containing silicate minerals, 249; of chert, 230; of clastic rocks, 228; of crystalline schists, 251; of diabase, Medford, Massachusetts, 218; of diabase, Venezuela, 222; of diabase, Stamford, Connecticut, 275; of diorite, Albemarle County, Virginia, 224; of dolomitic limestones, 250; of eruptive and sedimentary rocks, relative rapidity of, 271; of feldspathic quartzite, 251; of fine-grained homogeneous rocks, 250; of gneiss, Albemarle County, Virginia, 213; of granite of the District of Columbia, 206; of granite, Lake Huron, 275; of granite, Pike's Peak, 243; of grano-diorites, 274; of limestone, 232, 250; of limestones, process one of solution, 231; of peridotite, 225; of phonolite, 217; of pyroxenite, 225; of quartzite boulders on deserts, 256; of quartzite in the District of Columbia, 251; of rhyolite, 255; of soapstone, Albemarle County, Virginia, 226; of soapstone, Fairfax

County, Virginia, 227; of syenite, Little Rock, Arkansas, 214; rate of, 268; rate of, influenced by climate, 278; relative amount of material lost through, 284; surface contours due to, 257; ultimate product of, 388; unequal, of bedded rocks, 253.
Weathered stone implements, 273.
Websterite, 100.
Wehrlite, 97.
Weight of soils, 381.
Whirlwinds, effects of, 346.
White ants, effects on soils, 391.
Williamsite, 115.
Wind action, 153, 184, 292.
Wind action on Cape Cod, 297.

Wind action on Wyoming soils, 296.
Wind-blown sand polish, 257.
Wisconsin, rock-weathering in, 278.
Wurtzilite described, 126.

Zeolites, as conservators of potash, 374; as rock constituents, 31; at Plombières, 375; composition of, 32; formed in deep-sea bottoms, 375; in soils, 374; origin of, 31; products of hydro-metamorphism, 375.
Zeolitic matter in soils, 370.
Zircon syenite, 79.
Zonal structure, 37.
Zonal structure incident to weathering, 258.

www.ingramcontent.com/pod-product-compliance
Lightning Source LLC
Chambersburg PA
CBHW051852300426
44117CB00006B/358